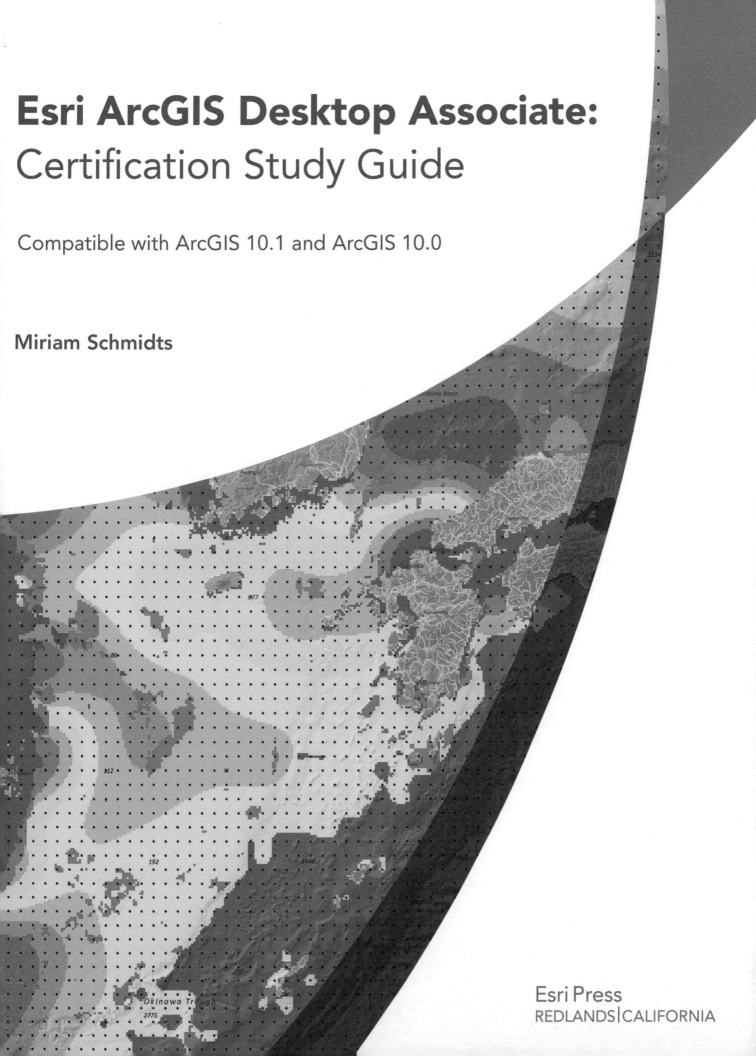

Esri ArcGIS Desktop Associate: Certification Study Guide

Compatible with ArcGIS 10.1 and ArcGIS 10.0

Miriam Schmidts

Esri Press
REDLANDS | CALIFORNIA

Cover image courtesy of Undersea Warfare Graphics.

Esri Press, 380 New York Street, Redlands, California 92373-8100
Copyright © 2013 Esri
All rights reserved. First edition 2013

Printed in the United States of America
17 16 15 14   2 3 4 5 6 7 8 9 10

The information contained in this document is the exclusive property of Esri unless otherwise noted. This work is protected under United States copyright law and the copyright laws of the given countries of origin and applicable international laws, treaties, and/or conventions. No part of this work may be reproduced or transmitted in any form or by any means, electronic or mechanical, including photocopying or recording, or by any information storage or retrieval system, except as expressly permitted in writing by Esri. All requests should be sent to Attention: Contracts and Legal Services Manager, Esri, 380 New York Street, Redlands, California 92373-8100, USA.

The information contained in this document is subject to change without notice.

US Government Restricted/Limited Rights: Any software, documentation, and/or data delivered hereunder is subject to the terms of the License Agreement. The commercial license rights in the License Agreement strictly govern Licensee's use, reproduction, or disclosure of the software, data, and documentation. In no event shall the US Government acquire greater than RESTRICTED/LIMITED RIGHTS. At a minimum, use, duplication, or disclosure by the US Government is subject to restrictions as set forth in FAR §52.227-14 Alternates I, II, and III (DEC 2007); FAR §52.227-19(b) (DEC 2007) and/or FAR §12.211/12.212 (Commercial Technical Data/Computer Software); and DFARS §252.227-7015 (DEC 2011) (Technical Data – Commercial Items) and/or DFARS §227.7202 (Commercial Computer Software and Commercial Computer Software Documentation), as applicable. Contractor/Manufacturer is Esri, 380 New York Street, Redlands, CA 92373-8100, USA.

@esri.com, 3D Analyst, ACORN, Address Coder, ADF, AML, ArcAtlas, ArcCAD, ArcCatalog, ArcCOGO, ArcData, ArcDoc, ArcEdit, ArcEditor, ArcEurope, ArcExplorer, ArcExpress, ArcGIS, arcgis.com, ArcGlobe, ArcGrid, ArcIMS, ARC/INFO, ArcInfo, ArcInfo Librarian, ArcLessons, ArcLocation, ArcLogistics, ArcMap, ArcNetwork, *ArcNews*, ArcObjects, ArcOpen, ArcPad, ArcPlot, ArcPress, ArcPy, ArcReader, ArcScan, ArcScene, ArcSchool, ArcScripts, ArcSDE, ArcSdl, ArcSketch, ArcStorm, ArcSurvey, ArcTIN, ArcToolbox, ArcTools, ArcUSA, *ArcUser*, ArcView, ArcVoyager, *ArcWatch*, ArcWeb, ArcWorld, ArcXML, Atlas GIS, AtlasWare, Avenue, BAO, Business Analyst, Business Analyst Online, BusinessMAP, CityEngine, CommunityInfo, Database Integrator, DBI Kit, EDN, Esri, esri.com, Esri—Team GIS, Esri—*The GIS Company*, Esri—The GIS People, Esri—The GIS Software Leader, FormEdit, GeoCollector, Geographic Design System, Geography Matters, Geography Network, geographynetwork.com, Geoloqi, Geotrigger, GIS by Esri, gis.com, GISData Server, GIS Day, gisday.com, GIS for Everyone, JTX, MapIt, Maplex, MapObjects, MapStudio, ModelBuilder, MOLE, MPS—Atlas, PLTS, Rent-a-Tech, SDE, SML, Sourcebook•America, SpatiaLABS, Spatial Database Engine, StreetMap, Tapestry, the ARC/INFO logo, the ArcGIS Explorer logo, the ArcGIS logo, the ArcPad logo, the Esri globe logo, the Esri Press logo, The Geographic Advantage, The Geographic Approach, the GIS Day logo, the MapIt logo, The World's Leading Desktop GIS, *Water Writes*, and Your Personal Geographic Information System are trademarks, service marks, or registered marks of Esri in the United States, the European Community, or certain other jurisdictions. CityEngine is a registered trademark of Procedural AG and is distributed under license by Esri. Other companies and products or services mentioned herein may be trademarks, service marks, or registered marks of their respective mark owners.

Ask for Esri Press titles at your local bookstore or order by calling 800-447-9778, or shop online at esri.com/esripress. Outside the United States, contact your local Esri distributor or shop online at eurospanbookstore.com/esri.

Esri Press titles are distributed to the trade by the following:

In North America:
Ingram Publisher Services
Toll-free telephone: 800-648-3104
Toll-free fax: 800-838-1149
E-mail: customerservice@ingrampublisherservices.com

In the United Kingdom, Europe, Middle East and Africa, Asia, and Australia:
Eurospan Group
3 Henrietta Street
London WC2E 8LU
United Kingdom
Telephone: 44(0) 1767 604972
Fax: 44(0) 1767 601640
E-mail: eurospan@turpin-distribution.com

Contents

Acknowledgments   ix
Using this book   x

Chapter 1: Understanding ArcGIS products and extensions   1
    ArcGIS products   2
    ArcGIS for Desktop extensions   6

Chapter 2: Working with geographic data storage formats   15
    Data formats   16

Chapter 3: Using the geodatabase   27
    Types of geodatabases   28
    Defining geodatabase tables   34
    Advanced geodatabase elements   37

Chapter 4: Managing geographic data   43
    Methods for adding data to the geodatabase   44
    Creating data from layers   55
    Layer source data   60

Chapter 5: Coordinate systems   67
    Types of coordinate systems   68
    Projection on the fly   73
    Data in different projected coordinate systems   75
    Data with an unknown coordinate system   77
    Managing data in different geographic coordinate systems   79

Chapter 6: Evaluating data   87
    Evaluating data for a task   88
    Data documentation   92

Chapter 7: Associating tables   101
    Types of table relationships   102
    Table associations   104
    Relationship classes   109

Chapter 8: Georeferencing and spatial adjustment 117
Georeferencing 118
Spatial adjustment 126

Chapter 9: Geocoding 133
Geocoding components 134
Address matching 139
The geocoding environment 142
Geocoding results 146

Chapter 10: Creating feature geometry 149
Feature templates 150
Feature construction tools 152
Segment construction methods 155
Creating features from existing features 159

Chapter 11: Updating feature geometries 165
Modifying feature shape 166
Dividing features into parts 172

Chapter 12: Editing attributes 179
Attribute editing methods 180
Field calculations 183

Chapter 13: Maintaining attribute integrity 191
Attribute editing with default values 192
Attribute editing with domains 194

Chapter 14: Editing with topology 201
Topology 202
Geodatabase topology 208

Chapter 15: Geoprocessing for analysis 223
Geoprocessing tools for analysis 224
Geoprocessing models for analysis 230
Using Python scripts for analysis 235

Chapter 16: Analyzing and querying tables — 241
   Extracting information from attribute tables — 242
   Attribute queries — 246
   Spatial queries — 248
   Queries in analysis — 250
   Working with selections — 252

Chapter 17: Performing spatial analysis — 257
   Proximity analysis — 258
   Overlay analysis — 262
   Statistical analysis — 266
   Temporal analysis — 268

Chapter 18: Organizing layers — 273
   Working with layers and data frames — 274

Chapter 19: Displaying layers — 283
   Vector layer symbology — 284
   Raster layer symbology — 288
   Managing the amount of data viewed in a map — 290

Chapter 20: Composing map layouts and graphs — 297
   Map layout — 298
   Graphs — 303

Chapter 21: Creating Data Driven Pages — 309
   The index layer — 310
   Enabling Data Driven Pages — 311
   Refining Data Driven Pages — 314

Chapter 22: Creating map text and symbols — 319
   Map text — 320
   Custom symbols — 329

Chapter 23: Preparing maps for publishing — 335
   Optimizing maps for the web — 336
   Sharing a map as a service — 337
   Preparing temporal data for a web map — 342

Chapter 24: Sharing maps and data   345
    Exporting and printing maps   346
    Printing maps   **349**
    Sharing maps and data through packaging   **353**
    Sharing map documents   **357**

Image and data credits   363
Installing the data and software   373
Data license agreement   379

Acknowledgments

Many people contributed to this book. First and foremost, I want to express my gratitude to Eileen Napoleon, my instructional designer. A very special thank you also to my colleagues Gregory Emmanuel, Beth Guse, and Eric Bowman, who prepared selected exercise workflows and data; to Carl Byers for his technical reviews; and to all the instructors and editors who helped with review and testing.

Many thanks to all the organizations that provided the data used to create the exercises in this book, especially the following:

The City of Austin provided data, including the lots and row basemap, street centerlines, address points, parcels, parks, historical landmarks, facilities, and aerial photography used in chapters 10, 11, 12, 13, and 17.

Hamilton County, Indiana, provided layers, including street centerlines, addresses, parcels, annexation areas, precincts, and hydrology line used in chapters 9, 14, 15, and 16.

Yellowstone National Park provided layers of the park boundary, greater Yellowstone area cities, points of interest, trails, campgrounds, and aerial photography used in chapters 2, 3, and 4.

Using this book

The Esri Technical Certification Program recognizes qualified individuals who are proficient in best practices for using Esri software. Achieving an Esri Technical Certification is a process of preparing for and taking an examination. Exams consist of 85 to 95 multiple choice questions for which candidates have 2 to 2 ½ hours to complete. Candidates may register and sit for exams at over 5,000 locations around the world through Pearson VUE, our testing partner. Obtaining an Esri Technical Certification is a great way to benchmark your skills against an established level of competency and demonstrate your knowledge in using Esri best practices. Learn more about the certification program at **training.esri.com/certification**.

This book is a practical guide intended to help you review the skills and knowledge measured by the ArcGIS Desktop Associate exam as outlined in the "Skills measured" section of the ArcGIS Desktop Associate web pages. Each skill is listed at the beginning of the chapter that addresses the related content. This book, in conjunction with the ArcGIS Help library, is intended as a comprehensive guide for preparing for the ArcGIS Desktop Associate exam. It is not by any means an all-inclusive, in-depth compendium of fundamental GIS topics.

As a qualified exam candidate, you should already have broad experience (at least two years) using ArcGIS software. You can use this book to refresh and reinforce your knowledge and skills. The chapters in this book build on your existing knowledge of fundamental ArcGIS concepts and workflows (prerequisite knowledge and skills are listed at the beginning of each chapter).

Chapter 1 gives you a conceptual overview of ArcGIS products and ArcGIS for Desktop extensions. All other chapters contain conceptual content and hands-on exercises that allow you to practice software skills. The 24 chapters of this book are independent of one another. For a comprehensive review, it is advisable to study all of the chapters. Alternatively, you may choose to focus on particular chapters to close any gaps in your knowledge. If you need to study a subject in more depth, each chapter provides references to relevant topics in the ArcGIS Help library in the Resources section at the end of each chapter.

Most exercises in each chapter are independent of one another. In a few cases, where exercises depend on the completion of another exercise in the same chapter, this will be noted in the introduction to the exercise. Exercises can be completed using ArcGIS Desktop 10 or ArcGIS 10.1 for Desktop software and do not require the use of any extensions. Appendix C describes the process for installing the exercise data provided on the DVD in the back of this book.

We encourage you to use this book as one step on your path to achieving certification. The best opportunity for success lies in the combination of study and hands-on experience.

chapter 1

Understanding ArcGIS products and extensions

ArcGIS products 2
- ArcGIS for Desktop
- ArcGIS for Server
- ArcGIS Online
- ArcGIS for Mobile
- ArcGIS Explorer
- Challenge 1

ArcGIS for Desktop extensions 6
- ArcGIS Spatial Analyst
- ArcGIS Geostatistical Analyst
- ArcGIS 3D Analyst
- ArcGIS Network Analyst
- ArcGIS Schematics
- ArcGIS Tracking Analyst
- ArcGIS Publisher
- Challenge 2

Answers to challenge questions 12

Key terms 13

Resources 13

Chapter 1: Understanding ArcGIS products and extensions

Welcome to the ArcGIS platform. The ArcGIS products complement each other to create a complete geographic information platform that lets you author, serve, and use geographic information (figure 1.1). In this chapter, you will get an overview of ArcGIS products that will help you identify the right product for a given task.

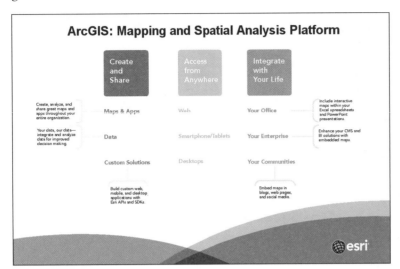

Figure 1.1 ArcGIS is a platform for designing and managing solutions through the application of geographic knowledge. It enables you to perform deep analysis, gain a greater understanding of your data, and perform high-level decision making. Esri.

Skills measured

Given a task, identify the necessary ArcGIS products and/or resources to complete the task.

ArcGIS products

ArcGIS is a complete platform that includes several products for managing, integrating, and sharing geographic data, performing spatial analysis, and displaying your results as professional-quality maps. To determine the correct product to complete a given task, you need to know what each of the different products is used for. Below you will find a brief description of the main products in the ArcGIS platform and some examples of when to use them.

ArcGIS for Desktop

ArcGIS for Desktop is used to author and edit maps and geospatial content. It includes two applications: ArcMap and ArcCatalog. ArcCatalog is also integrated into ArcMap as a Catalog window. Use these two applications for the following:

- Use ArcMap for visualizing and editing geographic data, performing GIS analysis, and creating professional-quality map products.
- Use ArcCatalog (or the ArcMap Catalog window) for browsing, managing, and documenting geographic data.

Examples:
- An analyst could use ArcGIS for Desktop geoprocessing tools to analyze the spatial distribution of 911 calls and visualize the results as a series of maps and graphs. The analyst could then run the analysis repeatedly on different data by building a geoprocessing model.
- You could use the ArcGIS for Desktop predefined templates and simple wizards to create quality maps (figure 1.2).
- A technician could use ArcGIS for Desktop to create and edit roads and highways. For new roads, the technician could use Coordinate Geometry (COGO) tools to create features based on survey measurements.
- As part of the GIS department in a larger organization, you could use ArcGIS for Desktop to prepare maps for publishing and share them as map services and map packages in ArcGIS Online.

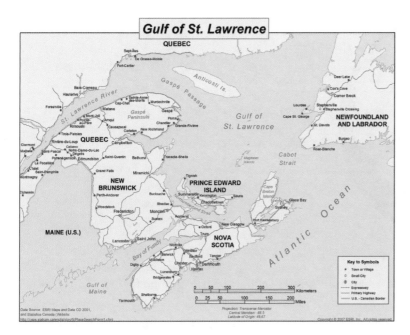

Figure 1.2 ArcGIS for Desktop provides a large library of predefined and customizable symbols, tools for advanced cartographic label placement and conflict detection (Maplex), and predefined map templates. Map by Esri; data from Esri Data & Maps, 2001, courtesy of ArcWorld Supplement and DMTI Spatial Inc.

ArcGIS for Server

ArcGIS for Server is used to create and distribute maps and geospatial services over the web. Use ArcGIS Server for the following:

- Publishing GIS information and maps as web services
- Accessing GIS services through web applications
- Managing data in an enterprise geodatabase

Examples:
- An employee of a city GIS department could use ArcGIS for Server to publish city map layers as GIS services and make them available to the general public via a web application.
- An employee of a weather agency could use ArcGIS for Desktop to create an animation showing the progression of hurricane paths from time-enabled data and then use ArcGIS for Server to publish the map as a GIS service that can be accessed through a web application.
- A Department of Emergency Management employee could use ArcGIS for Server to integrate statewide hazmat, transportation, and shelter status from several GIS services, along with a weather data feed into one central web application to support decisions that involve hazardous materials.
- A city employee could store city spatial data holdings in a central enterprise geodatabase that employees from different city departments could access and use to edit and maintain the data. Enterprise geodatabase functionality is part of the ArcGIS for Server license.

ArcGIS Online

ArcGIS Online helps organizations collaborate and manage maps and geospatial content in the cloud (figure 1.3). Organizations can set up their own custom ArcGIS Online website. Members of the organization use ArcGIS Online for the following:

- Creating and sharing maps that can be accessed by anyone through a browser, a mobile device, ArcGIS for Desktop, or a custom application
- Accessing maps, datasets, services, tools, and other geospatial content shared by others
- Managing geospatial content
- Sharing content publicly, with specific groups, or keeping it private within the organization

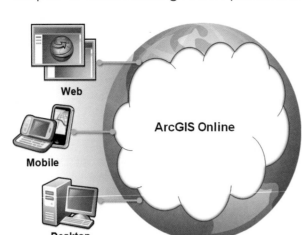

Figure 1.3 ArcGIS Online is part of the ArcGIS platform. It can be accessed through desktop, mobile, or web applications. Esri.

Examples:
- You could use the World Street Map basemap from ArcGIS Online as a background for displaying analysis results in ArcMap.
- You could create a map of median household income in different parts of the county, and then create a map package and post it to ArcGIS Online.
- A web developer could create an application that combines ArcGIS Online map services of renewable energy projects and solar radiation potential with an ArcGIS Online basemap, allowing users to calculate the solar energy potential of a building or area. The developer could then share the application on ArcGIS Online.
- A public health initiative employee could create a web map of health-food supermarkets using the ArcGIS Online built-in map viewer. Then the employee could share the web map by distributing the URL to members of the initiative's ArcGIS Online group.

ArcGIS for Mobile

ArcGIS for Mobile is used to access maps and geospatial content on mobile devices, including tablets and smartphones. It consists of ready-to-deploy applications for field data collection and inspections, plus Software Development Kits (SDK) for creating custom applications.

ArcGIS for Mobile consists of the following product suite:

- **ArcPad:** ArcPad is designed for GIS professionals who need mobile field mapping and advanced data collection software. ArcPad includes key GIS and GPS (Global Positioning Systems) capabilities for capturing, editing, and displaying geographic information quickly and easily.
- **ArcGIS for Windows Mobile:** ArcGIS for Windows Mobile helps organizations deliver GIS capabilities and data from centralized servers to a range of Windows-based mobile devices. You can use ArcGIS for Windows Mobile to deploy intuitive GIS applications to increase the accuracy and improve the currency of GIS data across your organization. Easy-to-use ArcGIS for Windows Mobile applications enable field crews who do not necessarily have any GIS experience to do field mapping, GIS editing, spatial queries, sketching, and GPS integration.
- **Apps for Smartphones and Tablets:** ArcGIS for Smartphones and Tablets is used for navigating maps, data collection, reporting and analysis using iOS, Android, or Windows Phone devices. It includes a free application that you can download from the Apple App Store, Google Play/Android Market, or Windows Store. Also provided are developer-focused SDKs that you can use to build custom applications.

Use ArcGIS for Mobile to do the following:

- Provide simple-to-use data collection applications for smartphone and tablet devices
- Synchronize directly with ArcGIS for Server to make data and map updates available to field staff and desktop users both in the office and in the field

Examples:
- You could use ArcPad on a Windows-based mobile device in the field to collect and edit spatial and attribute data for water and sewer mains, manholes, fire hydrants, pole inspections, and so on, using various customization tools such as scripts and applets. Back in the office, you could quickly check new data into a geodatabase to update the city's data holdings.
- City field workers who are not trained in GIS could use the out-of-the-box ArcGIS for Windows Mobile application to collect information about city street signs via a map service that accesses the GIS features of the map.
- A water department employee could use an ArcGIS for Windows Mobile application to view work orders, find the water valves to shut off for servicing water features, add notes for the user to the map about updates in the water features, and synchronize the information with the database in the office twice a day.
- You could use the ArcGIS for Windows Mobile SDK to create a data collection application to be used on a Windows 7 tablet or laptop PC and distribute it to field crews.
- A city resident without any GIS knowledge could use the city's custom reporting smartphone or tablet application that leverages ArcGIS technology to report civic issues and make service requests to the different city departments using your mobile phones.

ArcGIS Explorer

ArcGIS Explorer is a free viewer that can be used to do the following:

- View, query, and analyze spatial data
- Deliver authoritative data to a broad audience

ArcGIS Explorer includes two applications: ArcGIS Explorer Desktop and ArcGIS Explorer Online (figure 1.4).

- Use **ArcGIS Explorer Desktop** (downloadable) for exploring, visualizing, and sharing geographic information and author presentations.
- Use **ArcGIS Explorer Online**, a browser-based viewer, for creating web maps, performing queries, and creating dashboards and author presentations.

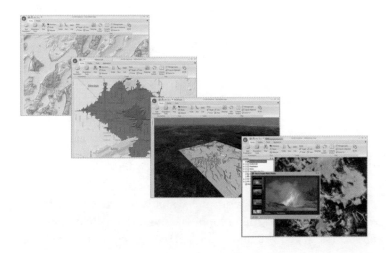

Figure 1.4 With ArcGIS Explorer Desktop you can access ready-to-use ArcGIS Online basemaps and layers, perform spatial analysis (e.g., visibility analysis, modeling, proximity search), fuse your local data with map services to create custom maps, and add photos, reports, videos, and other information to your maps. Esri.

Examples:

- An employee in the department of transportation could use ArcGIS Explorer Desktop to view and query data about road construction projects, query traffic data, and integrate external map services and basemaps to create presentations for a public meeting.

- A real-estate agent could use ArcGIS Explorer Online to show available properties to clients, along with socio-economic, transportation, school, and other family resource information in a one-mile buffer area.

- A group of disc golf enthusiasts could use ArcGIS Explorer Online to create a web map of disc golf courses in a county along with a description and driving directions to the individual courses. Then, they could share their web map with others using ArcGIS Online.

- A geography teacher could use ArcGIS Explorer Desktop in the classroom to view, query, and analyze maps as part of geography education.

Challenge 1

For each of the following scenarios, pick the right products to complete the task.

1. Which two ArcGIS products or applications should you use to create a new geoprocessing model? (Choose two.)

    a. ArcGIS Explorer Desktop
    b. ArcCatalog
    c. ArcGIS Mobile
    d. ArcMap
    e. ArcGIS for Server

2. Which ArcGIS product should you use to share GIS maps so they can be accessed from within a web browser?

    a. ArcGIS Explorer Online
    b. ArcPad
    c. ArcGIS for Server
    d. ArcGIS for Desktop

3. You need to add imagery to a map document so that it displays in the background of some analysis results. Which product should you use to find the imagery?

    a. ArcGIS for Desktop
    b. ArcGIS Online
    c. ArcGIS Explorer Online
    d. ArcGIS Mobile

4. You need to use advanced labeling tools for fitting the maximum number of labels on a map without conflict. Which product or application will provide this functionality?

    a. ArcMap
    b. ArcGIS Online
    c. ArcPad
    d. ArcGIS Explorer Online
    e. ArcGIS Mobile

ArcGIS for Desktop extensions

Let's briefly talk about special products that extend the functionality of ArcGIS for Desktop. ArcGIS for Desktop extensions are products that add specialized capabilities for advanced analysis and enhanced productivity to ArcGIS for Desktop. They give you tools for performing sophisticated tasks such as raster geoprocessing, 3D analysis, and network analysis. Below, you will find a summarized description of the most commonly used ArcGIS for Desktop extensions.

ArcGIS Spatial Analyst

ArcGIS Spatial Analyst is used for creating, querying, mapping, and analyzing cell-based raster data (figure 1.5).
Examples:
- You could use an elevation raster to calculate surfaces, for example, slope, aspect, and hillshade.
- You could find suitable locations for a new ski resort by reclassifying different input rasters according to their suitability and then combining them in a weighted overlay.
- You could derive a crime density surface from a point layer of crime incidents. Since some point locations represent more than one crime incident, a density surface will present a different pattern than the point layer.
- You could perform a cost-of-travel analysis between two locations by creating a cost-weighted distance raster and identifying the best path based on travel cost.

Figure 1.5 ArcGIS Spatial Analyst provides tools to derive slope, aspect, or hillshade rasters from an elevation raster. Esri.

ArcGIS Geostatistical Analyst

ArcGIS Geostatistical Analyst complements the functionality of Spatial Analyst. Most of the interpolation methods available in Spatial Analyst are represented in ArcGIS Geostatistical Analyst as well, but in Geostatistical Analyst there are many more statistical models and tools, and all their parameters can be manipulated to derive optimum surfaces.
Examples:
- A California Health Department employee could use the kriging interpolation method from Geostatistical Analyst to create a raster surface of ozone concentrations for the state of California from point measurements at different ozone monitoring stations (figure 1.6).

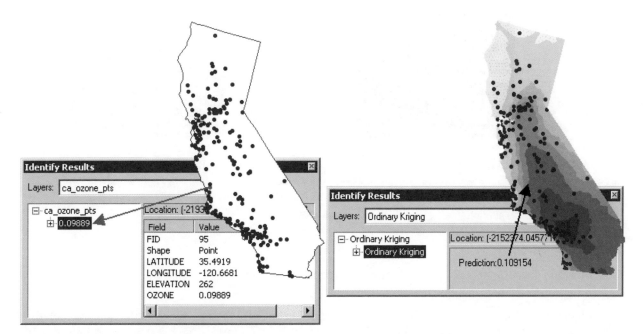

Figure 1.6 There are 193 ozone monitoring stations in California. The prediction map shows a predicted ozone value for every location in the state. Esri.

- From the same data, you could create a standard error map that quantifies the uncertainty of the predictions (figure 1.7).

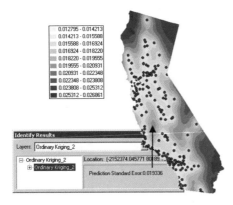

Figure 1.7 An error map quantifies the uncertainty of the predictions. The larger the standard error, the more uncertain the predictions. Esri.

- A scientist working for an environmental science university lab could use Geostatistical Analyst to determine radioactive soil contamination after the Chernobyl reactor accident based on point measurements of rainfall within a few days after the accident.

ArcGIS 3D Analyst

ArcGIS 3D Analyst is used for creating, visualizing, and analyzing three-dimensional data. ArcGIS 3D Analyst extends the functionality of ArcMap and ArcCatalog but also includes two viewing applications: ArcGlobe and ArcScene (figure 1.8).

- Use ArcGlobe for viewing and navigating large amounts of data on a globe surface.
- Use ArcScene for 3D viewing and analyzing datasets in a local area.

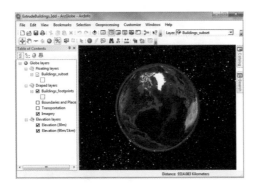 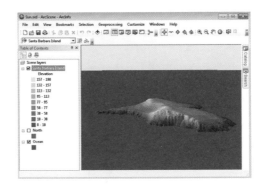

Figure 1.8 You can visualize very large amounts of 3D data in a global view using the ArcGlobe application, or you can view site-level data in a local coordinate system using the ArcScene application. Map by Esri; data from Esri Data & Maps, 2010, courtesy of Esri, i-cubed, USDA FSA, USGS, AEX, GeoEye, AeroGRID, Getmapping, IGP, DeLorme, NAVTEQ, TomTom, Intermap, AND, NRCAN, Kadaster NL, and the GIS User Community.

Examples:
- You could visualize data in 3D to enhance understanding of spatial relationships and build a 3D animation to share in a public meeting.
- You could use interactive tools in a 3D view to determine line of sight, measure heights in 3D, calculate 3D volumes, and determine the steepest path.

- You could use ArcGIS 3D Analyst to analyze how the shadow of a proposed building will affect the neighboring buildings.
- You could build surfaces of tree heights and crown cover from light detection and ranging (lidar) data to estimate the biomass in a forest area.

ArcGIS Network Analyst

ArcGIS Network Analyst is used for spatial analysis of transportation networks, such as street, pedestrian, and railroad networks that allow for travel in multiple directions.

Examples:
- A truck driver in a fleet of delivery vehicles could use ArcGIS Network Analyst to find the quickest route for a vehicle to drive from location A to location B and print travel directions.
- A fleet manager could create the most efficient routes for a fleet of vehicles for multiple customer locations, optimizing travel time and minimizing transportation costs (figure 1.9).

Figure 1.9 Three food delivery trucks at a distribution center are assigned grocery stores and routes to the stores that minimize transportation costs. Vehicle capacities, lunch breaks, and maximum travel time constraints are included in the analysis. Esri.

- A 911 dispatcher could determine which ambulances or patrol cars can respond quickest to an accident and locate the closest hospital from an accident location.
- A marketing analyst of a department store chain could define the service area that is covered by a department store and determine the optimal location for a new department store.

ArcGIS Schematics

ArcGIS Schematics is used for creating a representation of a network in a schematic diagram.

Examples:
- An electrical company employee could use ArcGIS Schematics to create a simplified representation of an electrical power network, intended to explain its structure and the way it operates (figure 1.10).

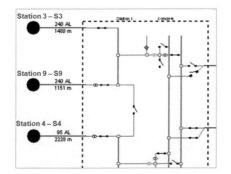

Figure 1.10 Schematic representation of an electrical network showing how three outside stations are connected to the network inside a substation. From the position of the switches, you can tell how electricity flows through this network. Esri.

- A system administrator could use ArcGIS Schematics to visualize a computer network in a schematic diagram.
- An airline could use ArcGIS Schematics to visualize resource dependencies of a flight schedule and the cockpit crew, the cabin crew, and the plane (figure 1.11).
- An employee of an organic food brand could use ArcGIS Schematics to visualize the path of food, for example, from the farm where it's grown, to the packing facility, to the distribution facility, to a warehouse, and finally, to the store and the customer.

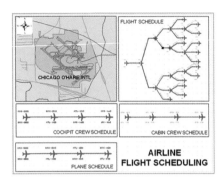

Figure 1.11 Schematic representation of resource dependencies of a flight schedule. Referring to these schematics, you can easily tell how a delay in the flight schedule would affect the other schedules. Esri.

ArcGIS Tracking Analyst

ArcGIS Tracking Analyst is used for visualizing, analyzing, and understanding spatial patterns and trends in your data in the context of time (figure 1.12).

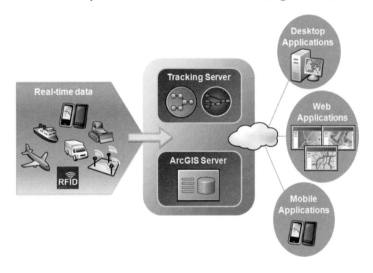

Figure 1.12 ArcGIS Tracking Analyst allows you to visualize and analyze the movement of resources. Esri.

Examples:
- A fleet manager could track the positions of a fleet of delivery trucks in real time. ArcGIS Tracking Analyst supports network connections to GPS units and other tracking devices so the fleet manager could create a connection to a tracking service or GPS device that can stream real-time temporal data.
- A National Weather Service employee could create a time-series animation of a hurricane, with the hurricane's center symbolized by wind speed and with a special symbol for the most current speed.
- An airport employee could analyze patterns in airplane delays by creating and summarizing charts that display the temporal distribution of the data.
- An employee of the Center for Disease Control could analyze the spread of a highly infectious disease to come up with the best strategy to stop it.

ArcGIS Publisher

ArcGIS Publisher is used by ArcGIS Desktop users to share and distribute their maps, globes, and data. ArcGIS Publisher creates Published Map Files (PMFs) (figure 1.13). PMF files are viewable through ArcGIS for Desktop products, including ArcReader, a free downloadable and customizable application that supports collaboration via markup tools. Since ArcGIS Publisher does not require an Internet connection, it is frequently used for distributing content inside and outside of an organization with secure control over data sources and viewer access. It's a cost-effective way to distribute maps and data to many different people.

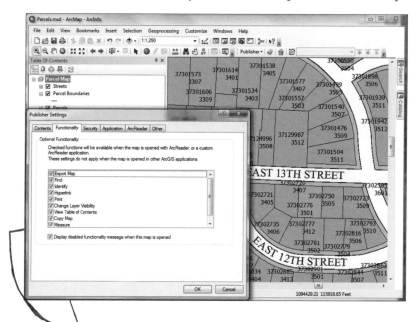

Figure 1.13 With ArcGIS Publisher you can share data and maps locally, over networks, or on the Internet, allowing colleagues to view and interact with maps simultaneously. Esri.

Examples:
- A government agency could use ArcGIS Publisher to share an interactive parcel map, protect it with a password, and package the required data for distribution to the public.
- The GIS manager in a large organization could use ArcGIS Publisher to distribute maps to all GIS groups within the company.
- A GIS consultant could use ArcGIS Publisher to distribute preauthored maps to their customers.
- A GIS developer in a consulting firm could use ArcGIS Publisher to distribute custom map viewers to their customers.

For a complete list of Esri products, refer to the product pages at **http://www.esri.com/products** and click the Alphabetical List of All Products link at the bottom of the page.

Challenge 2

For each of the following scenarios, pick the extension or product to complete the task.

1. Which ArcGIS for Desktop extension should you use to find the best location for a new ski resort based on the following criteria?
   - Must be on intermediate slopes, 15%–35%
   - Must be in deeper snow; over 23 inches is best
   - Must be in an area that gets more shade (so snow doesn't melt quickly)

      a. ArcGIS 3D Analyst
      b. ArcGIS Spatial Analyst
      c. ArcGIS Geostatistical Analyst
      d. ArcGIS Tracking Analyst
      e. ArcMap

Chapter 1: Understanding ArcGIS products and extensions

2. Which extension product should you use to find the shortest route for driving your car from home to the closest train station, taking the train to the central station downtown, and then walking to your workplace?

   a. ArcGIS Spatial Analyst
   b. ArcGIS Tracking Analyst
   c. ArcGIS Network Analyst
   d. ArcGIS Mobile

3. Which product should you use to visualize data with a time component?

   a. ArcGIS Network Analyst
   b. ArcGIS Spatial Analyst
   c. ArcGIS Tracking Analyst
   d. ArcGIS Publisher

Answers to challenge questions

Challenge 1
Correct answers shown in bold.

1. Which two ArcGIS products or applications should you use to create a new geoprocessing model? (Choose two.)

   a. ArcGIS Explorer Desktop
   b. ArcCatalog
   c. ArcGIS Mobile
   d. ArcMap
   e. ArcGIS for Server

2. Which ArcGIS product should you use to share GIS maps so they can be accessed from within a web browser?

   a. ArcGIS Explorer Online
   b. ArcPad
   c. ArcGIS for Server
   d. ArcGIS for Desktop

3. You need to add imagery to a map document so that it displays in the background of some analysis results. Which product should you use to find the imagery?

   a. ArcGIS for Desktop
   b. ArcGIS Online
   c. ArcGIS Explorer Online
   d. ArcGIS Mobile

4. You need to use advanced labeling tools for fitting the maximum number of labels on a map without conflict. Which product or application will provide this functionality?

   a. ArcMap
   b. ArcGIS Online
   c. ArcPad
   d. ArcGIS Explorer Online
   e. ArcGIS Mobile

Challenge 2
For each of the following scenarios, pick the extension or product to complete the task.

1. Which ArcGIS for Desktop extension should you use to find the best location for a new ski resort based on the following criteria?

- Must be on intermediate slopes, 15%–35%
- Must be in deeper snow; over 23 inches is best
- Must be in an area that gets more shade (so snow doesn't melt quickly)

   a. ArcGIS 3D Analyst
   b. ArcGIS Spatial Analyst
   c. ArcGIS Geostatistical Analyst
   d. ArcGIS Tracking Analyst
   e. ArcMap

2. Which extension product should you use to find the shortest route for driving your car from home to the closest train station, taking the train to the central station downtown, and then walking to your workplace?

   a. ArcGIS Spatial Analyst
   b. ArcGIS Tracking Analyst
   c. ArcGIS Network Analyst
   d. ArcGIS Mobile

3. Which product should you use to visualize data with a time component?

   a. ArcGIS Network Analyst
   b. ArcGIS Spatial Analyst
   c. ArcGIS Tracking Analyst
   d. ArcGIS Publisher

Key terms

ArcSDE geodatabase: A geodatabase stored in an RDBMS served to client applications using ArcSDE technology. An ArcSDE geodatabase can support long transactions and versioned workflows, be used as a workspace for geoprocessing tasks, and provide the benefits of a relational database such as security, scalability, backup and recovery, and SQL access.

Geoprocessing tool: An ArcGIS tool that can create or modify spatial data, including analysis functions (overlay, buffer, slope), data management functions (add field, copy, rename), or data conversion functions.

Map service: A type of web service that generates maps.

Web service: A software component accessible over the World Wide Web for use in other applications.

Web map: In ArcGIS Online, a web-based, interactive map that allows you to display and query the layers on the map. A web map contains one or more ArcGIS Server map services that are referenced to ArcGIS Online.

Resources

- ArcGIS Help 10.1 > ArcGIS Tutorials
  - ArcGIS Spatial Analyst
  - Geostatistical Analyst
  - ArcGIS Network Analyst
  - ArcGIS Tracking Analyst
  - Schematics

- ArcGIS Help 10.1 Help > Extensions
  - Spatial Analyst > Introduction
    - What is the Spatial Analyst extension?
    - A quick tour of Spatial Analyst
  - 3D Analyst
    - What is the ArcGIS 3D Analyst extension?
    - A quick tour of the ArcGIS 3D Analyst extension
  - Geostatistical Analyst > Introduction to Geostatistical Analyst
    - What is the ArcGIS Geostatistical Analyst extension?
    - A quick tour of Geostatistical Analyst
  - Network Analyst
    - What is the ArcGIS Network Analyst extension?
    - A quick tour of the ArcGIS Network Analyst extension
  - Schematics > Basics
    - What is Schematics?
    - A quick tour of Schematics
  - Tracking Analyst
    - What is Tracking Analyst?
    - A quick tour of Tracking Analyst
  - Publisher > Getting started with Publisher
    - An overview of ArcGIS Publisher
- ArcGIS Help 10.1 Help > Professional Library > ArcGIS Server
  - What is ArcGIS Server?
- ArcGIS for Desktop product pages at **http://www.esri.com/products/index.html**
- ArcGIS for Desktop extension product pages at **http://www.esri.com/products/index.html**
- Virtual Campus courses
  - 3D Visualization Techniques Using ArcGIS 10
  - 3D Analysis of Surfaces and Features Using ArcGIS 10
  - Using Lidar Data in ArcGIS 10
  - Learning ArcGIS Spatial Analyst
  - Introduction to ArcGIS 9 Geostatistical Analyst
- Recorded training seminars (free)
  - ArcGIS Explorer Desktop Quick Start Tutorial
  - Getting the Most Out of ArcGIS Explorer Online
  - Introduction to ArcGIS Schematics
  - Making and Sharing Maps with ArcGIS Online
  - Using Network Analyst in ArcGIS 10
  - Introduction to the ArcGIS API for iOS

chapter 2

Working with geographic data storage formats

Data formats 16
    Vector data
        Feature Classes
    3D features
    Exercise 2a: Connect to exercise data and turn on file extensions
    Challenge 1
    Raster data
        Raster pyramids
    Exercise 2b: Build raster pyramids
    Tabular data formats
    Challenge 2

Answers to chapter 2 questions 24

Answers to challenge questions 24

Key terms 25

Resources 25

Chapter 2: Working with geographic data storage formats

There is no GIS without geographic data. We work with geographic data every day. But what exactly is geographic data? Geographic data is measurements, counts, probabilities, descriptions, and classifications about phenomena on the earth's surface and their recorded locations.

Skills measured

- Compare and contrast the various types of datasets and data formats supported in ArcGIS and their uses.
- Explain raster pyramids and the appropriate circumstances for their use.

Prerequisites

- Hands-on experience working with geographic datasets in ArcGIS
- Knowledge of the vector and the raster data models
- Knowledge of storing data in a geodatabase

Data formats

Geographic data is organized into datasets, which are stored in various data formats. The geodatabase is the primary data format for storing data in ArcGIS. However, geographic datasets can also be stored in other formats.

Vector data

In the vector data model, geographic data is stored as point, line, and polygon features. Features consist of a geometry part that is defined by its x,y coordinates (the shapes that you see in the map) and an attribute part that is stored as a record in an attribute table. You can store point, line, and polygon features either inside the geodatabase or outside of it in different file formats.

If you store point, line, and polygon features in the geodatabase, they can have either single-part or multipart geometry. In a single-part feature, each geometry element in the map—each point, each line, or each polygon—corresponds to one record in the attribute table. In a multipart feature, multiple geometry elements correspond to the same record in the attribute table (figure 2.1). For example, in a US States feature class, the state of Hawaii is often represented as a multipart polygon feature. The islands are represented as multiple polygons on the map, but there is only one record in the attribute table for attributes such as state name, population, and so on.

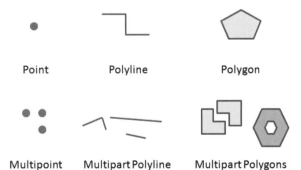

Figure 2.1 Multipart features include multipoints, multipart polylines, and multipart polygons. Polygons with a hole (doughnut polygons) are considered multipart as well. Created by the author.

Feature classes

Vector data is stored as a feature class: a collection of features—either point, line, or polygon features that share the same set of attributes and spatial reference (figure 2.2). For example, all the fire hydrants in a city area can be represented as a point feature class. World countries can be represented as a polygon feature class.

Figure 2.2 Inside a geodatabase, feature classes can be stored under the root level of the geodatabase (stand-alone) or inside of a feature dataset. Created by the author.

Tables 2.1 and 2.2 give you an overview of the data formats that are used in ArcGIS for Desktop to store feature classes.

Table 2.1 Data formats inside the geodatabase		
Data format	What it is	When to use it
Stand-alone feature class	A collection of features stores both geometry and attribute information about the features.	This format is recommended for storing datasets in ArcGIS for Desktop.
Feature class inside of a feature dataset	All feature classes in a feature dataset are stored in the same spatial reference.	The primary purpose for storing feature classes in a feature dataset is to provide the underlying structure for adding advanced geodatabase functionality such as topologies, geometric networks, network datasets, terrain datasets, or parcel fabrics.

Table 2.2 Data formats outside the geodatabase		
File format	What it is	When to use it

Shapefile	• A shapefile is a single feature class stored as a collection of at least three files (all three files are needed for the shapefile to draw):
	  • ShapefileName.shp—stores feature geometry
	  • ShapefileName.dbf—stores feature attributes
	  • ShapefileName.shx—stores the index of the feature geometry
		• Use when a lightweight data format is needed, e.g., to add to an ArcGIS Online web map or to ArcPad.
		• A shapefile is compatible with non-Esri GIS software packages.

Coverage	• A coverage is a collection of feature classes usually representing a single thematic layer (e.g., schools, streets, etc.). The main coverage feature classes are point, arc (lines), polygon, label, and tic. • Native data format to ArcInfo Workstation software, which preceded ArcGIS for Desktop.	Coverages are used mostly for legacy datasets.

Continued on next page

Chapter 2: Working with geographic data storage formats

Table 2.2 Data formats outside the geodatabase (continued)

File format	What it is	When to use it
Computer-aided design (CAD) datasets	• CAD files are drawing files created in CAD software (e.g., AutoCAD, Microstation, etc .). • ArcGIS for Desktop has direct-read capability for CAD files and interprets them as a collection of feature classes. • The information to define these feature classes is derived from the CAD file and held in memory, while the CAD file remains in its native format.	Use when working with CAD data in its native format; for example, when collaborating with a group or organization that creates or uses native CAD files.

For a complete list of all supported data formats, refer to the ArcGIS Help topic "About geographic data formats."

3D features

Points, lines, and polygons can also store data in 3D. Three-dimensional features are points, lines, or polygons that store a z-value as part of their geometry in addition to their x,y coordinates. Three-dimensional vector data also includes multipatch features. Multipatch features are a type of 3D feature used to represent the outer surface (or shell) of features (figure 2.3). 3D features and multipatches can be stored as shapefiles or geodatabase feature classes.

Figure 2.3 Example of 3D multipatch features. Created by the author from Esri online course, "Visualizing 3D Data."

Exercise 2a: Connect to exercise data and turn on file extensions

In order to visualize data in different data formats in the ArcGIS for Desktop Catalog window or the ArcCatalog application, you need to create a connection to the workspace where the data is stored. If you need to distinguish data in different file formats, it is also useful to turn on the display of file extensions.

1. Start ArcCatalog.
2. If necessary, establish a folder connection to your ...\DesktopAssociate folder.

Before you start working with the datasets you received, you will make sure that ArcCatalog displays the file extensions.

3. From the Customize menu, open ArcCatalog Options.

4. At the bottom of the General tab, turn off the Hide file extensions option.

☐ Hide file extensions.
☑ Return to last used location when ArcCatalog starts up.

5. In the Catalog tree, expand to the DesktopAssociate\Data\Chapter02 folder.

The Chapter02 folder contains examples of four different datasets in different file formats storing geographic data: a CAD feature dataset, a coverage, a raster dataset, and a shapefile. The next concept will cover raster datasets in more detail.

6. If you would like, take a few minutes to preview and explore the properties of the datasets.

7. Close ArcCatalog.

Challenge 1

For each of the following scenarios, choose the most appropriate data format for storing geographic data.

1. You are designing a database for a city. Which is the best data format to store street centerlines of the city area, ensuring that street features are properly connected?

   a. Shapefile
   b. Geodatabase feature class in a feature dataset
   c. Geodatabase mosaic dataset
   d. Coverage

2. You are preparing a point dataset of well locations that will be added to a web map in the ArcGIS Online map viewer. Which is the best data format to use?

   a. Geodatabase feature class
   b. CAD dataset
   c. Shapefile
   d. Geodatabase feature class in a feature dataset

3. You are creating a new feature class that will represent universities. One feature may have 55 points representing different buildings across a university campus. What should the user choose as the geometry type?

   a. Point features
   b. Dimension features
   c. Multipoint features
   d. Multipatch features

Raster data

Raster data is stored as raster datasets, also referred to as images if they have been produced by an optical or electronic device, such as a camera or a scanning radiometer. Raster datasets and images can be stored in many formats. The geodatabase provides components that are developed for storing raster datasets and images. There are also many raster formats outside a geodatabase. Each format has different properties and requirements, and supports specialized uses. Some offer high data compression, some handle color better than others, and some are designed specifically to store geographic data. It is beyond the scope of this book to cover all the image file formats, so we will focus on storage formats for raster datasets in general.

Tables 2.3 and 2.4 give you an overview of the main data storage formats in ArcGIS for Desktop for storing raster datasets.

Table 2.3 Data storage formats inside the geodatabase

Data format	What it is	When to use it
Geodatabase raster dataset	The format is a single homogeneous raster dataset stored under the root level of a geodatabase. • This format is organized into one or more bands.	Geodatabase raster datasets are useful for the following: • As a data source for many geoprocessing and analysis tools • For fast display at any scale and for saving space • For easy data management if you have only a few, relatively small rasters to manage • As an attribute of features, such as a picture of a house, associated with the corresponding building footprint polygon
Geodatabase raster catalog	This format is a collection of raster datasets in the geodatabase. (It predates the mosaic dataset. Still in use at ArcGIS 10.0 and 10.1.) • Raster datasets are organized as a geodatabase feature class. Each row in the attribute table represents a raster dataset. • The footprints of the raster datasets are stored as polygons. • Raster datasets can remain in their native format and coordinate system, and do not need to be stored in the geodatabase.	Use raster catalogs to display adjacent or overlapping raster datasets. • Cannot be used as a data source in geoprocessing. • Mostly used for archiving raster datasets.
Geodatabase mosaic dataset	This format is a collection of raster datasets or images that can be viewed either individually or as one on-the-fly mosaicked image. • Raster datasets can remain in their native format on disk or, if required, can be physically loaded into the geodatabase. • This format can be a heterogeneous collection of raster datasets with multiple formats, sources, cell sizes, number of bands, pixel depths, file sizes, and coordinate systems.	Use mosaic datasets for the following: • Managing, querying, and visualizing large collections of raster data • Controlling the display or the ordering of the rasters • On-the-fly processing (Processed as the rasters are accessed. The source pixels are not altered or converted.)

Data formats **21**

Table 2.4 Data storage formats outside the geodatabase		
File format	What it is	When to use it
Esri GRID, TIFF, MrSID, JPG, and many more	This format is a dataset or image stored in a valid raster or image file format.	Store your raster dataset as independent files in the following situations: • When you want to use a few rasters in their native format, typically as a series of raster files • When the datasets are not adjacent to each other or are rarely used on the same project

For a complete list of the supported file-based raster storage formats, refer to the ArcGIS Help topic "Supported raster dataset file formats."

Raster pyramids

An important property of a raster dataset is the presence or absence of raster pyramids (figure 2.4). Raster pyramids are reduced-resolution copies of a raster dataset that are used to improve display performance. At full extent only the coarsest resolution pyramid is drawn. As you zoom in, the software only retrieves the pyramid level at the resolution that is required for the display. When you build raster pyramids, ArcGIS creates downsampled versions of the original raster dataset, where each successive pyramid level is downsampled at a scale of 2:1. Raster pyramids are stored in a file with the same name as that of the raster data and have an OVR (overview file, or .ovr) file extension.

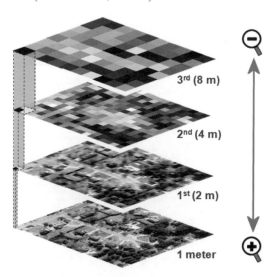

Figure 2.4 In this example, a raster with a 1-by-1-meter cell size has four levels of pyramids built. As you zoom out, notice that the cell size doubles at each level in the pyramid: the first level has a 2-meter resolution, the second level has a 4-meter resolution, and so on. Since each level in the pyramid is a degraded copy of the level below, it draws faster. Esri.

Question 1: What is the purpose of building raster pyramids?

Exercise 2b: Build raster pyramids

In this exercise, you will build raster pyramids for the Yellowstone.tif raster dataset. Yellowstone.tif represents an aerial photo of a volcanic area in Yellowstone National Park.

1. Start ArcMap with a blank map.

2. Open the Catalog window and navigate to your ...\DesktopAssociate\Data\Chapter02 folder.

3. In the catalog tree, select the Yellowstone.tif raster dataset.

Note: If a message appears asking you whether or not to build raster pyramids, click No.

4. Right-click the Yellowstone.tif image and choose Build Pyramids.

The Build Pyramids geoprocessing tool builds raster pyramids based on the Environment settings that have been defined.

5. Click Environments and expand Raster Storage.

To build raster pyramids, ArcGIS needs to resample the Yellowstone.tif raster dataset. Before you build the pyramids, you will change the resampling method.

6. Inspect the Pyramid resampling techniques.

There are three resampling methods available for creating the pyramid levels: nearest neighbor, bilinear, and cubic convolution.

Question 2: For what kind of raster dataset is it appropriate to use the nearest neighbor resampling technique, and for what kind of raster dataset should you use bilinear interpolation or cubic convolution?

Decision point: Which resampling technique(s) are appropriate for the pyramid levels in the Yellowstone.tif image?

To build pyramids for the Yellowstone.tif image you could use bilinear interpolation or cubic convolution.

7. For Pyramid resampling techniques, choose BILINEAR.

8. Make sure that the Build pyramids option is selected.

9. For the pyramid levels, enter 10.

10. Leave the default for the Pyramid compression type, and close the Environment settings.

11. Run the Build Pyramids tool.

It will take a few moments to build the raster pyramids.

12. When the Build Pyramids tool is finished, zoom in on the Yellowstone.tif image and observe its drawing performance.

Once raster pyramids have been built, they are accessed each time the dataset is viewed.

13. Close ArcMap.

Tabular data formats

In ArcGIS, tables are used to store attributes or other information about geographic features.

There are two types of tables used in ArcGIS: One is the attribute table, which is part of a feature class and stores attributes about point, line, and polygon features. Feature class attribute tables contain a Shape field that connects each record in the table to the corresponding point, line, or polygon. Another type of table is a stand-alone table, which is not part of a feature class and does not have a Shape field. Stand-alone tables can store attribute information about features and can also store location information as descriptive text, for example, as street addresses, postal codes, x,y coordinates, or location descriptions.

Similar to feature classes and raster datasets, stand-alone tables can be stored in the geodatabase or in other file formats (tables 2.5 and 2.6).

For a complete list of supported table formats, refer to the ArcGIS Help topic "Tabular data sources."

Table 2.5 Table formats inside the geodatabase		
Data format	What it is	When to use it
Geodatabase table	A table stored directly at the root level of a geodatabase (Nonspatial tables cannot be stored inside of a feature dataset.)	• Typically used in a table association with feature class attribute tables • Can create, read, and edit in ArcGIS for Desktop

Table 2.6 Table formats outside the geodatabase		
File format	What it is	When to use it
dBase, Microsoft Excel, Microsoft Access, text such as TXT, CSV, and many more	A table stored in a file-based data format. • ArcGIS for Desktop can read Microsoft Excel tables directly (read-only). • ArcGIS for Desktop must connect to Microsoft Access tables through an OLE DB connection (read-only).	• Use dBase format for creating and editing file-based tables in ArcGIS. • Use Microsoft Excel or Microsoft Access to collaborate with groups that use these formats.

Challenge 2

For each of the following scenarios, choose the most appropriate dataset and data format to store geographic data.

1. You need to store a single raster that will be used as a base layer for digitizing trails for a project. All project data is being stored in a geodatabase. Which is the best dataset to use?

    a. Mosaic dataset
    b. Raster catalog
    c. Geodatabase raster dataset
    d. Feature dataset

2. You are preparing a stand-alone table for storing owner information of land parcels that occasionally will be joined to a parcel polygon feature class but otherwise will be managed separately by a different group. All groups in the organization work in the same ArcSDE geodatabase.

    a. Excel table
    b. Geodatabase table
    c. Geodatabase feature class
    d. Comma delimited text file (.csv)

3. You need to store and manage 16 TB of TIFF images covering the area of an entire country. Which is the best data format to use?

    a. Raster dataset
    b. GRID
    c. Terrain dataset
    d. Mosaic dataset

Chapter 2: Working with geographic data storage formats

Answers to chapter 2 questions

Question 1: What is the purpose of building raster pyramids?
Answer: Use raster pyramids to increase display performance for large rasters. Raster pyramids are retrieved when the user changes scales. When studying an area at full resolution, no pyramids would be used.

Question 2: For what kind of raster dataset is it appropriate to use the nearest neighbor resampling technique, and for what kind of raster dataset should you use bilinear interpolation or cubic convolution?
Answer: The nearest neighbor resampling technique is the default and works for any type of raster dataset. However, it is recommended that you use nearest neighbor for discrete (nominal) data or raster datasets with color maps, such as land-use data, scanned maps, and pseudocolor images. Bilinear interpolation should be used for continuous data such as elevation. The use of cubic convolution is usually limited to aerial photography.

Answers to challenge questions

Correct answers shown in bold.

Challenge 1

1. You are designing a database for a city. Which is the best data format to store street centerlines of the city area, ensuring that street features are properly connected?

    a. Shapefile
    b. Geodatabase feature class in a feature dataset (Storing the street centerlines as a geodatabase feature class in a feature dataset allows you to build a topology that will check for line features that are not connected.)
    c. Geodatabase mosaic dataset
    d. Coverage

2. You are preparing a point dataset of well locations that will be added to a web map in the ArcGIS Online map viewer. Which is the best data format to use?

    a. Geodatabase feature class
    b. CAD dataset
    c. Shapefile (The ArcGIS Online map view allows you to create a web map combining a basemap with your own data from a zipped shapefile.)
    d. Geodatabase feature class in a feature dataset

3. You are creating a new feature class that will represent universities. One feature may have 55 points representing different buildings across a university campus. What should the user choose as the geometry type?

    a. Point features
    b. Dimension features
    c. Multipoint features (A multipoint feature is a single feature with a single record, but is made up of a number of different points. Using a point feature class would result in 55 separate records for the university. Dimension features are a special kind of geodatabase annotation that is used to show lengths or distances, and they are not appropriate for representing features. Multipatch features are used to represent three-dimensional features.)
    d. Multipatch features

Challenge 2

1. You need to store a single raster that will be used as a base layer for digitizing trails for a project. All project data is being stored in a geodatabase. Which is the best dataset to use?

   a. Mosaic dataset
   b. Raster catalog
   c. Geodatabase raster dataset (Since you are storing only one raster, the raster dataset is the best option. Since all the other project data is stored in a geodatabase, it's best to store the raster dataset in the same geodatabase.)
   d. Feature dataset

2. You are preparing a stand-alone table for storing owner information of land parcels that occasionally will be joined to a parcel polygon feature class but otherwise will be managed separately by a different group. All groups in the organization work in the same ArcSDE geodatabase.

   a. Excel table
   b. Geodatabase table (Storing the owner information table in the same ArcSDE geodatabase as the parcels allows you to create a temporary table association between the parcels feature class and the table.)
   c. Geodatabase feature class
   d. Comma delimited text file (.csv)

3. You need to store and manage 16 TB of TIFF images covering the area of an entire county. Which is the best data format to use?

   a. Raster dataset
   b. GRID
   c. Terrain dataset
   d. Mosaic dataset (A mosaic dataset is well-suited for storing and managing large amounts of raster data.)

Key terms

Data format: The structure used to store a computer file or record.

Raster: Spatial data that defines space as an array of equally sized cells arranged in rows and columns, and composed of single or multiple bands. Each cell contains an attribute value. Unlike a vector structure, which stores coordinates explicitly, raster coordinates are contained in the ordering of the matrix.

Vector: A coordinate-based data model that represents geographic features as points, lines, and polygons. Each point feature is represented as a single coordinate pair, while line and polygon features are represented as ordered lists of vertices. Each vector feature is associated with attributes.

Resources

- ArcGIS Help 10.1 > Geodata > Data types
  - Introduction
    - About geographic data formats
  - Coverage
    - What is a coverage?
  - CAD
    - > Introduction >

Chapter 2: Working with geographic data storage formats

- What is CAD data?
  - Fundamentals of CAD data >
- Elements of a CAD drawing
  - How CAD data is organized
  - Shapefiles
    - What is a shapefile?
  - Rasters and images >
    - Introduction
- What is raster data?
  - Rasters and images >
    - Fundamentals of raster data
- Raster data organization
  - Rasters and images >
    - Supported raster data
- List of supported image and raster data formats
  - Rasters and images >
    - Properties of raster data>
- Raster pyramids
  - Raster pyramids
  - Rasters and images >
    - Building and managing a raster database
- How raster data is stored and managed
  - Tables
    - Tabular data sources
  - Tables
    - Creating and editing tables
    - Creating tables
- ArcGIS Help 10.1 > Geodata > Geodatabases > An overview of the geodatabase
  - Table basics
  - Feature class basics
  - Raster basics

chapter 3

Using the geodatabase

Types of geodatabases 28
   Connecting to an ArcSDE geodatabase
   Exercise 3a: Create a geodatabase
   Building geodatabase schema
   Exercise 3b: Create a stand-alone feature class
   Feature classes inside a feature dataset
   Exercise 3c: Create a feature class in a feature dataset

Defining geodatabase tables 34
   Exercise 3d: Create a stand-alone table
   Specifying fields in an ArcSDE geodatabase
   Challenge 1

Advanced geodatabase elements 37
   Managing lidar data (applies to ArcGIS 10.1 and higher)
   Challenge 2

Answers to chapter 3 questions 40

Answers to challenge questions 40

Key terms 41

Resources 41

Chapter 3: Using the geodatabase

The geodatabase is the native data structure for ArcGIS and is the primary data format used for storing, editing, and managing geographic data in ArcGIS. Why is that, you might ask? Probably the most compelling reason is that the geodatabase lets you create geodatabase elements that allow for modeling real-world behavior with your GIS data.

There are different types of geodatabases that support different data editing workflows and storage needs. However, regardless of which type of geodatabase you choose, the settings for building the structure of the geodatabase and the geodatabase elements you can build to store data will be the same.

Skills measured

- Given a scenario, determine an appropriate name and location for a file geodatabase.
- Given a scenario, determine appropriate specifications for creating a new geodatabase feature class (e.g., geometry type, feature attributes, coordinate system, inside or outside a feature dataset, z-enabled, m-enabled).
- Explain how to specify a coordinate system in ArcGIS for Desktop when creating a new or editing an existing feature class or feature dataset.
- Given a scenario, determine appropriate specifications for creating a new geodatabase feature dataset (e.g., coordinate system, location).
- Given a scenario, determine appropriate specifications for creating a new geodatabase table (e.g., location, name, fields, field types).

Prerequisites

- Knowledge of the main geodatabase elements used to store data in a geodatabase
- Hands-on experience working with data stored in a geodatabase

Types of geodatabases

There are three main types of geodatabases (personal, file, ArcSDE) that have been designed to support different data storage needs and data-editing workflows (figure 3.1). While personal and file geodatabases are considered single-user geodatabases, ArcSDE geodatabases are considered multi-user geodatabases, allowing multiple editors to edit the same data at any given time.

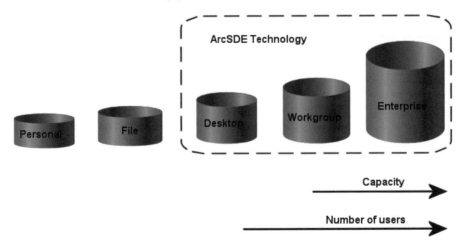

Figure 3.1 The three types of geodatabases (personal, file, and ArcSDE) mainly differ in their storage capacity and in the number of concurrent users and editors. ArcSDE geodatabases require the use of ArcSDE technology. Created by the author.

Tables 3.1 and 3.2 summarize when to use each of the three main geodatabase types.

Table 3.1 Single-user geodatabases

Type of geodatabase	What it is	When to use it
Personal geodatabase	• Original ArcGIS for Desktop database format • A single Microsoft Access file (.mdb) • 2 GB storage limit per database, effective size limits are much lower, between 250–500 MB • Supports one editor per geodatabase at the same time	• For single-user projects and small workgroups with smaller datasets • For working on a Windows platform only
File geodatabase	• Improved ArcGIS for Desktop database format • Folder in a file system that holds binary files • 1 TB storage limit per dataset (can be configured to 256 TB), no storage limit per database • Supports many readers, and one editor per feature dataset, stand-alone feature class, or table	• For project databases and small workgroups with moderate to larger size datasets • Improved desktop format: recommended over the personal geodatabase • Can support multiple editors: design a workflow that ensures that editors edit different feature datasets, stand-alone feature classes, or tables at the same time

Table 3.2 ArcSDE geodatabases

Type of geodatabase	What it is	When to use it
Enterprise	• Datasets stored in a relational database management system (RDBMS), such as Oracle, Microsoft SQL Server, IBM DB2, IBM Informix, or PostgresSQL • No storage limit	• Unlimited users and editors • For larger organizations with a large number of users, storing large datasets, potentially on a central server • For versioned workflows, e.g., for isolating multiple edit sessions, automatic archiving, or historical queries
Workgroup	• Datasets stored in Microsoft SQL Express (free download) • Storage up to 10 GB	• Ten concurrent users, all can edit at the same time • For small to medium-sized organizations

Continued on next page

Table 3.2 ArcSDE geodatabases (continued)		
Type of geodatabase	What it is	When to use it
Desktop	• Datasets stored in Microsoft SQL Express (free download) • Storage up to 4 GB	• Three concurrent users, only one can edit at the time • For using versioned workflows on a local drive: e.g., a user would like to check out a version of a feature class from an enterprise geodatabase to a local drive, take the data to the field and edit it, and then check the edited data back in

For more information on geodatabase types, refer to the ArcGIS Help topic "Types of geodatabases."

Connecting to an ArcSDE geodatabase

The workflow for connecting to an ArcSDE geodatabase differs slightly depending on the type of geodatabase you connect to:

- To connect with an ArcSDE enterprise geodatabase, use the Add Database Connection tool from the Database Connections node in ArcCatalog or the Catalog window. After you enter database information, such as the instance, username, and password, a spatial database connection file (file extension .sde) is created in your user profile. The .sde connection file can then be placed in any computer's user profile or placed on a centralized server for all users to access.

- To create a connection to an ArcSDE Desktop or Workgroup geodatabase, you use the Add Database Server tool from the Catalog's Database Servers node. Once you have established a connection to a database server, you can attach a Desktop or Workgroup geodatabase. If you choose to save the connection to the attached Desktop or Workgroup geodatabase, ArcGIS will create a connection file in your user profile.

Exercise 3a: Create a geodatabase

Suppose you are creating a geodatabase for a data collection project in Yellowstone National Park. Eventually, you will want to store a copy of the geodatabase on several laptop computers. Every project participant will edit their own copy of the data, but everyone will work on feature classes. You don't know how large the final geodatabase will be.

Decision point: What type of geodatabase is appropriate for this project?

A file geodatabase is the most appropriate geodatabase type for the project. Since the geodatabase will be stored locally on laptop computers and only one person at a time needs to edit the data, a single user geodatabase is appropriate. Since you don't want to reach the 2 GB storage limit of a personal geodatabase, the file geodatabase is the best choice for a single user geodatabase. For new geodatabases, the file geodatabase is also recommended over a personal geodatabase because of its more efficient storage format and better performance.

1. Start ArcMap and open the Create File GDB geoprocessing tool.

Hint: The tool is located in the Data Management tools and the Workspace toolset.

Decision point: What would be an appropriate location for creating a file geodatabase?

You can create a file geodatabase in any folder on your file system. Since all the data used for this exercise resides in the ...\DesktopAssociate\Chapter03 folder, it is best to create the file geodatabase in the same location.

2. For File GDB Location, click the Browse button and navigate to your ...\DesktopAssociate folder.

3. Click the Chapter03 folder, and then click Add.

For project geodatabases, it is best to use a descriptive name.

4. For the File GDB Name, type **ProjectMaster**.

Question 1: What are some naming restrictions for a file geodatabase?

By default, the new file geodatabase will be created in the current version of ArcGIS. Optionally, you could create a file geodatabase that is compatible with previous versions of ArcGIS.

5. For File GDB Version (optional), accept the default of CURRENT.

6. Click OK to create the file geodatabase.

Question 2: Can you think of another way to create a file geodatabase?

7. If you are planning to do the next exercise, leave ArcMap open.

Building geodatabase schema

Geodatabase schema is the structure or design of the geodatabase. It includes the structure of its feature datasets, feature classes, and tables; the fields in each table; and the relationships between fields and tables. In other words, the term schema describes the structure of a feature dataset, feature class, or table—including their field definitions—or the structure of the geodatabase as a whole.

When you create the schema of a geodatabase, you start by defining the schema of its components. For example, for a feature class, you define its geometry type; spatial reference; x-, y-, and z-tolerances; storage configuration; and attribute field definitions.

Exercise 3b: Create a stand-alone feature class

You can create a feature class by using the Create Feature Class geoprocessing tool or through ArcCatalog, which you will use in this exercise. However, no matter which method you choose, the parameters for creating the new feature class will be the same.

Dependency: You need to complete exercise 3a before running this exercise. If you have not completed exercise 3a, please do so before starting this exercise.

1. If necessary, start ArcMap and navigate to your ...\DesktopAssociate\Chapter03 folder.

2. In the Catalog window, create a new feature class in the ProjectMaster.gdb and name it **Rivers**.

Hint: Right-click the ProjectMaster.gdb geodatabase and point to New.

The geometry of the new feature class will determine what type of features can be stored in the feature class.

3. Click the Type drop-down arrow and examine the available geometry types.

Decision point: What geometry type is most appropriate for storing rivers at the scale and extent of the entire Yellowstone National Park?

___

Line features are the most appropriate geometry type for rivers. (**Note:** If you wanted to store rivers to be used at a larger scale, you could also use polygon geometry.)

4. Set the geometry type to Line Features.

Decision point: One of the specifications that you received for your new feature class is to enable it for storing elevation values. Which type of geometry property values should you include in the coordinates?

___

In order to enable the feature class for storing 3D features, you have to include z-values.

5. At the bottom of the panel, check Coordinates include z-values to store 3D data.

The next panel prompts you to choose the coordinate system that will be used for the x,y coordinates of the data in the new feature class.

Note: A spatial reference describes where features are located in the real world. It includes a coordinate system for the x-,y-, and z- coordinate values as well as tolerance values and extents for the x-, y-, and z-values. For more information about coordinate systems, refer to chapter 5 of this book.

You can select the coordinate system from a list of predefined coordinate systems, import the coordinate system from an existing dataset, or search for a coordinate system (in ArcGIS 10.1).

6. Import the coordinate system from the ...\DesktopAssociate\Chapter03 folder and select **YellowstoneRivers.shp**.

Note: In ArcGIS 10.0, click the Import button. In ArcGIS 10.1, click the Add Coordinate System button and choose Import.

The North American Datum 1983 (NAD83) UTM Zone 12 North projected coordinate system will be assigned for the x,y coordinate system of the new feature class.

In the next panel, you will specify a vertical coordinate system. Since you don't have any existing data sources available to import the vertical coordinate system, you will set the vertical coordinate system manually.

7. If you are working in ArcGIS 10.1, search for **NAD 1983**. If you are working in ArcGIS 10.0, continue with step 8.

8. Expand the Vertical Coordinate Systems/North America folder, and then select NAD1983 for the vertical coordinate system.

9. The next panel prompts you to specify the x- and y-tolerance values.

Decision point: What are the recommended x- and y-tolerance values to obtain good results for most cases?

___

It is recommended to keep the default tolerance, which is 0.001 meters (or 1 millimeter), or the equivalent in the coordinate system's units, for example, 0.003281 feet.

10. Accept the default x-, y-, and z-tolerance values. Also accept the default resolution and domain extent.

The next panel prompts you for a configuration keyword.

11. Accept the Default configuration keyword.

The next panel lets you define the attribute fields for the new feature class.

12. Import the fields from the ...\DesktopAssociate\Chapter03\YellowstoneRivers.shp.

13. The field names and data types of these fields have been copied from the corresponding fields in the shapefile.

14. **Finish creating the feature class.**

The new Rivers feature class does not contain any features yet. The attribute table contains only the attribute fields, but no records. In chapter 4 of this book, you will review how to add data into an existing feature class.

15. **If you are planning to do the next exercise, leave ArcMap open.**

> ### *Feature classes inside a feature dataset*
> Use feature datasets to store feature classes that are thematically and topologically related. Topological relationships among feature classes in a feature dataset include coincidence, connectivity, adjacency, or containment. For example, streams in a particular watershed are connected to rivers; therefore, streams and rivers are topologically related.
>
> Storing feature classes in a feature dataset provides the underlying structure for adding advanced geodatabase elements, such as topologies, geometric networks, network datasets, terrain datasets, or parcel fabrics.

Exercise 3c: Create a feature class in a feature dataset

Suppose you want to create a new feature dataset to store transportation data and then create several feature classes inside it.

1. **If necessary, start ArcMap and navigate to your ...\DesktopAssociate\Chapter03 folder in the Catalog window.**

2. **Create a new feature dataset in the ProjectMaster.gdb and name it Transportation.**

Hint: Right-click the ProjectMaster.gdb geodatabase and point to New.

In the next panel, you define the x,y coordinate system of the new feature dataset. You will specify the NAD 1983 UTM Zone 12N coordinate system for the new feature dataset.

3. **Select the NAD 1983 UTM Zone 12N coordinate system from the Projected Coordinate Systems\UTM\ NAD1983 folder.**

You will not specify a vertical coordinate system for this feature dataset.

4. **Click Next.**

The next panel prompts you to specify the x- and y-tolerance values for the new feature dataset.

5. **Accept the default values for the x-, y-, z-, and m-tolerances. Also accept the default resolution and domain extent.**

6. **Finish creating the feature dataset.**

To create feature classes inside of the Transportation feature dataset, you will use the Create Feature Class geoprocessing tool, which has the advantage that you can run it in batch mode, creating all three feature classes by running the tool only once.

7. **Open the Create Feature Class geoprocessing tool in batch mode.**

Hint: In ArcGIS 10.1, open the Search window and locate the Create Feature Class geoprocessing tool. Right-click the Create Feature Class tool and choose Batch. In ArcGIS 10.0, navigate to the System Toolboxes,

Data Management Tools toolbox, Feature Class toolset. Then right-click the Create Feature Class tool and choose Batch.

8. **Fill out the dialog box to create three new feature classes in the ...\DesktopAssociate \Chapter03\ ProjectMaster.gdb\Transportation feature dataset. Name them the following:**
   - Roads
   - BusRoutes
   - BusStops

Hint: Right-click the cell under Feature Class Location and choose Browse. Then navigate to the ProjectMaster/Transportation feature dataset. To specify the feature class name, type it in under Feature Class Name. Click the plus sign button to add another row to the table.

Decision point: What will be the geometry types for roads, bus routes, and bus stops?

Roads and bus lines will be polylines (line features with more than two vertices), and bus stops will have point geometry.

9. **Assign POLYLINE for the geometry type of the Roads and BusRoutes feature classes. Assign POINT geometry to the BusStops feature class.**

Hint: Click the cell under Geometry Type twice to display the geometry type drop-down arrow.

10. **Notice that you have the option to specify a template feature class.**

Choosing a template feature class will allow you to import the attribute fields for the new feature classes from existing datasets. You don't have any templates for the new feature classes, so you will not import the attribute fields for now.

11. **Scroll to the right and examine the remaining columns.**

12. **Leave the default DISABLED value in the Has M and Has Z cells.**

You do not have to specify a coordinate system for your new feature classes. They will be automatically created in the coordinate system of the Transportation feature dataset.

13. **Run the Create Feature Class batch tool.**

14. **When the tool is finished, open the properties of one of the new feature classes in the Catalog window and activate the XY Coordinate System tab.**

The feature classes inherited the NAD 1983 UTM Zone 12N coordinate system from the feature dataset.

15. **Click the Fields tab.**

The feature classes were created with only the OBJECTID, Shape, and Shape_Length fields, which are automatically generated system fields. To specify additional user-defined fields, you would type in the field names and set their types in this tab.

16. **Close the feature class Properties.**

17. **If you are planning to do the next exercise, leave ArcMap open. Otherwise, close ArcMap.**

Defining geodatabase tables

The geodatabase stores information about geographic features in two types of tables: spatial tables and nonspatial tables. The feature class attributes tables that you created in the previous exercise are examples of spatial tables. They contain a Shape field that defines the geometry of the features and is an integral part of the feature class. Nonspatial tables do not contain a Shape field and are always stored directly under the root level of the geodatabase.

To specify user-defined fields in both spatial and nonspatial tables, you select a field data type. This will determine which type of data that field will store.

Table 3.3 lists the most common data types supported for fields in the geodatabase. The ranges listed are for file and personal geodatabases. Ranges differ slightly in ArcSDE geodatabases.

Table 3.3 Data types supported in geodatabase fields

Data type	Specific range, length, or format	Size (bytes)	Applications
Short integer	−32,768 to 32,767	2	Numbers without fractions within specific range, coded values
Long integer	−2,147,483,648 to 2,147,483,647	4	Numbers without fractions within specific range
Single-precision floating point number (Float)	Approx. −3.4E38 to 1.2E38*	4	Numbers with fractions within specific range
Double-precision floating point number (Double)	Approx. −2.2E308 to 1.8E308*	8	Numbers with fractions within specific range
Text	Up to 64,000 characters	Varies	Names or other textual qualities
Date	mm/dd/yyyy hh:mm:ss AM/PM	8	Date and/or time
BLOB (Binary Large Object)[†]	Varies	Varies	Images or other multimedia
Raster	Varies	Varies	Images or pictures
GUID	36 characters, enclosed in curly brackets	16 or 38	Customized applications requiring global identifiers

[†] A large block of data, such as an image, a sound file, or geometry, stored in a database. The database cannot read the BLOB's structure and only references it by its size and location.

For a complete list of supported field data types, refer to the ArcGIS Help topic "Geodatabase field data types."

Exercise 3d: Create a stand-alone table

Suppose you want to create a stand-alone table to store information about visitors of Yellowstone National Park. The park administration will use this table for planning purposes.

Decision point: Where in the Yellowstone geodatabase should you create the nonspatial table?

You will create the nonspatial table at the root level of the geodatabase. Nonspatial tables cannot be stored in a feature dataset.

1. In the Catalog window, right-click the ProjectMaster.gdb geodatabase and create a new table.

2. **Create a table with the following specifications:**
   - Name: VisitorInformation
   - Alias: Visitor Information
   - Configuration Keyword: Default
   - Fields:
     - OBJECTID
     - VISITORS_TOTAL
     - NOTES

You will create two fields in the new table: one for the total number of visitors in the park per year and another one for notes about the visitors in the park (for example, whether the number increased or decreased compared to the previous year). You will start with the field for the yearly number of total visitors in the park.

Hint: Press the Tab key after typing in the field name.

3. Notice that Text is the default data type for the VISITORS_TOTAL and NOTES fields.

Decision point: In 2010, more than 3.4 million people visited Yellowstone National Park. What is an appropriate data type for the VISITORS_TOTAL field?

Since the number of visitors will be a whole number, it is best to use an integer field. Since the total number of visitors per year will most likely be above 32,767, long integer will be the appropriate data type.

Decision point: The NOTES field will store notes about visitors in the park. What is an appropriate data type for the NOTES field?

To store notes about visitors in the park, text is the appropriate data type.

4. Set the data type of the VISITORS_TOTAL and the NOTES fields to the appropriate data types.

Text fields are created with a default length of 50 characters, which means that a maximum of 50 characters can be stored in the field. You will increase the length of the NOTES field to 300 characters.

5. In the Field Properties area of the NOTES field, change the field's length to 300.

6. Click Finish.

The new table is added to the root level of the ProjectMaster.gdb geodatabase.

7. Close ArcMap.

Specifying fields in an ArcSDE geodatabase

If you're specifying numeric fields for a table in a file or personal geodatabase, you need only specify the data type. If you're specifying numeric fields for a multiuser ArcSDE geodatabase, you should also specify the precision, which is the maximum length of the field, and scale, which is the maximum number of decimal places. For example, if you specify a float field with a precision of 4 and a scale of 2, the field will accept 12.34. If you try to enter 12.345 into the field, an error message will display, as this exceeds the maximum number of digits and decimal places allowed. On the other hand, if you specify a float with a precision of 5 and a scale of 3, you will be able to enter 12.345.

Table 3.4 lists data types and their possible precision and scale values for ArcSDE geodatabases.

Table 3.4 ArcSDE data types		
Data type	Precision (field length)	Scale (decimal places)
Short Integer	1–5 (Oracle, SQL Server, PostgreSQL); 5 (DB2, Informix)	0
Long Integer	6–10 (Oracle and PostgreSQL); 6–9 (DB2, Informix, and SQL Server)	0
Float	1–6	1–6
Double	7+	0+

Challenge 1

For each of the following scenarios, choose the most appropriate data storage option.

1. You are creating a new field in a geodatabase feature class attribute table to store the latitude and longitude values of the polygon centroids in a polygon feature class. Which data type should you use?

   a. Long integer
   b. Text
   c. Float
   d. BLOB

2. You are creating a new ArcSDE geodatabase table with a precision of 6 and a scale of 2. Which of the numbers below can be stored in that field?

   a. 63.1234
   b. 6312.34
   c. 6312.1234
   d. 6312.345

Advanced geodatabase elements

A key aspect of the geodatabase is that it allows you to model real-world behavior in your GIS data. For example, you can model how a taxi driver maneuvers through a street network or the adjacency of parcels without any gap or overlap.

Chapter 3: Using the geodatabase

One way to implement behavior in the geodatabase is by creating elements in the geodatabase that manage the spatial and attribute integrity of the data and make the data "behave" similar to the real-world entities that they represent. These elements are special datasets that extend the capabilities of feature classes, feature datasets, and tables. Table 3.5 below lists some these geodatabase datasets.

Table 3.5 Geodatabases datasets			
Dataset	What it is	When to use it
Geometric network	A collection of point and line features that comprises a connected system of edges and junctions • Model network systems with directional (one directional or indeterminate) flow, e.g., utility networks or hydrologic networks	• Use if the elements move through the network without making decisions, e.g., for water, gas, electricity, or river networks • Use to perform tracing operations, e.g., upstream or downstream tracing, or model problems on the network by disabling some parts and finding alternate paths through the network
Network dataset	A collection of topologically connected network elements (edges, junctions, and turns) that are derived from simple line feature classes • Models transportation networks such as road or subway systems with undirected flow • Network type used by the ArcGIS Network Analyst extension	• Use if the elements that move through the network make decisions, e.g., to analyze route and travel time of a delivery truck • Use to model multimodal networks with different transportation modes, e.g., find the shortest route for driving your car to the closest train station, taking the train, and then walking to your workplace • Use to model 3D networks, e.g., hallways and staircases in a multistory building
Parcel fabric	A dataset that models the spatial relationships for lot lines, parcel corners, and parcel polygons as an interconnected fabric • Parcels in a fabric are defined by polygon features, line features, and point features • Stores survey information for subdivisions and parcel plans	• Use for managing and editing a spatially integrated set of features, such as land parcels covering an area • Use to preserve parcel history
Terrain dataset	A multiresolution surface, based on a triangulated irregular network (TIN)	• Use to model surfaces of point, line, and polygon features with z-values • Use to manage and integrate massive point collections of 3D data, for example, billions of elevation points collected through lidar (light detection and ranging) sensors

Table 3.5 illustrations from Esri.

> ## *Managing lidar data*
> ## *(applies only to ArcGIS 10.1 and higher)*
>
> Lidar (light detection and ranging) data is becoming more common and more affordable, and therefore more frequently used in GIS workflows. Lidar data usually comes as a collection of LAS files, which in ArcGIS 10.1 can be managed in ArcGIS for Desktop while remaining in its native format. There are two general ways to manage a collection of LAS files and treat them as a single unit: you can either reference them into a geodatabase mosaic dataset, or you can combine them into a file-based LAS dataset.
>
> A mosaic dataset allows you to create a raster-based view of the LAS files and manage them similarly to how you would manage imagery: you will be able to see geographically where you have lidar data, identify which LAS files cover a specific area, and download them, if necessary.
>
> A LAS dataset is a file outside the geodatabase and provides vector-based access to LAS files. That is, you can interact with a collection of LAS files either as individual points or as a TIN-based surface. LAS datasets will allow you to do the following:
>
> - View lidar points directly or as a high-resolution surface in 3D
> - Analyze lidar surfaces, such as the first-return surfaces or bare-earth surfaces
> - Edit LAS class code attributes for individual points
> - Provide a backdrop for feature extraction or digitizing new features

Challenge 2

For each of the following scenarios, choose the most appropriate type of geodatabase dataset to store the data in.

1. You need to find the shortest route from an accident location to the nearest hospital. Which dataset should you use to do this?

    a. Geometric network
    b. Triangulated irregular network
    c. Network dataset
    d. Geodatabase line feature class

2. You have a 3D point dataset in which the z-values represent the first return from lidar data. What is the best data format to model a vegetation canopy surface of these point features?

    a. LAS dataset
    b. Network dataset
    c. Terrain dataset
    d. Triangulated irregular network

3. You need to trace a water network upstream from a broken water main to find the valve that needs to be shut off for the repairs. Which dataset is best for doing this?

    a. Geometric network
    b. Triangulated irregular network
    c. Network dataset
    d. Geodatabase line feature class

Answers to chapter 3 questions

Question 1: What are some naming restrictions for a file geodatabase?
Answer: Since a file geodatabase is really just a folder on your file system, you can give it any name that is allowed by your operating system. For example in Windows (using the NTFS file system), the characters / ? < > \ : * | " and any character that you type with your Ctrl key are illegal. Under Windows using the NTFS file system, folder names may be up to 256 characters long. For more information on Windows naming conventions, refer to the Microsoft Developer Network website at **http://msdn.microsoft.com** and search for Naming Files, Paths, and Namespaces. Also, for project geodatabases, it is best to use a descriptive name.

Question 2: Can you think of another way to create a file geodatabase?
Answer: An alternative way to create a file geodatabase in ArcCatalog is to navigate to the folder where you want to create it, right-click the folder, point to New, and then click File Geodatabase. When you create a new file geodatabase in ArcCatalog, it is automatically created in the current version of ArcGIS for Desktop with a default name of New File Geodatabase.gdb, which you can change.

Answers to challenge questions

Correct answers shown in bold.

Challenge 1

1. You are creating a new field in a geodatabase feature class attribute table to store the latitude and longitude values of the polygon centroids in a polygon feature class. Which data type should you use?

   a. Long integer
   b. Text
   c. Float (You would use Float because the latitude and longitude values are numbers with decimal places.)
   d. BLOB

2. You are creating a new ArcSDE geodatabase table with a precision of 6 and a scale of 2. Which of the numbers below can be stored in that field?

   a. 63.1234
   b. 6312.34 (The precision of 6 determines that you can store up to 6 digits in the field. The scale of 2 allows you to store 2 decimal places.)
   c. 6312.1234
   d. 6312.345

Challenge 2

1. You need to find the shortest route from an accident location to the nearest hospital. Which dataset should you use to do this?

   a. Geometric network
   b. Triangulated irregular network
   c. Network dataset (A network dataset allows you to model transportation networks and find the route between two locations.)
   d. Geodatabase line feature class

2. You have a 3D point dataset in which the z-values represent the first return from lidar data. What is the best data format to model a vegetation canopy surface of these point features.

   a. LAS dataset (A terrain dataset allows you to model an elevation surface from lidar data, but a LAS dataset provides the highest resolution surface data.)
   b. Network dataset
   c. Terrain dataset
   d. Triangulated irregular network

3. You need to trace a water network upstream from a broken water main to find the valve that needs to be shut off for the repairs. Which dataset is best for doing this?

   a. Geometric network (A geometric network allows you to model one directional flow between two locations.)
   b. Triangulated irregular network
   c. Network dataset
   d. Geodatabase line feature class

Key terms

Geodatabase: A database or file structure used primarily to store, query, and manipulate spatial data. Geodatabases store geometry, a spatial reference system, attributes, and behavioral rules for data. Various types of geographic datasets can be collected within a geodatabase, including feature classes, attribute tables, raster datasets, network datasets, topologies, and many others. Geodatabases can be stored in IBM DB2, IBM Informix, Oracle, Microsoft Access, Microsoft SQL Server, and PostgreSQL relational database management systems, or in a system of files, such as a file geodatabase.

Feature class: In ArcGIS, a collection of geographic features with the same geometry type (such as point, line, or polygon), the same attributes, and the same spatial reference. Feature classes can be stored in geodatabases, shapefiles, coverages, or other data formats. For example, highways, primary roads, and secondary roads can be grouped into a line feature class named "roads." In a geodatabase, feature classes can also store annotation and dimensions.

Feature dataset: In ArcGIS, a collection of feature classes stored together that share the same spatial reference; that is, they share a coordinate system and their features fall within a common geographic area. Feature classes with different geometry types may be stored in a feature dataset.

Schema: The structure or design of a geodatabase or geodatabase object, such as a feature dataset, feature class, or table.

Resources

- ArcGIS Help 10.1 > Geodatabases > An overview of the geodatabase
  - What is the geodatabase?
  - A quick tour of the geodatabase
  - Types of geodatabases
- ArcGIS Help 10.1 > Geodatabases > Defining the properties of data in a geodatabase > Geodatabase table properties
  - ArcGIS field data types

- ArcGIS Help 10.1 > Geodata > Data types
  - Geometric networks
    - What are geometric networks?
  - LAS dataset
    - What is a LAS dataset?
  - Network datasets
    - What is a network dataset?
  - TIN
    - What is a TIN surface?
- ArcGIS Help 10.1 > Administering geodatabases > Geodatabases in SQL Server > Connecting to a geodatabase in SQL Server
  - Geodatabase Connections in ArcGIS Desktop
- ArcGIS Help 10.1 > Extensions> Network Analyst > Guide books > Network Analyst tutorial: Exercise 1L Build a network dataset
- Michael Zeiler. *Modeling Our World: The ESRI Guide to Geodatabase Concepts*, second edition. (Redlands, CA: Esri Press, 2010). Chapter 1 can be downloaded for free, **http://esripress.esri.com/display/index.cfm?fuseaction=display&websiteID=178&moduleID=29**.
- Esri product page: Geodatabases—Data Storage

chapter 4

Managing geographic data

Methods for adding data to the geodatabase   44
    Copy and paste
    Export/Import
        Exercise 4a: Export multiple shapefiles into the geodatabase
        Exercise 4b: Export a shapefile into a feature dataset
    Export to XML
        Exercise 4c: Export a geodatabase to XML
    Export/Import nonspatial data
    Loading data into a geodatabase feature class
        Exercise 4d: Load features using the Simple Data Loader
        Exercise 4e: Load features using the Append tool
        Exercise 4f: Load features using the Object Loader
    Snapping
    Challenge

Creating data from layers   55
    Exporting layers
        Exercise 4g: Export a subset of a layer
        Exercise 4h: Export the attribute table of a layer
    Layer files and layer packages
        Exercise 4i: Set layer properties and create a layer package

Layer source data   60
        Exercise 4j: Change data sources for a layer

Answers to chapter 4 questions   62

Answers to challenge questions   63

Key terms   64

Resources   64

Chapter 4: Managing geographic data

Exporting plays a major role in data management. You can export a feature class to change its name and storage format—for example, export from shapefile to geodatabase format or vice versa. Or, you can export from geodatabase to XML format to share geodatabase data with others. When you add feature classes as layers to ArcMap, you can export layers in the coordinate system of the data frame, or you can export a subset of the layer.

You will now zero in on three different aspects of data management: First, you will look at the data management tools that help you add data to the geodatabase. You will also explore how ArcMap layers can support data management, and then look into managing the relationship between layers and their source data.

Skills measured
- Given a scenario, determine appropriate specifications (selections, input dataset, format, coordinate system, output location, output format) for exporting spatial and nonspatial data.
- Given a scenario, determine appropriate methods and specifications for updating layer data sources.
- Identify layer properties that are potentially affected by changing the data source.
- Given a scenario, determine appropriate loading process, tools, and environment for loading data into a feature class or table and validating the success of the data load.

Prerequisites
- Knowledge of common data types and storage formats used in ArcGIS
- Knowledge of the main elements stored in a geodatabase
- Hands-on experience working with layers in ArcMap

Methods for adding data to a geodatabase

There are a variety of methods for adding data to the geodatabase. You can add existing data from data formats such as CAD or shapefile into the geodatabase; you can move data between different types of geodatabases, for example, between ArcSDE and file geodatabases; or you can create new data from scratch by digitizing it in ArcMap.

The methods below give you an overview of the four different methods of adding existing data to the geodatabase:
- Copy and paste
- Export/Import
- Export/Import to/from XML
- Load into feature class/table

Copy and paste
Copy and paste is one of the most straightforward methods to bring data into a geodatabase. Provided that their geometries match, you can quickly transfer them from one feature class to another by copying and pasting features inside of an edit session. You can also copy and paste at the dataset level. Provided that a feature class, table, or raster dataset is already in a geodatabase, you can copy and paste it to another geodatabase. Table 4.1 summarizes the different copy and paste methods.

Export/Import
The export and import methods create a new feature class, table, or raster dataset in a geodatabase or a new feature class in a feature dataset. Export and import are equivalent methods to bring data into the geodatabase. For example, to bring a shapefile or CAD feature class into the geodatabase, you can either export them to geodatabase format or import to the geodatabase. Alternatively, you can bring datasets into the geodatabase using the geoprocessing tools from the Conversion Tools toolbox, To Geodatabase toolset. These geoprocessing tools include specific tools for tasks such as bringing CAD or raster data into the geodatabase. They honor geoprocessing environment settings such as output workspace, output coordinate system, and output extent, and can also be used in geoprocessing models.

Table 4.2 summarizes the export/import methods.

Table 4.1 Methods for copying and pasting feature classes, tables, and raster datasets

Method	Outcome
Copy/Paste feature(s)	Adds features to an existing feature class
Copy/Paste feature class, table, or raster dataset	Creates a new feature class, table, or raster dataset in a geodatabase

Table 4.2 Methods for exporting and importing feature classes, tables, and raster datasets

Method	Outcome	Condition	When to use
Export or import a feature class, table, or raster dataset	Creates a new feature class, table, or raster dataset in a geodatabase or a new feature class in a feature dataset.	When exporting or importing into a feature dataset, note the following: • Source data will be automatically projected into the projected coordinate system of the feature dataset • Geographic coordinate system will not automatically be transformed (You must perform a geographic transformation before exporting/importing.)	• Use when there is no preexisting geodatabase feature class, table, or raster dataset • Use for converting entire datasets from file formats into geodatabase format • Use to bring a stand-alone geodatabase feature class into a feature dataset
Conversion geoprocessing tools	Creates a new feature class, table, raster dataset in the geodatabase	When bringing a feature dataset, note the following: • Source data will be automatically projected into the projected coordinate system of the feature dataset • Geographic coordinate system will not automatically be transformed (You must perform a geographic transformation before exporting/importing.)	Use to convert datasets from file formats into geodatabase format, e.g., to convert CAD feature classes, shapefiles, coverages, raster datasets, dBase tables, and XML files to geodatabase format

Both the export/import and the Conversion geoprocessing tools can be run either on a single source dataset or in batch mode to convert multiple feature classes, rasters, and tables at the same time.

In the next set of exercises, you return to the Yellowstone geodatabase project. Recall that in the previous chapter, you created a file geodatabase for Yellowstone National Park. Now you will bring the data into your Yellowstone geodatabase that you obtained from the park administration and the surrounding counties.

Exercise 4a: Export multiple shapefiles into the geodatabase

You will start by converting some shapefiles to geodatabase feature classes by exporting them to the Yellowstone geodatabase.

1. Start ArcMap with a blank map. In the Catalog window, navigate to your ...\DesktopAssociate\ Chapter04\NewData folder.

2. The NewData folder contains a number of shapefiles with data about Yellowstone National Park.

3. Right-click the Park Area.shp, point to Export, and click To Geodatabase (multiple).

4. The Feature Class to Geodatabase (multiple) tool opens with the ParkArea.shp already added to the list of input feature classes. You will add Trails.shp as additional input.

5. For Input Features, add the ...\Chapter04\NewData\Trails.shp.

6. For the Output geodatabase, add the ... \Chapter04\Yellowstone.gdb geodatabase.

7. Run the tool.

8. When the export process completes, expand Yellowstone.gdb and add the new ParkArea and the Trails geodatabase feature classes as layers to the map.

Tip: To add feature classes as layers to ArcMap, you can just drag them from the Catalog window into the map.

9. Open the attribute tables of both layers and scroll to the right end.

When you brought the feature classes into the geodatabase, a Shape_Length field was added to both attribute tables. In addition, a Shape_Area field was added to the ParkArea feature class. The Shape_Length and Shape_ Area fields are added to the attribute table when a feature class is stored in the geodatabase and automatically updated when line or polygon geometry is edited.

10. Close the table window.

If a feature class is already in a geodatabase, you can copy and paste (or drag and drop) it into another geodatabase instead of exporting or importing it. You will copy and paste the Campgrounds feature class from a personal geodatabase into the Yellowstone file geodatabase.

11. In the Chapter04 folder, expand the Camps.mdb personal geodatabase.

12. Right-click the Campgrounds feature class and choose Copy.

13. Right-click the Yellowstone geodatabase and choose Paste.

14. Click OK to transfer the Campgrounds feature class into the Yellowstone geodatabase.

15. If you are planning to do the next exercise, leave ArcMap open. Otherwise, close ArcMap.

So far, you brought shapefiles into the geodatabase and stored them directly at the root level of the geodatabase. Feature classes at the root level of the geodatabase can have any coordinate system. However, if you export or import feature classes into a geodatabase feature dataset, they will be automatically projected into the coordinate system of the feature dataset. Also, if the feature dataset has a different geographic coordinate system, keep in mind that a geographic transformation must be performed manually prior to bringing in the feature classes.

Exercise 4b: Export a shapefile into a feature dataset

In the previous exercise, you exported shapefiles to geodatabase format. Now, you will export the MajorRoads.shp shapefile into the Transportation feature dataset.

1. If necessary, start ArcMap, and navigate to your ...\DesktopAssociate\Chapter04 folder.

2. In the Yellowstone.gdb geodatabase, expand the Transportation feature dataset.

Methods for adding data to a geodatabase   **47**

The Transportation feature dataset contains empty feature classes for bus lines and bus stops. Remember, you created these feature classes in the previous chapter.

   3. **In the ...Chapter04\NewData folder, open the properties of the MajorRoads.shp. Click the XY Coordinate System tab.**

Hint: Right-click MajorRoads.shp and choose Properties.

Major roads are in the NAD 1983 geographic coordinate system.

   4. **Close the feature class properties.**

   5. **Open the properties and the XY Coordinate System tab of the Yellowstone.gdb\Transportation feature dataset and inspect its coordinate system information.**

The coordinate system of the Transportation feature dataset is NAD 1983 (North American Datum 1983) UTM Zone 12N projected coordinate system. When you export the MajorRoads shapefile into the Transportation feature dataset, ArcGIS will automatically project it into the projected coordinate system of the feature dataset. You will review geographic and projected coordinate systems in the next chapter.

   6. **Close the Feature Dataset Properties.**

Next, you will export the MajorRoads.shp shapefile to the Yellowstone.gdb/Transportation feature dataset.

Question 1: Why do you store major roads in the Transportation feature dataset in the Yellowstone geodatabase?

   7. **Right-click the MajorRoads.shp, point to Export, and choose To Geodatabase (single).**

   8. **Run the tool with the following parameters:**
- Output Location: ...Chapter04\Yellowstone.gdb\Transportation
- Output Feature Class: MajorRoads.

   9. **Open the Properties of the new ...Yellowstone.gdb\Transportation\MajorRoads feature class. The x,y coordinate system is NAD_1983_UTM_Zone_12N. This is the same coordinate system as one of the Transportation feature dataset.**

  10. **Close the feature class properties.**

Question 2: The geographic coordinate system of both the original MajorRoads shapefile and the Transportation feature dataset is NAD 1983. What additional operation would you need to perform to export a shapefile to a feature dataset if their geographic coordinate systems did not match?

  11. **If you are planning to do the next exercise, leave ArcMap open. Otherwise, close ArcMap.**

Export to XML

So far in this lesson you brought data from file formats into geodatabase format. However, suppose that your data is in geodatabase format, and you would like to share data with others. Exporting your geodatabase, or parts of it, to Extensible Markup Language (XML) format is a method to exchange data between geodatabases.

    Note: Other methods for sharing data include exporting ArcMap layers to layer packages and map packages.

    Depending on whether you want to export entire feature classes or tables, or subsets of feature classes or tables, there are two types of XML documents you can export to: XML workspace documents and XML recordset documents. Both types of XML documents can then be shared and imported or loaded into other geodatabases.

    An XML workspace document is an XML specification for a geodatabase that can hold all or a subset of the contents of a geodatabase. When you export feature datasets, feature classes, and tables to an XML workspace document, any dependent data is exported. For example, if you export a feature class that participates in a geodatabase topology, all the other participating feature classes in the topology are exported to the XML workspace document. Or, if you export a feature class that is connected to another feature class or table by a relationship class, the relationship class and the related items are exported to the XML workspace document.

An XML recordset document is an XML specification for holding subsets of feature classes or tables. You can select a subset of features from a feature class or rows from a table and export them into an XML recordset document. Unlike exporting to an XML workspace document, when you export selected features or records to an XML recordset document, no additional geodatabase-related information is exported to the output file. For example, topologies or relationship classes are not exported into an XML recordset document.

Table 4.3 summarizes the XML export methods.

Table 4.3 XML export methods

Method	Outcome	Condition	When to use
Export to/Import from XML workspace document	• Creates new feature classes. • Dependent data is exported.	Requires an ArcGIS for Desktop Standard or Advanced license.	• Use to exchange entire geodatabases, datasets, or the dataset schema only. • Can be used to share a geodatabase schema or to copy the contents, rules, objects, and behaviors to another geodatabase.
Export to/load XML recordset document	• Exporting and loading XML recordset document adds features to an feature class or records to a table. • Dependent data is NOT exported.	• Requires an ArcGIS for Desktop Standard or Advanced license. • Loading data from an XML recordset document requires an ArcMap edit session.	• Use to exchange datasets, parts of a dataset, or dataset schema only. • Use the XML recordset document to exchange rows from a single feature class or table as simple features or records.

For more information on XML methods, refer to the ArcGIS Desktop Help topic "Geodatabase XML."

Exercise 4c: Export a geodatabase to XML

Suppose you want to share the contents of the Yellowstone geodatabase with a colleague or a client. You will export the geodatabase elements to Extensible Markup Language (XML).

1. If necessary, start ArcMap and navigate to the ...\DesktopAssociate\ Chapter04 folder.

2. Right-click Yellowstone.gdb, point to Export, and choose XML Workspace Document.

On the first panel of the Export XML Workspace Document wizard, you can specify whether to export all the data (plus the schema) or the schema only. Schema here refers to the structure of the data: feature class structure, coordinate systems, table fields, domains, topologies, and more. Also, you can choose whether to export to binary format or not.

3. Verify that the options to export the Data, Binary, and Export Metadata are selected.

4. Save the output XML file to ...\Chapter04 folder and name it Yellowstone.xml.

5. Click Next.

The panel lists the geodatabase elements that can be exported.

6. Leave all the boxes checked and click Finish.

7. When the tool is finished, refresh the ...\DesktopAssociate\Chapter04 folder in the Catalog window, and verify that the Yellowstone.xml has been created.

Hint: To refresh the Chapter 4 folder in the Catalog tree, select it and press the F5 key on your keyboard.

The Yellowstone.xml workspace document can now be uploaded to a server or attached to an e-mail, and then ultimately imported into another geodatabase.

8. If you are continuing to the next exercise, leave ArcMap open. Otherwise, close ArcMap.

> ### *Export/Import nonspatial data*
> Exporting and importing nonspatial data from and into a geodatabase works very similarly to exporting and importing spatial data. If you would like to store tabular data with your spatial data in a geodatabase, you can import it from, for example, other geodatabase or dBase tables, CSV files, or Microsoft Excel worksheets. Similarly, you can export geodatabase tables to dBase format or to other geodatabases. For more information on exporting and importing tables, refer to the ArcGIS Help topics "Table to Geodatabase (Conversion)" and "Table to Table (Conversion)."

Loading data into a geodatabase feature class

If you want to add features to an existing geodatabase feature class or table, use the loading method. You use the loading method for transferring features from a shapefile, coverage, or CAD feature class into a geodatabase feature class and also for transferring features among geodatabase feature classes. Besides loading from an XML recordset document, there are three other loading tools available in ArcGIS: the Simple Data Loader, the Object Loader, and the Append tool.

You use the Simple Data Loader or the Append tool to add features to a simple feature class or for loading additional records into tables. Simple feature classes are point, line, or polygon feature classes that are not network feature classes, that do not have feature-linked annotation, or that are not related to other objects through a relationship class.

You use the Object Loader to load data into complex feature classes that have a geometric network, a relationship with messaging, or feature-linked annotation. In addition to loading features, the Object Loader can apply validation rules, such as attribute domains or network connectivity rules, update feature-linked annotation with the newly loaded features, or snap the loaded features to the existing ones based on the current snapping environment.

When you load data from multiple data sources using the Simple Data Loader or the Object Loader, the schema of the source datasets and the target feature class must be identical.

An alternative to using the Simple Data Loader is the Append tool. Input datasets can be point, line, or polygon feature classes, tables, rasters, raster catalogs, annotation, or dimensions feature classes. The Append tool honors the geoprocessing environment settings such as output workspace, output coordinate system, and output extent, and it can also be used in geoprocessing models. The Append tool handles schema differences with its TEST and NO_TEST modes. In TEST mode, the schema of the input datasets and the target must match. In NO_TEST mode, input and target schemas do not have to match.

Table 4.4 summarizes the loading tools available in ArcGIS.

Table 4.4 Loading tools in ArcGIS

Method	Outcome	Condition	When to use
Simple Data Loader	Adds features to an existing simple feature class/table	Target feature class must exist. Feature class can be empty or already contain features	Use for simple features that do not participate in networks, relationships, or feature linked annotation. Use for loading the following: • Features from a shapefile, coverage, or CAD feature class into a simple geodatabase feature class • Records from a dBase, INFO, or geodatabase table to a geodatabase table • Multiple data sources at the same time
Object Loader	Adds features to an existing feature class • Option to validate features as they load • Features that violate rules will be selected • Option to snap features to existing features	Must be used in an ArcMap edit session. Existing feature class can be empty or may already contain features	Use for loading the following: • Features from a shapefile, coverage, or geodatabase feature class • Records from a dBase, INFO, or geodatabase table Data into a geometric network, feature classes in a relationship, or feature classes that have feature-linked annotation. Use if there is a possibility you need to undo changes. Use to load from multiple data sources at the same time
Append tool	Adds features to an existing feature class	Target feature class must exist	Use to append multiple input datasets into an existing target dataset. Use NO_TEST when loading from one or more data sources whose schema doesn't match the schema of the target dataset. Can be used for loading data as part of a geoprocessing model

For more information on loading tools, refer to the ArcGIS Help topics "About loading data into existing feature classes and tables," "Loading data in the Catalog tree," "About loading data in ArcMap," and "Append (Data Management)."

Exercise 4d: Load features using the Simple Data Loader

You will load river features from the YellowstoneRivers shapefile into the YellowstoneRivers feature class.

1. If necessary, start ArcMap, and navigate to and expand the ...\DesktopAssociate\Chapter04\Yellowstone.gdb geodatabase.

2. Add YellowstoneRivers geodatabase feature class to the map.

Hint: Make sure that you add the YellowstoneRivers feature class from the Yellowstone.gdb geodatabase, not the YellowstoneRiver.shp shapefile.

Question 3: Suppose the YellowstoneRivers feature class was stored in an ArcSDE geodatabase instead of a file geodatabase. How would you access a feature class from an ArcSDE geodatabase to add it as a layer to the map?

The YellowstoneRivers feature class currently does not contain any features.

3. In the Catalog window, right-click the YellowstoneRivers feature class, point to Load, and choose Load Data.

The Simple Data Loader wizard opens.

4. Click Next.

5. For the Input Data, add ...Chapter04\YellowstoneRivers.shp.

6. Click Next.

You have the option to load features directly into a particular subtype in the geodatabase.

7. Accept the default option not to load features into a subtype.

8. Click Next.

The next panel summarizes the Target fields in the YellowstoneRivers feature class and the matching Source Fields in the YellowstoneRivers shapefile.

9. Click Next.

This panel allows you to create a query to load only a subset of the source data.

10. Load all of the source data and click Next.

The last panel shows you a summary of the data load operation.

11. Click Finish.

12. When the loading is finished, in the table of contents zoom to the extent of the Yellowstone Rivers layer.

The layer now contains the river features that you just loaded.

13. If you are continuing to the next exercise, leave ArcMap open. Otherwise, close ArcMap.

Exercise 4e: Load features using the Append tool

Next, you will load features from the YellowstoneLakes coverage into the YellowstoneLakes feature class.

1. If necessary, start ArcMap, and navigate to your ...\DesktopAssociate\Chapter04 folder and expand the yellowstlakes coverage.

2. Add the yellowstlakes\polygon feature class to the map.

3. If you get a geographic coordinate system warning, close it.

When you use the Append tool, it will append features to the target feature class in the coordinate system of the data frame.

4. Next, add YellowstoneLakes feature class from the Yellowstone geodatabase to the map.

5. If you get a geographic coordinate system warning, close it.

The YellowstoneLakes feature class does not contain any features yet.

The yellowstlakes polygon feature class contains lakes of Yellowstone National Park.

6. Open the tables and compare the schema of yellowstlakes polygon and YellowstoneLakes.

Decision point: Do the schemas of the yellowstlakes polygon and YellowstoneLakes feature classes match?

The yellowstlakes polygon feature class has fields such as AREA, PERIMETER, YELLOWSTLAKES# and YELLOWSTLAKES-ID that are not present in the YellowstoneLakes feature class. Therefore, the schemas don't match.

Since the yellowstlakes polygon feature class has additional fields that are not present in the YellowstoneLakes feature class, you will use the Append tool with the NO_TEST option to add the lake features from yellowstlakes to the YellowstoneLakes geodatabase feature class.

7. From the Data Managements Tools toolbox, General toolset, open the Append tool.

8. Run the tool with the following parameters:
- Input Datasets: yellowstlakes polygon
- Target Datasets: YellowstoneLakes
- Schema Type: NO_TEST

9. When the Append tool is finished, zoom to the extent of the YellowstoneLakes layer.

10. Open the YellowstoneLakes attribute table.

The features from yellowstlakes have been loaded into the YellowstoneLakes feature class. For matching fields such as the FCODE or FCODE_DESC fields, the attribute values have been loaded; nonmatching fields such as YELLOWSTLAKES# and YELLOWSTLAKES-ID have been dropped.

11. Close the table window.

12. If you are continuing to the next exercise, leave ArcMap open. Otherwise, close ArcMap.

Exercise 4f: Load features using the Object Loader

Next, you will use the Object Loader to add rivers from the surrounding counties to the YellowstoneRivers feature class.

Dependency: You need to complete exercise 4d before running this exercise. If you have not completed exercise 4d, please do so before starting this exercise.

1. If necessary, start ArcMap, and navigate to your …\DesktopAssociate\Chapter04\NewData folder.

2. Expand the Rivers folder and add all five river shapefiles to the map.

Hint: In ArcGIS 10.1, click the Toggle Contents Panel button in the top right of the Catalog window to display the panel underneath the tree view. In ArcGIS 10.0, click the Show Next View button in the top right of the Catalog window to display the panel underneath the tree view.

In the bottom panel, you can select all five shapefiles (holding down the Ctrl key) and drag them into the map display.

The shapefiles contain rivers from the counties surrounding Yellowstone National Park.

3. **Also add YellowstoneRivers geodatabase feature class to the map.**

Remember, in exercise 4d, you have loaded features into the YellowstoneRivers feature class.

Decision point: Do the schemas (field definitions) of these river shapefiles match?

Yes, they match. All five shapefiles have the same field definitions. They have the same field names and the same data types. Therefore you can load them at the same time using the Object Loader.

Question 4: Suppose that you did not want to visually compare the schema of the two tables. Which geoprocessing tool could you use to automatically compare the schema feature classes?

To use the Object Loader, you first need to add it to the ArcMap interface.

4. **From the Customize menu, choose Customize Mode. Click the Commands tab. In the list of Categories, select Data Converters.**

5. **Drag the Load Objects command to any toolbar in the ArcMap interface. Then close the Customize dialog box.**

You will use the Object Loader with an option that snaps the new features to the existing ones. You will now set up your snapping environment.

> ### *Snapping*
> Snapping is an editing operation in which features are moved to coincide with nearby features if they are within a specified distance (snapping tolerance). Snapping allows you to automatically connect features to each other. The settings for snapping are located on the Snapping toolbar. You will learn more about snapping in chapter 10 of this book.

6. **Display the Editor and Snapping toolbars.**

7. **On the Snapping toolbar, make sure that End Snapping is enabled. Disable all other snap types, as shown in the following graphic.**

8. **Click Snapping and choose Options. For the Snapping tolerance, type in 500 and click OK.**

Note: Since the snapping tolerance is measured in pixels and some monitor resolutions are very high, you set a high snapping tolerance to make sure the new river features will snap to the existing ones.

9. **On the Editor toolbar, click Editor and Start Editing. Choose the Yellowstone.gdb file geodatabase as the workspace to edit. Click Continue on the warning message.**

Now you are ready to start the loading process.

10. **Click Load Objects to open the Object Loader.**

11. For the Input Data, add the …Chapter04\NewData\Rivers\ Rvs_Fremont_Cnty.shp. Then, one by one, add the other four shapefiles from the Rivers folder. Click Next.

12. Make sure that YellowstoneRivers is the Target.

13. Continue to click Next, accepting all defaults until you reach the panel asking if you want the input features to be snapped based on the current snapping environment.

14. Check Yes to snap the input features to the existing features, as shown in the following graphic.

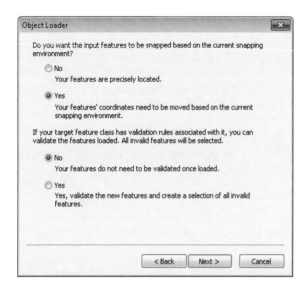

15. Click Next, and then click Finish in the Summary panel.

16. On the Editor toolbar, click Editor, choose Stop Editing, and click Yes to save your edits.

17. If you are continuing to the next exercise, leave ArcMap open. Otherwise, close ArcMap.

Challenge

For each of the following scenarios, choose the best method for adding data to the geodatabase.

1. You want to add new GPS waypoints from three different shapefiles to a geodatabase feature class with 10,000 existing features. The shapefiles have the same schema, which also matches the schema of the target geodatabase feature class. You want to load the data running a single tool. (Choose two.)

    a. Simple Data Loader
    b. Copy/Paste features
    c. Import
    d. Append tool (with TEST schema type)
    e. Export to XML

2. You want to bring building footprints from a shapefile into a geodatabase (no preexisting feature class) and retain the existing schema.

    a. Simple Data Loader
    b. Export to geodatabase (single)
    c. Object Loader
    d. Copy/Paste feature class

3. You want to transfer three street centerlines from an ArcSDE (multiuser) geodatabase feature class into a feature class in a file geodatabase.

    a. Simple Data Loader
    b. Export to geodatabase (single)
    c. Export to XML
    d. Copy/Paste feature class

4. You want to add water lines from a shapefile to a geodatabase feature class that is part of a geometric network. You'd like to snap the new water lines to the existing features, honoring the current snapping environment.

    a. Simple Data Loader
    b. Append tool (with NO_TEST schema type)
    c. Object Loader
    d. Copy/Paste feature class

5. You want to bring adjacent polygons from a geology coverage into a geodatabase feature class. In the coverage the polygons have a numeric attribute fields called Geology# and Geology-ID.

    a. Simple Data Loader
    b. Copy/Paste features
    c. Append tool (with NO_TEST schema type)
    d. Export to XML

6. You want to send several geodatabase feature classes to a contractor and have them import the feature classes into their geodatabase.

    a. Export to XML
    b. Simple Data Loader
    c. Copy/Paste features
    d. Export/Import feature class

Creating data from layers

ArcMap layers reference datasets on disk and specify how to display that data using layer properties. For example, the symbology and label properties, transparency settings, and definition queries control how a layer displays.

If a layer's source data is in a file or personal geodatabase, you can access it directly just by navigating in the catalog tree to the storage location. However, to access datasets in an ArcSDE geodatabase and add them as layers to a map, you must connect to the geodatabase first and then navigate to the source data through the connection file.

Exporting layers

Once a feature class has been added as a layer to ArcMap, you can export the layer and create a new dataset on disk (in geodatabase or shapefile format). Exporting a layer gives you various opportunities for data management. You can export either the entire layer or a subset of it, by exporting the selected features only. Also, you can export either in the native coordinate system of the source data or in coordinate system of the data frame (see chapter 5, exercise 5c). If you wanted to separate the attribute table from the geometry of a layer, you could export the attribute table to a stand-alone table.

Chapter 4: Managing geographic data

Exercise 4g: Export a subset of a layer

Exporting only the selected features in a layer is a quick way to create a subset of a feature class.

1. **If necessary, start ArcMap. In the Catalog window, navigate to the ...\DesktopAssociate\Chapter04\ NewData folder.**

2. **Add the Gya_Cities.shp to the map and zoom to the extent of the Gya_Cities layer.**

3. **If necessary, add the ...Yellowstone.gdb\Park_Area feature class.**

You only need to export cities within 50 miles of the park boundary to your geodatabase. You will select and export a subset of the GYA_cities layer to a new feature class.

4. **Using Select By Location, select features from Gya_Cities that are within a distance of 50 miles of the ParkArea layer. If necessary, use the following graphic for reference.**

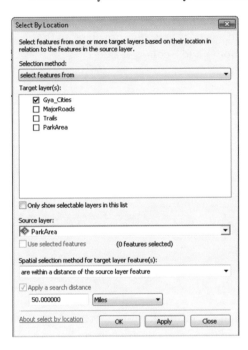

5. **Open the Gya_Cities attribute table.**

278 out of the 857 city features are selected. However, the city of Idaho Falls, a larger city in the vicinity of Yellowstone National Park, is currently not selected. Since you want to include this city in your set of cities to export, you will manually add Idaho Falls to the selection.

6. **Right-click the NAME field and choose Sort Ascending.**

7. **Scroll down and locate Idaho Falls (FID 150) in the NAME field.**

8. **Holding down the Ctrl key, click the gray box at left edge of the Idaho Falls record to add the record to the selection.**

Now you will export the selected city features to a geodatabase feature class.

Hint: Do **NOT** use the Export command from the Table Options menu. This will export just the selected records of the table, not the point geometries that you see in the map. You will use the Table Options Export command in the next step and see the difference.

9. **Close the Gya_Cities attribute table.**

10. **In the ArcMap table of contents, right-click Gya_Cities layer, point to Data, and choose Export Data.**

You have the option to export the features either in their native coordinate system or in the coordinate system of the data frame.

11. Export the Selected features using the coordinate system of the layer's source data and save the output feature class as **YellowstoneCities** in the … Chapter04\Yellowstone.gdb.

12. Click Yes to add the exported data as a layer to the map.

13. Remove the Gya_Cities layer from the map.

14. Zoom to the extent of the YellowstoneCities layer.

YellowstoneCities contain just the cities within 50 miles of the park (plus the city of Idaho Falls).

15. If you are planning to do the next exercise, leave ArcMap open. Otherwise, close ArcMap.

Exercise 4h: Export the attribute table of a layer

You can also export the attribute table of a feature class to a stand-alone table.

1. If necessary, start ArcMap, and navigate to your …\DesktopAssociate\Chapter04\NewData folder.

2. Add the Peaks.shp to the map.

3. Open the Peaks attribute table.

The Peaks attribute table contains information about the names, elevation, and the x,y coordinates of the summits of the Yellowstone peaks.

4. From the Table Options button in the upper left corner of the table window, choose Export.

5. Save the Output table as **YellowstonePeaks** in the … Chapter04\Yellowstone.gdb geodatabase.

6. Click Yes to add the new table to the current map.

When the stand-alone table was added to ArcMap, the table of contents switched to List By Source view. In List By Source the layers are grouped by the location of their source data.

7. In the table of contents, open the new YellowstonePeaks table.

The YellowstonePeaks table contains all the fields from the Peaks attribute table, except for the Shape field, which makes it a stand-alone table.

8. Close the table window.

9. If you are continuing to the next exercise, leave ArcMap open. Otherwise, close ArcMap.

> ### *Layer files and layer packages*
>
> So far, you have worked with layers inside an ArcMap document (.mxd). However, to share a layer and use it in other map documents, you can also save a layer, either as a layer file (.lyr) or as a layer package (.lpk).
>
> A layer file (.lyr) stores a layer's properties, but not the source dataset referenced by the layer. Similar to a layer in an ArcMap document (.mxd), a layer file (.lyr) needs access to the source data.
>
> A layer package (.lpk) includes both the layer properties and the dataset referenced by the layer. With a layer package, you can save and share everything about the layer—its symbolization, labeling, field properties, and the data. Layer packages are another great way to share geographic data.

58 Chapter 4: Managing geographic data

Exercise 4i: Set layer properties and create a layer package

You will import the symbology for MajorRoads from a layer file, set up labels, and then export MajorRoads to a layer package.

Dependency: You need to complete exercise 4b before running this exercise. If you have not completed exercise 4b, please do so before starting this exercise.

1. If necessary, start ArcMap, and navigate to your ...\DesktopAssociate\Chapter04 folder. If necessary, add the MajorRoads from the ...\Yellowstone.gdb\Transportation feature dataset to the map.

You will import the symbology definition for the Major Roads layer from a layer file.

2. Open the layer properties of MajorRoads and click the Symbology tab.

You can import layer symbology from ArcGIS layer files (.lyr) or any layers in the current map document, as long as they have the same geometry type.

3. On the Symbology tab, import the symbology from your ...Chapter04\Roads.lyr layer file.

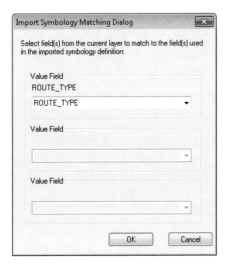

4. Make sure ROUTE_TYPE is set for the Value field, and then click OK to import the symbology definition.

5. In the Layer Properties, click Apply.

The MajorRoads layer is symbolized with Unique values symbology based on the ROUTE_TYPE field. However, labels were not included in the layer file. You will label the US highways with the highway number and a shield label.

6. In the Layer Properties, click the Labels tab. Check to label features in this layer.

7. Set the following label properties:
   - Method: Define classes of features and label each class differently
   - Label field: HWY_NUM
   - Label styles: US Route

You will label only the US highways, not the state and the park roads.

8. Click the SQL Query button and build the following query: "ROUTE_TYPE" = 'US'

9. In the Layer Properties, click Apply.

The Major Roads layer has been labeled with highway shields.

10. Click OK to close the layer properties.

Now, you will now export the MajorRoads layer to a layer package.

11. Right-click MajorRoads and choose Create Layer Package.

12. Save the MajorRoads.lpk layer package to a file in the …\DesktopAssociate\Chapter04 folder.

Workflow for ArcGIS 10.1:

13. Click Analyze.

You receive three errors in the Prepare window: First, a layer description is required for packaging. Also, the Tags and the Summary are missing in the Item Description.

14. Right-click the error "A layer description is required for packaging," and then choose Change General Layer Properties.

15. For the Description, type in **Major roads in the greater Yellowstone area**. Then click OK.

16. In the Layer Package dialog box, click Item Description.
    - For the Summary, type in **Major Roads in the greater Yellowstone area**.
    - For Tags, type in **Roads, Transportation, Yellowstone, Wyoming, USA**
    - Click Analyze again.

Workflow for ArcGIS 10.0:

13. Click Validate.

You receive one error in the Prepare window: A layer description is required for packaging.

14. Right-click the error "A layer description is required for packaging," and then choose Change General Layer Properties.

15. For the Description, type in **Major Roads of Major roads in the greater Yellowstone area**. Then click OK.

16. Click Validate again.

For both ArcGIS 10.1 and 10.0: There are no more errors listed in the Prepare window.

17. Click Share. Click OK on the message you receive.

The layer properties and source data are compressed into the .lpk file.

18. Remove the MajorRoads layer from the map.

19. Navigate to your …\DesktopAssociate\Chapter04 folder, right-click the MajorRoads.lpk, and choose Unpack.

A new MajorRoads layer is added to the map.

20. Open the layer properties of the new MajorRoads layer (that you added from the layer package) and activate the Source tab.

By default, layer packages unpack to the Documents\ArcGIS\Packages in your user profile.

21. Close the layer properties.

Question 5: How is a layer package different from an XML workspace or XML recordset document?

22. If you are continuing to the next exercise, leave ArcMap open. Otherwise, close ArcMap.

Layer source data

Occasionally, you may want to reset the data source of a layer to a different dataset, for example, because you received more current source data. Suppose that you changed the source data of a layer to a different dataset, and the new source dataset has a different schema, that is, different attribute field definitions. What would happen to the layer properties that reference fields in the attribute table?

Any layer property that references one or more fields in the source data could be affected by a change in the source data. Table 4.5 lists the problems in the layer properties of a vector layer that could potentially be caused by a change in the source data.

Table 4.5 Problems caused by changes in source data in vector layers

Tab	Property	Problem caused by a change in source data	Solution
General	Scale ranges	Existing scale ranges may not be appropriate for the new data source.	Adjust visible scale ranges.
Source	Path to the source data	The layer might not be displayed if the path to the source data cannot be resolved.	Reset path to the source data.
Display	Display expressions	If the fields used in display expressions change, they may not display properly.	Re-create display expression with new fields.
Symbology	Categories Quantities Charts Multiple Attributes	If the new data source does not contain the value fields used for symbolizing the layer, it may fail to draw.	Reset value fields in the symbology options.
Fields	Visible fields in the attribute table	If the new data source contains fields that were not present in the original data source, these fields are visible by default.	If desired, turn off additional fields in the Fields tab.
Definition query	Definition queries	If the new data source contains different attribute fields, definition queries may break and the layer may fail to draw.	Re-create definition queries with new fields.
Labels	Label fields Label expressions SQL queries	If the new data source contains different attribute fields than the original data source, label fields, label expressions, and label SQL queries may cause the labels not to display.	• Reset label fields to existing fields in the new source data. • Re-create label expressions and SQL queries with new fields.
Joins & Relates	Joins and relates	If the new data source does not contain the key field used for a join or a relate, the table association will break.	Re-create joins and relates using new fields as key fields.
Time	Time field(s) may be different.	Time series animations fail to display.	Re-create time series animation with new time field(s).

Exercise 4j: Change data sources for a layer

Next, you will experiment with different source data for the Major Roads layer. Suppose you extracted highways from Data and Maps for ArcGIS, which you now want to use as an alternative to the major roads layer's source data.

Dependency: You need to complete exercise 4i before running this exercise. If you have not completed exercise 4i, please do so before starting this exercise.

1. Add Highways.shp from the ...\DesktopAssociate\Chapter04 folder as a layer to the map.
2. Zoom to the extent of the Highways layer.

Highways has a larger extent than the Major Roads layer.
Next, you will compare the attribute tables of the two layers.

3. Open the attribute tables of both Major Roads and Highways.

Hint: The two tables appear as two tabs in the table window. To view the two tables right next to each other, click the Highways tab and then drag the MajorRoads tab over one of the blue arrows that appear in the table window.

Some of the attribute fields match between the two tables, but some of the fields are different. For example, the Route_TYPE and the Shield fields are present in the MajorRoads table, but not in the Highways table.

4. Close the table window.
5. Turn off the Highways layer.

You will change the source data of the MajorRoads layer to reference the Highways shapefile and observe the impact on the layer properties of MajorRoads.

6. Open the layer properties of MajorRoads.
7. In the Source tab, click Set Data Source.
8. Set the data source of the layer to the ...Chapter04\Highways.shp. Then click OK.

You receive an error message that one or more layers failed to draw. The MajorRoads layer does not draw on the map.

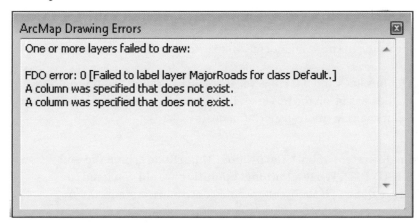

Decision point: Why do you think the MajorRoads layer does not draw? What can you do to fix the problem?

The symbology of the MajorRoads layer was based on ROUTE_TYPE as a value field. Since the Highways layer does not contain the ROUTE_TYPE field, ArcMap cannot draw the layer based on the current settings in the layer properties. You will set single symbol symbology to make the layer draw.

9. Close the message.

10. Open the layer properties of MajorRoads.

11. In the Symbology tab, set single symbol symbology and click Apply.

12. Move the Layer properties dialog box to the side of your screen so that can see the map.

The MajorRoads layer draws now, but you still receive a drawing error message.

Decision point: What other layer property did you set for the MajorRoads that may have also been affected by the change in the source data?

The Label properties have been affected. The SHIELD field that you used as a label field is not present in the new source data.

13. Click the Labels tab.

14. Set HWY_SYMBOL as the Label field and click Apply.

The road features still cannot be labeled because of the SQL Query that is based on the ROUTE_TYPE field.

15. Click the SQL Query button and clear the query expression.

16. Click OK, and then click Apply.

Now the Highways layer draws.

17. Click OK to close the Layer Properties of MajorRoads. Also close the error window.

Question 6: Suppose that the source data for all layers in a map document has been moved to a different storage location. Rather than updating the data sources one by one, what is a more efficient method to update the data sources of all of the layers in a map document?

Question 7: What is the most efficient way to update the data sources in multiple map documents?

18. Close ArcMap.

Answers to chapter 4 questions

Question 1: Why do you store major roads in the Transportation feature dataset in the Yellowstone geodatabase?
Answer: Although you don't have data for bus lines and bus stops yet, it is safe to assume that bus routes will run along the major roads and that bus stops will be on the bus lines. Because of the topological relationships among those layers, it is best to store major roads in the Transportation feature dataset.

Question 2: The geographic coordinate system of both the original MajorRoads shapefile and the Transportation feature dataset is NAD 1983. What additional operation would you need to perform to export a shapefile to a feature dataset if their geographic coordinate systems did not match?
Answer: Before you export a shapefile into a feature dataset with a different geographic coordinate system, you must perform a geographic transformation. Using the Project geoprocessing tool, you would transform the shapefile to the geographic coordinate system of the feature dataset. The Project tool will generate a new shapefile, which you can then export to the geodatabase feature dataset.

Question 3: Suppose the YellowstoneRivers feature class was stored in an ArcSDE geodatabase instead of a file geodatabase. How would you access a feature class from an ArcSDE geodatabase to add it as a layer to the map?
Answer: To add layers from ArcSDE geodatabase data sources, you navigate through the connection file to the ArcSDE geodatabase to the source dataset.

Question 4: Suppose that you did not want to visually compare the schema of the two tables. Which geoprocessing tool could you use to automatically compare the schema feature classes?
Answer: The Feature Compare tool compares the geometry, attributes, spatial reference, and schema of two input feature classes.

Question 5: How is a layer package different from an XML workspace or XML recordset document?
Answer: Layer packages are compressed files of layers and their source data, including the specifications for how to display the source data. XML workspace or recordset documents are XML specifications of geodatabase content, such as feature classes, tables, or subsets of them. They don't include information about how to display that data.

Question 6: Suppose that the source data for all layers in a map document has been moved to a different storage location. Rather than updating the data sources one by one, what is a more efficient method to update the data sources of all of the layers in a map document?
Answer: To update the layer sources for all layers in a map document, you can right-click a map document in ArcCatalog (or the Catalog window) and choose Set Data Sources. Another option is to right-click one layer in the ArcMap and choose Data, Repair Data Source. If the source data for all of the layers in a data frame have been moved to the same location, updating one layer will then update the data sources of all layers in the data frame.

Question 7: What is the most efficient way to update the data sources in multiple map documents?
Answer: You can use Python to perform bulk updates of your data sources. For more information about updating data sources in bulk, refer to the ArcGIS for Desktop Help document "Updating and fixing data sources with arcpy.mapping."

Answers to challenge questions

Challenge

Correct answers shown in bold.

1. You want to add new GPS waypoints from three different shapefiles to a geodatabase feature class with 10,000 existing features. The shapefiles have the same schema, which also matches the schema of the target geodatabase feature class. You want to load the data running a single tool. (Choose two.)

    a. **Simple Data Loader**
    b. Copy/Paste features
    c. Import
    d. **Append tool (with TEST schema type)**
    e. Export to XML

2. You want to bring building footprints from a shapefile into a geodatabase (no preexisting feature class) and retain the existing schema.

   a. Simple Data Loader
   b. Export to geodatabase (single)
   c. Object Loader
   d. Copy/Paste feature class

3. You want to transfer three street centerlines from an ArcSDE (multiuser) geodatabase feature class into a feature class in a file geodatabase.

   a. Simple Data Loader
   b. Export to geodatabase (single)
   c. Export to XML
   d. Copy/Paste feature class

4. You want to add water lines from a shapefile to a geodatabase feature class that is part of a geometric network. You'd like to snap the new water lines to the existing features honoring the current snapping environment.

   a. Simple Data Loader
   b. Append tool (with NO_TEST schema type)
   c. Object Loader
   d. Copy/Paste feature class

5. You want to bring adjacent polygons from a geology coverage into a geodatabase feature class. In the coverage the polygons have a numeric attribute fields called Geology# and Geology-ID.

   a. Simple Data Loader
   b. Copy/Paste features
   c. Append tool (with NO_TEST schema type)
   d. Export to XML

6. You want to send several geodatabase feature classes to a contractor and have them import the feature classes into their geodatabase.

   a. Export to XML
   b. Simple Data Loader
   c. Copy/Paste features
   d. Export/Import feature class

Key terms

Layer: A set of properties, including a reference to a data source (such as a shapefile, coverage, geodatabase feature class, or raster) and a symbology definition. Layers can also define additional properties, such as which features from the data source are included. Layers can be stored in map documents (.mxd) or saved individually as layer files (.lyr). Layers are conceptually similar to themes in ArcView 3.x.

Resources

- ArcGIS Desktop Help > Geodata > Geodatabases > Adding datasets and other geodatabase elements
  - An overview of adding datasets to the geodatabase
- ArcGIS Desktop Help > Geodata > Geodatabases > Adding datasets and other geodatabase elements > Exporting data
  - A quick tour of exporting data
  - About exporting selected features and records

- ArcGIS Desktop Help > Geodata > Geodatabases > Adding datasets and other geodatabase elements > Loading data
  - About loading data into existing feature classes and tables
  - Loading data in the Catalog tree
  - Loading data in ArcMap >
    - About loading data in ArcMap
  - Loading features and records from an export file
    - Loading an XML recordset document in the Catalog tree
    - Loading an XML recordset document in ArcMap
- ArcGIS Desktop Help > Desktop > Mapping > Working with layers
  - What is a layer?
  - A quick tour of map layers
  - Managing layers > Working with layers
    - Saving layers and layer packages
- ArcGIS Desktop Help > Desktop > Mapping > Working with layers > Displaying layers > Common layer display tasks
  - Importing symbology from another layer
- ArcGIS Desktop Help > Desktop > Mapping > Working with layers > Interacting with layer contents
  - Exporting features
- ArcGIS Desktop Help > Desktop > Geoprocessing > Tool reference > Conversion toolbox > To Geodatabase toolset
  - Table to Geodatabase
  - Table to Table
- ArcGIS Desktop Help > Desktop > Geoprocessing > Tool reference > Data Management toolbox > General toolset
  - Append
- ArcGIS Desktop Help > Desktop > Geoprocessing > ArcPy > Mapping module
  - Updating and fixing data sources with arcpy.mapping
- ArcGIS Desktop Help > Desktop > Geoprocessing > Tool reference > Data Management toolbox > Distributed geodatabase tooset
  - Export XML Workspace document
  - Import XML Workspace document

chapter 5

Coordinate systems

Types of coordinate systems 68
    Geographic coordinate systems
    Projected coordinate systems
    Challenge 1
    Local coordinate systems
    Exercise 5a: Examine the coordinate system information of a feature class
    Vertical coordinate systems

Projection on the fly 73
    Exercise 5b: Project layers on the fly

Data in different projected coordinate systems 75
    Valid input data types for the Project tool
    Exercise 5c: Project feature classes

Data with an unknown coordinate system 77
    Determining the coordinate system of a dataset
    Exercise 5d: Define the coordinate system of a feature class
    Defining an incorrect coordinate system

Managing data in different geographic coordinate systems 79
    Geographic transformations
    Exercise 5e: Set a transformation in the data frame
    NADCON transformation
    Exercise 5f: Project with a transformation
    Challenge 2

Answers to chapter 5 questions 83

Answers to challenge questions 83

Key terms 85

Resources 85

Chapter 5: Coordinate systems

Does the following situation sound familiar? You add your data layers to ArcMap and they display nicely one by one—just not together in the same location! Some of the data layers draw in the correct place. Some layers seem to line up, but there is a gap between them. And then, there are one or two layers that are completely off—they display in an entirely different part of the world, or maybe in the middle of the ocean (where they're not supposed to be!).

Coordinate systems give us a common frame of reference to define where geographic features are located and also how they are positioned relative to each other. Misalignments are often caused by coordinate system issues, such as data being in a different coordinate system or data not having a defined coordinate system. To understand those issues and distinguish them from misalignments caused by scale and accuracy issues, you will review the two types of coordinate systems found in geographic data: geographic and projected coordinate systems. Then you will see how ArcMap data frames deal with different coordinate systems and apply some techniques for troubleshooting (or preventing) common coordinate system issues.

Skills measured

- Compare and contrast geographic coordinate systems and projected coordinate systems and determine the appropriate coordinate system to use for a given purpose or task.
- Inspect the coordinate system of an ArcGIS layer, dataset, and data frame.
- Differentiate between changing a dataset's assigned coordinate system and reprojecting the data.
- Explain how to reproject data in ArcGIS for Desktop using project and export tools.
- Explain how data frames inherit coordinate systems in ArcGIS for Desktop.
- Explain how to implement a transformation for a data frame in ArcGIS for Desktop.

Prerequisites

- Knowledge of these terms:
  - coordinate system
  - longitude and latitude
  - datum
  - spheroid
  - projection
- Hands-on experience working with coordinate systems in ArcGIS

Types of coordinate systems

Geographic coordinate systems

All GIS datasets use coordinate systems to define their location on earth. The coordinates in a GIS dataset can be stored in a geographic, projected, or local coordinate system.

A geographic coordinate system (GCS) uses angles of longitude and latitude to define locations on earth on a three-dimensional model of the earth (figure 5.1). The terms geographic coordinate system (GCS) and datum are often used interchangeably. However, technically speaking, a datum is only one part of a GCS. Besides the datum and a spheroid, a geographic coordinate system includes the units for longitude and latitude, and the prime meridian that has been used for measuring longitude. For example, the North American Datum of 1983 (NAD83) geographic coordinate system is based on the GRS80 (Geodetic Reference System 1980) spheroid, the NAD83 datum, and the use of degrees as units to measure longitude and latitude. As many other geographic coordinate systems, NAD83 uses Greenwich, England, as its prime meridian, which acts as its 0 degrees longitude. Other common geographic coordinate systems used in North America that are identified by the datum are the North American Datum of 1927 (NAD27) and the World Geodetic System of 1984 (WGS84).

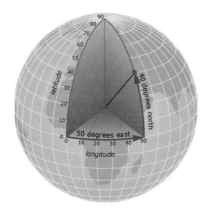

Figure 5.1 Many geographic coordinate systems (GCS) use the meridian through Greenwich, England, as 0 degrees longitude and the equator as 0 degrees latitude. Esri.

Projected coordinate systems

Geographic coordinate systems are good for storing data but not very good for measuring distances. This is because the angles of latitude and longitude can locate exact positions on the surface of the earth, but angles do not provide consistent distance measurements. Think about it: 30 degrees of longitude along the equator and 30 degrees of longitude close to the north pole represent much different distances.

To measure distances, we use projected coordinate systems (PCS), because they define locations on the earth in a flat (planar), two-dimensional surface. The coordinates in a projected coordinate system, also known as a Cartesian coordinate system, represent distances from an origin (0,0 point) along two perpendicular axes: a horizontal x-axis representing east-west, and a vertical y-axis representing north-south (figure 5.2).

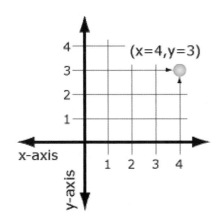

Figure 5.2 The two perpendicular axes in a Cartesian coordinate system have the same unit of length, for example, feet or meters. A Cartesian coordinate system specifies locations as a pair of numerical coordinates, which represent distances from the intersection point ((0,0) point) of the two axes. Esri.

Every projected coordinate system is based on a geographic coordinate system. To apply a projected coordinate system, we have to convert the three-dimensional surface of the GCS into a two-dimensional one. This mathematical operation is commonly referred to as a map projection. When a dataset is projected, the latitude and longitude values are converted into x,y values in meters or feet.

When datasets covering large geographic areas are projected (e.g., continental or world data), four properties of the data are distorted: shape, area, direction, and distance. When datasets covering small areas such as a neighborhood or a city are projected, these distortions are often negligible. Projections are categorized according to the spatial properties they preserve: some projections preserve shape (conformal projections), while others preserve area (equal area projections), distance (equidistant projections), or direction (true direction or azimuthal projections). Only one of these spatial properties can be preserved with a given projection; the other ones will be distorted. Some projections preserve a second spatial property reasonably well. For example, conformal projections preserve shape, but they also preserve direction fairly well. Equal area projections preserve area but can also preserve distance reasonably well. Alternatively, some projections distort all four of these spatial properties.

In addition to the projection and the underlying geographic coordinate system, a projected coordinate system includes the location of the origin, which is the 0,0 point for measuring the x,y coordinates, and their units, usually meters or feet. One commonly used projected coordinate system for the United States is the State Plane Coordinate System. It divides each state into different zones. Each zone has its own origin and uses a projection and adjusted projection parameters designed to minimize distortion. Other commonly used projected coordinate systems used for North America and the world are Universal Transverse Mercator (UTM, conformal), Albers Equal Area (true area), and Robinson (compromise).

Table 5.1 summarizes the components of geographic and projected coordinate systems and gives you some guidance about when to use each one.

Table 5.1 Properties of geographic and projected coordinate systems

Coordinate system	Components	When to use
Geographic	• Angular unit of measure (usually degrees, sometimes radians, microradians, grads, or gons) • A prime meridian (most often Greenwich, England) • A datum (based on a spheroid)	• To store datasets in a central database, allowing users to apply the projected coordinate system they need • To make a quick map • When shape, area, distance, or direction don't need to be preserved • When spatial queries based on location and distance won't be performed

Projected	• Two-dimensional units of measure (meters or feet)
	• Origin (0,0 point)
	• Map projection
	• Underlying geographic coordinate system
		• To make a map in which you need to preserve shape, area, distance, or direction (e.g., density or navigation maps)
		• To accurately measure distances and calculate areas on a map
		• To make small-scale maps such as national and world maps
		• For GIS analysis:
		  • To perform spatial queries based on location and distance
		  • To ensure that spatial analysis tools calculate accurate directions, distances, and areas
		• For GIS editing:
		  • To create accurate geometries for new features, especially when entering bearings and distances
		  • To maintain accurate geometries when editing existing features
		• When your organization mandates using a particular projected coordinate system for all maps.

For more information on geographic and projected coordinate systems, refer to the ArcGIS Help topics, "What are geographic coordinates systems" and "What are projected coordinate systems."

Challenge 1

For each of the following scenarios, determine which type of coordinate system to use.

1. You want to make a thematic map of population density in different counties.

    a. Geographic
    b. Projected

2. You want to create the boundary lines of land parcels from the bearings and distances of provided survey measurements.

    a. Geographic
    b. Projected

3. You want to store the epicenters of earthquakes in a geodatabase that can be accessed through a website that allows users to download the data.

    a. Geographic
    b. Projected

4. You want to collect samples of ozone concentrations in different cities worldwide at different times of the year.

    a. Geographic
    b. Projected

5. You want to use weather data in a map that will be used to plan the most efficient airplane routes between different continents of the world.

    a. Geographic
    b. Projected

6. You want to make a quick map to evaluate if a dataset has a sufficient amount of spatial detail (scale) and make sure it covers your study area.

    a. Geographic
    b. Projected

Local coordinate systems

Local coordinate systems are planar coordinate systems that are often used when creating computer-aided design (CAD) files. An arbitrary point is selected on the ground, often at a street intersection, property boundary corner, a survey monument, or other point. That point will become the 0,0 point for survey measurements. In the CAD program, bearings and distances for parcel data would then be drawn based on survey measurements from that point location.

Source: Margaret M. Maher, *Lining Up Data in ArcGIS* (Redlands, CA: Esri Press, 2010, 148)

Exercise 5a: Examine the coordinate system information of a feature class

For the exercises in this chapter, suppose you are creating a map that will be used by project planners of various organizations to find potential sites for cultural heritage and nature conservation projects. The extent of your database is the continents of Australia and Oceania, which include Australia, New Zealand, and Papua New Guinea, as well as the islands of the South Pacific Ocean, including the Melanesia and Polynesia groups. Parts of South Asia will be visible in the map as well.

Chapter 5: Coordinate systems

First, you will add a Continents feature class to observe how the coordinate system of the feature class affects how the continents are drawn in the map.

1. Start ArcMap with a blank map.
2. In the Catalog window, navigate to your ...\DesktopAssociate\Chapter05 folder and expand the Oceania geodatabase.
3. In the Oceania geodatabase, right-click the Continents feature class and click Properties.

Decision point: In which tab of the feature class properties do you find the coordinate system information?

The coordinate system information of a feature class is found in the XY Coordinate System tab.

4. Click the XY Coordinate System tab and examine the information listed for the current coordinate system.

Decision point: Which geographic and/or projected coordinate system are continents stored in?

Decision point: What is the prime meridian of the WGS 1984 geographic coordinate system?

Continents are stored in the WGS 1984 geographic coordinate system. They are not projected into any projected coordinate system. The WGS 1984 geographic coordinate system uses the meridian of Greenwich, England, as a prime meridian.

5. Add Continents to ArcMap.

The Continents layer contains the continents of Asia, Australia, and Oceania.

Decision point: Notice that the eastern tip of Asia and the easternmost islands of Oceania are drawn at the left edge of the map. Why are the continents drawn disconnectedly in the map?

Some features in the Continent feature class cross the 180-degree meridian, which is approximately the meridian of the international dateline. Since the features are stored in degrees of longitude and latitude, the feature parts east of the 180-degree meridian have lower longitude values (i.e., negative values) than the feature parts west of the 180-degree meridian. Using Greenwich, England, as a zero degrees starting line, ArcMap draws the features at lower longitudes to the left of the ones with the higher longitude values. You will fix this problem later in this chapter.

6. In the ArcMap table of contents, right-click the Continents layer and click Properties.

Decision point: Where in the Layer Properties dialog box do you find the information about the coordinate system of a layer?

The coordinate system of a layer is listed in the Source tab.

7. Click the Source tab and locate the coordinate system of the Continents layer. Read the coordinate system information. Then close the Layer Properties dialog box.

Decision point: Which other ArcMap component contains coordinate system information?

Coordinate system information is also a property of the ArcMap data frame.

8. **Right-click the Layers data frame and click Properties.**

9. **In the Coordinate System tab, examine the coordinate system of the data frame.**

When you added Continents to the data frame, its coordinate system was automatically set to the WGS 1984 geographic coordinate system to match the Continents layer. The first layer you add to a data frame defines its coordinate system.

10. **Close the data frame Properties dialog box.**

11. **If you are planning to do the next exercise, leave ArcMap open. Otherwise, close ArcMap.**

> ### *Vertical coordinate systems*
>
> Local coordinate systems are planar coordinate systems that are often used when creating CAD files. An arbitrary point is selected on the ground, often at a street intersection, property boundary corner, a survey monument, or other point. That point will become the 0,0 point for survey measurements. In the CAD program, bearings and distances for parcel data would then be drawn based on survey measurements from that point location.
>
> The geographic and projected coordinate systems discussed so far in this chapter pertained to the coordinate values that define locations on earth in horizontal direction. If the coordinates of a dataset, in addition to the x- and y-values, also store a z-value as part of their geometry, you have to define a vertical coordinate system for the z-values. A vertical coordinate system defines whether the z-values represent heights (elevations) or depths, the origin for height or depth values (usually sea level or mean low water), and the units of measure for the elevation values (meters or feet). Especially if you want to display a dataset in combination with other 3D datasets, it is important to define a vertical coordinate system.

Projection on the fly

Geographic and projected coordinate systems are not only a property of a dataset but also a property of an ArcMap data frame. There is no default coordinate system for a data frame in ArcMap. When you open a new map without any layers, the data frame has no coordinate system. To set a coordinate system for the data frame, you can search for the desired coordinate system in the data frame properties, narrow down your search using a spatial filter, and select the appropriate one from a list of geographic and projected coordinate systems. Also, you can add a layer with the desired coordinate system as the first layer to the data frame. The first layer that you add to a data frame defines its coordinate system.

ArcMap displays all layers in a data frame in the coordinate system that has been set for that data frame. This process is called projection on the fly. Projection on the fly makes it possible to align data from different coordinate systems in the same data frame. When you add a layer to a data frame that has a different coordinate system than the layer, the layer is projected on the fly into the coordinate system of the data frame. This means that ArcMap displays a dataset stored in coordinate system A as if it were in coordinate system B. The coordinate system of the source data (stored on disk) is not changed. For example, if the data frame is in a NAD 1983 Albers Equal Area Conic projected coordinate system, and you add a layer that is in a NAD 1983 Universal Transverse Mercator (UTM) projected coordinate system, the layer will be projected on the fly to match the Albers Equal Area Conic. For layers with a different geographic coordinate system than the data frame, see the "Managing different geographic coordinate systems" section later in this chapter.

Exercise 5b: Project layers on the fly

You will now prepare an ArcMap data frame for making a map of Australia and Oceania. Since you anticipate that project planners will need to evaluate the area of potential project areas, you will apply an equal area projection to the data frame and observe how layers are projected on the fly.

1. If necessary, start ArcMap with a blank map and navigate to your ...\DesktopAssociate \Chapter05 folder.

2. If necessary, add Oceania.gdb\Continents to ArcMap.

3. Right-click the Layers data frame and click Properties.

4. **ArcGIS 10.1 only:** In the Coordinate System tab, set the spatial filter to the Outline of the Continents layer. (If you are working in ArcGIS 10.0, skip to step 5).

Hint: In the upper left corner of the Coordinate System tab, click the Spatial Filter button and choose set Spatial Filter.

5. **ArcGIS 10.1:** Search for equal area.
   ArcGIS 10.0: Under Select a coordinate system, expand the Predefined folder.

6. Expand the Projected Coordinate Systems and World folders.

You will set the Cylindrical Equal Area coordinate system for the data frame. Since the map of Australia and Oceania will be centered around the 180 degree meridian, you will modify the projection to use 180 degree as the central meridian.

7. **ArcGIS 10.1:** Right-click Cylindrical Equal Area (world) and choose Copy and Modify.
   ArcGIS 10.0: Click Cylindrical Equal Area (world), then click the Modify button.

8. In the Projected Coordinate System Properties dialog box, change the Central_Meridian value to 180.

9. Name the modified coordinate system **World_Cylindrical_Equal_Area_180** as shown in the graphic below. Then click OK.

Since you will use the Cylindrical Equal Area_180 in the remaining exercises in this chapter, you will add it to your Favorites file.

10. **ArcGIS 10.1:** Right-click Cylindrical Equal Area_180 and choose Add To Favorites.

    ArcGIS 10.0: Under <custom>, Click the Cylindrical Equal Area_180, then click the Add To Favorites button. Click Apply.

Cylindrical Equal Area_180 is now available under Favorites for later use.

Question 1: What are other methods to assign a coordinate system to a data frame?

11. **Apply the modified coordinate system to the data frame and click OK to close the data frame properties.**

The Continents layer is projected on the fly into the modified Cylindrical Equal Area projected coordinate system and draws correctly around the 180 degree central meridian.

12. **From the ...\DesktopAssociate \Chapter05\Oceania.gdb, add Pop_places to the map.**

Pop_places represents populated places in Australia and Oceania.

13. **Open the data frame properties, then click the Coordinate System tab.**

14. **Click the plus sign next to the Layers folder and expand all items listed inside.**

The Layers folder lists the coordinate systems of the layers in the data frame.

Decision point: What is the coordinate system of the Pop_places layer?

The Pop_places layer is stored in the World_Mercator projected coordinate system.

In the map, the Pop_places layer is projected on the fly to match the World_Cylindrical_Equal_Area_180 projected coordinate system of the data frame. Therefore, it aligns with the Continents layer.

Note: The World_Mercator projected coordinate system of the Pop_places layer has also been modified to use 180 degrees as its central meridian.

15. **Click World_Mercator in the Layers folder.**

Selecting a coordinate system in the Layers folder assigns the coordinate system of a layer to the data frame.

16. **Move the Data Frame Properties dialog box to the right side of your screen so that you can see the map. Click Apply. Leave the Data Frame Properties open.**

The World Mercator projected coordinate system has been assigned to the data frame and all layers in the data frame are projected on the fly. The Mercator projection distorts the size and shape of large objects, such as Asia. Note how the shape of the continents changed when you applied the World Mercator coordinate system.

ArcGIS 10.1 only: You can also search for the well-known ID (WKID) number of a coordinate system. **(If you are working in ArcGIS 10.0, skip to step 20.)**

17. **In the Data Frame Properties dialog box, search for 54030.**

54030 is the World Robinson projected coordinate system's WKID number.

18. **Select Robinson (world) and click Apply. Leave the Data Frame Properties open.**

World Robinson is a compromise projection, which distorts all four of these spatial properties (shape, area, direction, and distance). However, since you want to preserve the area of your features, you will change the data frame coordinate system back to World Cylindrical Equal Area_180.

19. **In the Data Frame Properties dialog box, clear the search box and click Search.**

20. **From the Favorites folder, select World_Cylindrical_Equal_Area_180.**

21. **Click OK to close the Data Frame Properties dialog box.**

You reverted the coordinate system of the data frame back to World Cylindrical Equal Area 180 projected coordinate system.

22. **If you are planning to do the next exercise, leave ArcMap open. Otherwise, save the map document and close ArcMap.**

Data in different projected coordinate systems

Projection on the fly displays the layers in the projected coordinate system of the data frame, without changing the coordinate system of the source datasets stored on disk. However, projection on the fly is intended to facilitate mapmaking and cartographic development but should not be used when analyzing or editing your GIS data. To obtain correct analysis results (e.g., accurate area and distance results) or accurate geometries when editing, you should (re-) project the datasets stored on disk (change the projection of the source data).

> ### *Valid input data types for the Project tool*
> Geodatabase feature classes, feature datasets, and shapefiles are valid input for the Project tool. Coverages, raster datasets, and raster catalogs are not supported as input to this tool. To project raster datasets, you use the Project Raster tool.

Exercise 5c: Project feature classes

Projecting will physically recalculate the coordinates of a dataset into a different projected and/or geographic coordinate system. You will now explore different methods for projecting datasets into different coordinate systems.

Dependency: You need to complete exercise 5b before running this exercise. If you have not completed exercise 5b, please do so before starting this exercise.

1. If necessary, open the map document that you saved at the end of exercise 5b

2. From the …\DesktopAssociate \Chapter05\Oceania.gdb, add Countries to the map.

Decision point: What is the coordinate system of the Countries feature class?

___

3. From the Data Management Tools toolbox, Projections and Transformations, Feature toolset, open the Project tool.

4. **ArcGIS 10.1: Run the tool with the following parameters:**
   - Input Dataset or Feature Class: Countries
   - Output Dataset or Feature Class: …\DesktopAssociate \Chapter05\Oceania.gdb**OceaniaCountries**.
   - Output Coordinate system: Favorites\World_Cylindrical_Equal_Area_180

   ArcGIS 10.0: Run the tool with the following parameters:
   - Input Dataset or Feature Class: Countries
   - Output Dataset or Feature Class: …\DesktopAssociate \Chapter05\Oceania.gdb**OceaniaCountries**.
   - Output Coordinate system:
     - Click Select and add the Projected Coordinate Systems\World\Cylindrical Equal Area (world).prj.
     - Click the Modify button and change the central meridian to 180 (as you did in the previous exercise).
     - Change the name of the modified coordinate system to World_Cylindrical_Equal_Area_180 (as you did in the previous exercise).
     - Click OK to close all dialog boxes.

5. Open the Properties of the OceaniaCountries feature class and confirm that the coordinate system is World_Cylindrical_Equal_Area_180.

Question 2: Can you think of another geoprocessing tool that allows you to project features from one coordinate system to another?

Next, you will project the source dataset of the Continents layer by exporting it using the coordinate system of the data frame. Remember in the previous exercise you set the coordinate system of the data frame to the World_Cylindrical_Equal_Area_180 projected coordinate system.

6. In the ArcMap table of contents, right-click the Continents layer, point to Data, and choose Export Data.

7. Check the option to use the same coordinate system as the data frame.

8. Save the Output feature class as …\DesktopAssociate\Chapter05\Oceania.gdb**OceaniaContinents**.

9. Click OK to export the data and Yes to add the layer to the map.

10. Open the Properties of the OceaniaContinents feature class and confirm that its current coordinate system is World_Cylindrical_Equal_Area_180.

Using the Project tool and exporting a layer in "using the same coordinate system of the data frame" are both valid methods for projecting datasets into different projected coordinate systems.

11. If you are planning to do the next exercise, leave ArcMap open. Otherwise, close ArcMap.

Data with an unknown coordinate system

All GIS data has a geographic coordinate system and might have a projected coordinate system (if projected). In order to use a dataset in ArcGIS, the information about its coordinate system must be "known" to ArcGIS. We "tell" ArcGIS about the coordinate system of a dataset by defining it; we create coordinate system documentation in the dataset's properties that ArcGIS can read. If the coordinate system of a dataset is defined, ArcMap finds that information and is then able to display the dataset in the correct location. For example, if the coordinates of a feature class are stored in a NAD 83 Lambert Conformal Conic, this coordinate system must be defined in the feature class properties so that ArcMap can interpret the coordinates as NAD 83 Lambert Conformal Conic.

Most GIS data comes with its coordinate system information defined. However, shapefiles in particular, which store their coordinate system information as an associated (.prj) file, sometimes lack the coordinate system definition, because the file is missing. When a dataset's coordinate system is "unknown" to ArcGIS (undefined), projection on the fly cannot happen. Also, you cannot reproject the dataset without the initial coordinate system being defined.

When you add a layer with an undefined coordinate system, ArcMap gives you a warning message, and the coordinate system information in the Layer Properties (Source tab) will say <Undefined>. However, ArcMap will still attempt to draw the layer by interpreting the "unknown" coordinates as if they are in the coordinate system of the data frame, which may or may not be correct. If the interpretation is incorrect, the layer will not align with other layers in the data frame. To resolve this misalignment, you have to find out what the coordinate system of the dataset is and define it using the Define Projection tool.

Determining the coordinate system of a dataset

If the coordinate system of a dataset is undefined, you first want to find out whether it is in a geographic or projected coordinate system. To find out, add the dataset as a layer to ArcMap and examine the dataset's coordinate extent (in the Layer Properties, Source tab). If the bounding coordinates are between −180 and 180 from left to right, and between 90 and −90 from top to bottom, it is very likely that the dataset is in a geographic coordinate system (then coordinates are longitude/latitude values). If the bounding coordinates in each direction are in the thousands, or even millions (of meters or feet), the dataset is in a projected coordinate system.

Then, you can apply a spatial filter and search for the geographic and projected coordinate systems that are appropriate for the coordinate extent (in the Data Frame Properties, Coordinate System tab). At this point, you will need to make an educated guess about which geographic or projected coordinate system the dataset might be in. It might help to check the metadata of the dataset, make phone calls, e-mail the person or place that provides the data, or check their website. Test your educated guess by setting the coordinate system of an ArcMap data frame to that coordinate system, add a reference layer with a known coordinate system to that data frame, and then add the questionable layer and see if it aligns with the reference layer. Once the "undefined" dataset aligns with the reference dataset, bingo! You have found its correct coordinate system. (Remember, ArcMap interprets the "unknown" coordinates as if they were in the coordinate system of the data frame.)

Exercise 5d: Define the coordinate system of a feature class

You will take a look at the misalignment caused by an undefined coordinate system in the source data of a layer and then fix the problem by defining the coordinate system.

Dependency: You need to complete exercise 5b before running this exercise. If you have not completed exercise 5b, please do so before starting this exercise.

1. If necessary, open the map document that you saved at the end of exercise 5b.

2. From the ...\DesktopAssociate\Chapter05 folder, add Hydro_lines shapefile to the map.

You receive an Unknown Spatial Reference warning message, telling you that Hydro_lines is missing spatial reference information. ArcMap will try to draw it, but it cannot project it on the fly.

3. Click OK to dismiss the warning message.

4. In the table of contents, right-click the Hydro_lines layer and choose Zoom to Layer.

Hydro_lines represents rivers and streams in Australia and Oceania. It draws in a different geographic area and does not align with the other layers in the data frame.

5. Use the Identify tool to identify different locations in the Hydro_lines layer.

In the Identify dialog's Locator area, you will see that the Hydro_lines features fall at longitudes around –179 degrees and latitudes around 0 degrees. This makes it very likely that Hydro_lines is stored in a geographic coordinate system.

6. Symbolize Hydro_lines with a very bright color (e.g., Ginger Pink) and set the Width field of the symbol to a thick line (e.g., 5 pt).

7. Zoom to the full extent.

8. Notice that at this extent, Hydro_lines dataset draws as a small dot near the 180 degrees meridian, near some islands in the ocean.

Hydro_lines originated from the Data and Maps for ArcGIS. Therefore, it is very likely that it is stored in the WGS 1984 geographic coordinate system. You will define the coordinate system of the Hydro_lines shapefile as WGS 1984.

9. From the Data Management Tools toolbox, Projections and Transformations toolset, open the Define Projection tool.

10. Run the tool with the following parameters:
    - Input Dataset or Feature Class: Hydro_lines
    - Coordinate System: Geographic Coordinate Systems\World\WGS 1984

11. **Zoom to the extent of the Hydro_lines layer.**

The Hydro_lines layer now lines up with the Continents.

12. **If you are planning to do the next exercise, leave ArcMap open. Otherwise, close ArcMap.**

> ## *Defining an incorrect coordinate system*
> Defining an incorrect coordinate system for a layer with an "unknown" coordinate system will result in a misalignment between the layer and other reference layers. For example, incorrectly defining the coordinates of a Yellowstone National Park dataset that is really in NAD 1983 UTM, Zone 12N coordinates as NAD 1983 State Plane Wyoming West will result in a shift of about 700 miles. Depending on how different the incorrect and the correct coordinate systems are, the shift can be even more drastic.

Managing data in different geographic coordinate systems

Once the geographic and/or projected coordinate system of a dataset is defined, projection on the fly can happen in the data frame, or you can reproject the dataset as needed.

However, if the layer that you are projecting on the fly has a different geographic coordinate system than the data frame, you have to set a geographic transformation in the data frame properties. Similarly, you have to set a geographic transformation in the Project tool if you transform a dataset to a different geographic coordinate system (or to a projected one with a different underlying geographic coordinate system).

Geographic transformations

A geographic transformation defines the mathematical operation for converting the coordinates of a dataset from one geographic coordinate system to another. There may be more than one transformation method available for converting between any two geographic coordinate systems. Each transformation method is designed for a particular area and each has a different accuracy. For example, there are eight transformation methods for converting between the NAD 1983 and WGS 1984 geographic coordinate systems. For data located in Alaska, you use NAD_1983_To_WGS_1984_2, whereas for data located in Hawaii you use NAD_1983_To_WGS_1984_3. Both NAD_1983_To_WGS_1984_4 and _5 apply to the lower 48 contiguous United States, but NAD_1983_To_WGS_1984_4 is an older transformation and should no longer be used (See ArcGIS Help topic "Choosing an appropriate transformation"). NAD_1983_To_WGS_1984_6, _7, and _8 apply to different provinces of Canada. NAD_1983_To_WGS_1984_1 applies to the entire North American continent, but its accuracy is lower than the one of the other seven (Maher, Lining Up Data, 26).

When choosing a transformation method for your datasets, it's important that you are consistent and use the same transformation method every time. To find out about the appropriate transformation methods in a given area, you may also use a spatial filter to restrict the search in the data frame properties. If there is no direct transformation method available between two geographic coordinate systems, the software will find an appropriate combination of transformation methods.

Suppose you have two layers that are not located in the lower 48 contiguous United States. One of them is in a NAD 1927 UTM coordinate system, and the other one is in NAD 1983 UTM. You will see a north-south shift between the two layers if you do not set a transformation in the data frame properties. Without a transformation set, projection on the fly cannot happen properly, which results in a slight shift between layers (Maher, *Lining Up Data*, 47).

For a complete list of all supported geographic transformations worldwide, refer to the geographic_transformations.pdf in the Documentation folder in your ArcGIS install location (e.g., C:\Program Files (x86)\ArcGIS\Desktop10.1\Documentation\geographic_transformations.pdf) or search for Knowledgebase article 21327 in the Esri Support Center at **support.esri.com**. For more information on geographic transformations, refer to the ArcGIS Help topics "Choosing an appropriate transformation" and "Geographic transformation methods."

Exercise 5e: Set a transformation in the data frame

You will take a look at the shift caused by a layer that is in a different geographic coordinate system than the data frame. You will then fix the problem by setting the transformation in the data frame properties.

Dependency: You need to complete exercises 5b and 5d before running this exercise. If you have not completed exercises 5b and 5d, please do so before starting this exercise.

1. Start ArcMap, if necessary, and navigate to the Hydro_lines_1972 shapefile in your ...\ DesktopAssociate \Chapter05 folder.

2. Add Hydro_lines_1972 to ArcMap.

You receive a warning message that Hydro_lines_1972 has a different geographic coordinate system than the data frame.

Decision point: What is the coordinate system of the Hydro_lines_1972 shapefile?

WGS 1972 is an older version of the World Geodetic System, which was published in 1972.

3. Close the warning message.

Hydro_lines_1972 represents the same rivers and streams in Australia and Oceania that you worked with in the previous exercise.

4. Zoom in closely to any Hydro_lines feature in southern Australia (approximately to a scale of 1:2,000). Turn off the display of all layers except for the Hydro_lines and the Hydro_lines_1972 layers.

Notice the shift between the Hydro_lines and the Hydro_lines_1972 layers.

5. Zoom to the extent of the Hydro_lines_1972 layer.

6. Open the properties of the Layers data frame.

Decision point: What is the geographic coordinate system of the data frame?

The data frame has been assigned the World_Cylindrical_Equal_Area_180 projected coordinate system, which is based on the WGS 1984 geographic coordinate system.

7. In the Coordinate System tab, click the Transformations button.

8. In the Geographic Coordinate System Transformations dialog box, set the geographic transformation as shown in the following graphic. Then click OK.

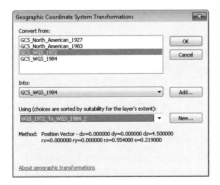

9. Move the Data Frame Properties dialog box to the right side of your screen so that you can see the map. Click OK.

Managing data in different geographic coordinate systems

When you apply the transformation to the data frame, the shift between the Hydro_lines and the Hydro_lines_1972 layers disappears and the layers line up with each other.

10. If you are planning to do the next exercise, leave ArcMap open. Otherwise, close ArcMap.

The transformation that you set in the data frame will affect only the layers in this data frame. The source data of the layers remain in their native geographic coordinate system. To change the geographic coordinate system of the source data, you project them with a transformation.

> ### *NADCON transformation*
> By default, there is no geographic transformation applied, neither in the ArcMap data frame properties nor in the Project tool. You must select a geographic transformation, or no transformation is applied (Maher, *Lining Up Data*, 112). There is one exception: The NAD_1927_To_NAD_1983_NADCON transformation is automatically loaded when you add at least one layer in NAD 1927 and another layer in NAD 1983 to the same data frame. This transformation is used in the lower 48 contiguous United States. If your data is located in a different area, such as Canada or Alaska, you must change the NADCON transformation method to the correct one.

Exercise 5f: Project with a transformation

You will now project a shapefile of rivers and streams in Australia and Oceania using a transformation in the Project tool.

Dependency: You need to complete exercise 5b before running this exercise. If you have not completed exercise 5b, please do so before starting this exercise.

1. Start ArcMap, if necessary, and navigate to the Hydro_lines_1972 shapefile in your ...\DesktopAssociate\Chapter05 folder.

2. If necessary, add Hydro_lines_1972 to ArcMap. Close the warning message.

3. From the Data Management Tools toolbox, Projections and Transformations toolset, Features toolset, open the Project tool.

4. **ArcGIS 10.1: Run the tool with the following parameters:**
   - Input Dataset or Feature Class: Hydro_lines_1972
   - Output Dataset or Feature Class ...\DesktopAssociate\Chapter05\Oceania.gdb**Hydro_lines**.
   - Output Coordinate system: Favorites\World_Cylindrical_Equal_Area_180
   - Geographic Transformation (optional): WGS1972_To_WGS1984_2

   ArcGIS 10.0: Run the tool with the following parameters:
   - Input Dataset or Feature Class: Hydro_lines_1972
   - Output Dataset or Feature Class: ......\DesktopAssociate \Chapter05\Oceania.gdb**Hydro_lines**.
   - Output Coordinate system:
     - Click Select and add the Projected Coordinate Systems\World\Cylindrical Equal Area (world).prj.
     - Click the Modify button and change the central meridian to **180**.

- Change the name of the modified coordinate system to **World_Cylindrical_Equal_Area_180**.
- Click OK to close all dialog boxes. Leave the Project tool open.
- Geographic Transformation (optional): WGS1972_To_WGS1984_2

You transformed the Hydro_lines feature class from the WGS1972 to the WGS 1984 geographic coordinate system.

Note: Exporting a layer using the same coordinate system of the data frame will NOT apply a transformation to the source data of the layer. You have to use the Project tool to transform a feature class into a different geographic coordinate system.

5. Open the properties of the …\Oceania.gdb\Hydro_lines feature class and confirm that its geographic coordinate system is WGS1984.

6. Close the data frame properties.

7. Close ArcMap.

Challenge 2

For each of the following scenarios, choose the right method to solve the coordinate system issues.

1. You add a shapefile as a layer to a new and empty data frame in ArcMap. You find that the coordinate system information in the data frame says "Unknown." What do you have to do?

   a. Define the coordinate system of the data.
   b. Apply a geographic transformation.
   c. Project the data.
   d. Use projection on the fly.

2. Your client has requested vector data in WGS 1984 Web Mercator. Your data is in WGS 1984 World Mercator. What do you have to do?

   a. Define the coordinate system of the data.
   b. Apply a geographic transformation.
   c. Project the data.
   d. Use projection on the fly.

3. The GCS of your data frame is NAD 1983 and the GCS of the incoming vector data is WGS 1984. What do you have to do?

   a. Define the coordinate system of the data.
   b. Apply a geographic transformation.
   c. Project the data.
   d. Use projection on the fly.

4. After you added the first layer to a data frame, you find out that your map needs to be in a different coordinate system than the first layer. What is a valid method for changing the spatial reference of the data frame?

   a. In ArcMap, right-click the data frame and choose Set Spatial Reference.
   b. In ArcMap, right-click the data frame and select Properties and use the Coordinate System tab.
   c. In ArcCatalog, right-click the map document and select Properties to change the coordinate system.
   d. In ArcMap, use the Define Projection tool to change the coordinate system.

5. A layer with no spatial reference or coordinate system is the first layer added to an Untitled.mxd (new map). What will ArcMap assign to the data frame as a spatial reference?

   a. The coordinate system will be defaulted to WGS 1984.
   b. The coordinate system will be inherited from the last opened map document (.mxd) on the user's computer.
   c. The coordinate system will be set to Unknown until the coordinate system is manually set for the data frame.
   d. The coordinate system will remain unknown until another layer with a known coordinate system is added to the data frame.

Answers to chapter 5 questions

Question 1: What are other methods to assign a coordinate system to a data frame?
Answer: In addition to searching and manually selecting it from a list of predefined coordinate systems, you can import the coordinate system of the data frame from a dataset, select it from one of the layers in the data frame (see step 13 in exercise 5b), or, if you are a coordinate system expert, define a your own new coordinate system for the data frame.

Question 2: Can you think of another geoprocessing tool that allows you to project features from one coordinate system to another?
Answer: You can also use the Copy feature tool from the Data Management Tools toolset, Features toolset to project features from one coordinate system to another. How would you do that? Since all geoprocessing tools honor the Environment settings, you could run the tool with the desired coordinate system as the Output Coordinate system Environment Setting.

Answers to challenge questions

Challenge 1

Correct answers shown in bold.

1. You want to make a thematic map of population density in different counties.

   a. Geographic
   b. Projected (For a population density map, it is important to show true area. Therefore, you should use an equal area projected coordinate system.)

2. You want to create the boundary lines of land parcels from the bearings and distances of provided survey measurements.

   a. Geographic
   b. Projected (When creating parcel boundary lines from survey measurements it is important to create accurate geometries. A projected coordinate system will allow you to do that.)

3. You want to store the epicenters of earthquakes in a geodatabase that can be accessed through a website that allows users to download the data.

   a. Geographic (Since users who download the data may project the data into the projected coordinate system that they need, it is appropriate to store the data in a geographic coordinate system.)
   b. Projected

4. You want to collect samples of ozone concentrations in different cities worldwide at different times of the year.

   a. Geographic (Each ozone sample will be collected at a particular point location, which may be recorded by GPS. A geographic coordinate system is appropriate for exactly defining and storing the sample location.)
   b. Projected

5. You want to use weather data in a map that will be used to plan the most efficient airplane routes between different continents of the world.

   a. Geographic
   b. Projected (To use the weather data in a map that will be used for navigation purposes, you need to use a projected coordinate system.)

6. You want to make a quick map to evaluate if a dataset has a sufficient amount of spatial detail (scale) and make sure it covers your study area. (You will review the concept of scale in chapter 6).

   a. Geographic (To make a quick map for quality control, a geographic coordinate system is appropriate.)
   b. Projected

Challenge 2

1. You add a shapefile as a layer to a new and empty data frame in ArcMap. You find that the coordinate system information in the data frame says "Unknown." What do you have to do?

   a. Define the coordinate system of the data.
   b. Apply a geographic transformation.
   c. Project the data.
   d. Use projection on the fly.

2. Your client has requested vector data in WGS 1984 Web Mercator. Your data is in WGS 1984 World Mercator. What do you have to do?

   a. Define the coordinate system of the data.
   b. Apply a geographic transformation.
   c. Project the data.
   d. Use projection on the fly.

3. The GCS of your data frame is NAD 1983 and the GCS of the incoming vector data is WGS 1984. What do you have to do?

   a. Define the coordinate system of the data.
   b. Apply a geographic transformation.
   c. Project the data.
   d. Use projection on the fly.

4. After you added the first layer to a data frame, you find out that your map needs to be in a different coordinate system than the first layer. What is a valid method for changing the spatial reference of the data frame?

   a. In ArcMap, right-click the data frame and choose Set Spatial Reference.
   b. In ArcMap, right-click the data frame and select Properties and use the Coordinate System tab.
   c. In ArcCatalog, right-click the map document and select Properties to change the coordinate system.
   d. In ArcMap, use the Define Projection tool to change the coordinate system.

5. A layer with no spatial reference or coordinate system is the first layer added to an Untitled.mxd (new map). What will ArcMap assign to the data frame as a spatial reference?

   a. The coordinate system will be defaulted to WGS 84.
   b. The coordinate system will be inherited from the last opened map document (.mxd) on the user's computer.
   c. The coordinate system will be set to Unknown until the coordinate system is manually set for the data frame.
   d. The coordinate system will remain unknown until another layer with a known coordinate system is added to the data frame.

Key terms

Coordinate system: A reference framework consisting of a set of points, lines, and/or surfaces, and a set of rules, used to define the positions of points in space in either two or three dimensions. The Cartesian coordinate system and the geographic coordinate system used on the earth's surface are common examples of coordinate systems.

Map projection: A method by which the curved surface of the earth is portrayed on a flat surface. This generally requires a systematic mathematical transformation of the earth's graticule of lines of longitude and latitude onto a plane. Some projections can be visualized as a transparent globe with a light bulb at its center (though not all projections emanate from the globe's center) casting lines of latitude and longitude onto a sheet of paper. Generally, the paper is either flat and placed tangent to the globe (a planar or azimuthal projection) or formed into a cone or cylinder and placed over the globe (cylindrical and conical projections). Every map projection distorts distance, area, shape, direction, or some combination thereof.

Geographic transformation: A systematic conversion of the latitude-longitude values for a set of points from one geographic coordinate system to equivalent values in another geographic coordinate system. Depending on the geographic coordinate systems involved, the transformation can be accomplished in various ways. Typically, equations are used to model the position and orientation of the "from" and "to" geographic coordinate systems in three-dimensional coordinate space; the transformation parameters may include translation, rotation, and scaling. Other methods, including one used in transformations between NAD 1927 and NAD 1983, use files in which the differences between the two geographic coordinate systems are given for a set of coordinates; the values of other points are interpolated from these.

Resources

- ArcGIS Help 10.1 > Guide Books > Map projections >
  - What are map projections?
  - Projection basics for GIS professionals
- ArcGIS Help 10.1 > Guide Books > Map projections >Geographic coordinate systems
  - What are geographic coordinate systems?
- ArcGIS Help 10.1 > Guide Books > Map projections >Projected coordinate systems
  - What are projected coordinate systems?
- ArcGIS Help 10.1 > Guide Books > Map projections > Geographic transformations
  - Choosing an appropriate transformation
  - Geographic transformation methods
- ArcGIS Help 10.1 > Guide Books > Map projections >Vertical coordinate systems
  - What are vertical coordinate systems?

- ArcGIS Help 10.1 > Desktop > Editing> Fundamentals of editing > About the editing environment
  - About editing data in a different projection (projecting on the fly)
- ArcGIS Help 10.1 > Geodata > Data Types > Shapefiles
  - Defining a shapefile's coordinate system
- Margaret M. Maher, *Lining Up Data in ArcGIS: A Guide to Map Projections* (Redlands, CA: Esri Press, 2010).
- Knowledge base article 21327 at the Esri Support Center at **support.esri.com**: How To: Select the correct geographic (datum) transformation when projecting between datums

chapter 6
Evaluating data

Evaluating data for a task 88
   Exercise 6a: Evaluate datasets
  Challenge

Data documentation 92
  Item description
   Exercise 6b: Edit the item description of a feature class
  Metadata
   Exercise 6c: Examine metadata
  Metadata templates

Answers to chapter 6 questions 96

Answers to challenge questions 97

Key terms 98

Resources 98

Chapter 6: Evaluating data

Evaluating data and choosing the right datasets for a project is the crucial first step to a successful GIS project. But how do you decide whether or not a dataset is appropriate for the project? For example, some datasets have been created with just enough spatial detail to be used at a continental or country scale, while others have been created with more detail to be used at regional scales. Also, what attributes do you need? Even if a dataset seems to have the right geometry and the right attributes, you might still ask yourself how old and how reliable the dataset is. Metadata, the documentation of a dataset, may help a lot to answer these questions. Proper documentation of your data may be crucial when it comes to evaluating whether or not to use a dataset for a given task. You will review the different metadata styles available in ArcGIS and, most importantly, what information will you find in the metadata.

Skills measured
- Given a dataset and a task, assess the appropriateness of the dataset for the task.
- Compare and contrast the different methods to document data in ArcGIS Desktop.

Prerequisites
- Knowledge of vector and raster data models
- Knowledge of what metadata is
- Knowledge of scale

Evaluating data for a task

When you evaluate a dataset to be used in a project, you anticipate the tasks you will need to perform on the dataset. For example, will it be used for mapping, for queries, for analysis, or for multiple tasks? You may ask yourself the following: Does it cover your study area? Does it have sufficient spatial detail and the right attributes? How old is the data? Ideally, a dataset already has all the properties you need to perform all the tasks, but oftentimes you may need to modify or update a dataset to make it really appropriate for a task.

There are a few data properties to evaluate when choosing a dataset for a task, as presented in table 6.1.

Table 6.1 Choosing a dataset (continued)		
Properties	**Considerations**	**Possible approaches**
Data model: vector or raster	• Are you modeling features with discrete boundaries or continuous phenomena? • Are most of your other data holdings vector data or raster data?	Discrete raster datasets can be converted to vector datasets, and vice versa (using Conversion tools).
Geometry type (vector data): point, line, or polygon	The geometry type used to represent a real-world entity is largely determined by the scale at which the datasets will be used. For example, at a large scale, you would represent a city as a polygon, while at a small scale, you would abstract it as a point.	Geometry types can be converted, using the following tools and methods: • Geoprocessing tools: Points to Line, Features to Polygon, Polygon to Lines. • By calculating centroids (points) from polygons. • Creating Thiessen polygons from points.

Continued on next page

Table 6.1 Choosing a dataset (continued)

Properties	Considerations	Possible approaches
Scale: amount of spatial detail	• In vector datasets, scale determines the amount of spatial detail. • In raster datasets, cell size determines the amount of spatial detail (spatial resolution). • A vector dataset should be used at scales similar to the scale at which it has been captured: at larger scales, boundaries appear jagged in the map, and analysis results may be inaccurate, if data was collected at a more generalized scale.	• If a dataset has too much spatial detail you can generalize it (using Cartography tools > Generalization tools, or Data Management > Generalization tools). • If a dataset has too little spatial detail for the task, you would most likely not use it.
Spatial accuracy and resolution	• Spatial accuracy is a measure of "closeness to reality," i.e., how closely a dataset represents features in the real world. • Spatial resolution is the smallest allowable separation between two coordinate values (x-, y-, z-, or m-values) in a feature class.	• Running a geodatabase topology may improve the spatial accuracy of a dataset (see chapter 14). • A lack of resolution is not fixable.
Attributes	• The dataset must contain the attribute fields you need for symbolization, labeling, queries, or analysis. • Attribute fields and values must be in the correct format.	• Existing attributes can be combined/parsed to create new attribute values. • Additional attributes can be added to an attribute table through joins, relates and relationship classes.

Data currency	• Acceptable time frame from which data can be used
	• Whether or not a dataset is outdated depends on the following:
	  • Level of currency required
	  • Amount of change that has occurred over time / space
		• In general, current data should be used.
		• In some cases, outdated datasets can be updated (e.g., by digitizing new features).

Data completeness	All the features in the study area and their attribute must be present in the data.	In some cases, incomplete datasets can be completed (e.g., by digitizing new features, assigning attribute values).
Data consistency	Does the combination of features and attributes make sense? • Logical consistency: consistent combination of attribute values (e.g., consistent address numbering along a street) • Physical consistency: consistent spatial relationships among features (e.g., consistent placement of water meters at the end of a service line)	If inconsistencies are not too severe, inconsistent dataset can be fixed by editing features and attributes.

Continued on next page

Table 6.1 Choosing a dataset (continued)

3D-enabled or not	A dataset can be used for 3D visualization and analysis if it stores the following: • Z-values as part of its geometry (3D features). • Elevation or height as an attribute.	You can convert 2D features to 3D. (See the Desktop Help topic: "Converting 2D features to 3D features.")
Coordinate system	A dataset needs to be in the correct geographic and/or projected coordinate system required for the task.	A dataset can be projected and/or transformed into the required coordinate system.
Data credibility / reliability of the data source	• Presence and completeness of metadata. • Authoritative data are most reliable. • Unknown data sources pose a risk.	Acquire metadata if data source is unknown.

Exercise 6a: Evaluate datasets

For the exercises in this chapter, suppose you are creating a map that will be used by project planners of various organizations to find potential sites for cultural heritage and nature conservation projects. The extent of your database is the continents of Australia and Oceania, which include Australia, New Zealand, and Papua New Guinea, as well as the islands of the South Pacific Ocean, including the Melanesia and Polynesia groups. Parts of South Asia will be visible in the map as well.

The map will be used at different scales. At a continental scale, it may provide an overview of different project areas, whereas at a regional scale, it will let the user analyze different project areas in more detail. An intern extracted various data layers from world data on the Data and Maps for ArcGIS media kit and has added them to a map document. Now it's your task to evaluate the data and decide which datasets to use for the map.

1. **ArcGIS 10.1: Start ArcMap and open the …\DesktopAssociate \Chapter05\OceaniaData.mxd.**

   ArcGIS 10.0: Start ArcMap and open the …\DesktopAssociate \Chapter05\OceaniaData_100.mxd.

You will evaluate which polygon feature class to use for the coastlines and the land masses.

At a continental scale (around 1:50,000,000), the coastline of Australia is clearly defined by the Countries layer.

2. **Open the attribute table of Continents and Countries and examine the attributes.**

Decision point: Most likely, project planners will use the map as a starting point to find out about local authorities and laws. Which of the two layers is more appropriate for this purpose?

The Countries layer seems more appropriate because it contains information about the government and status of the different islands.

3. **Zoom to the Samoa bookmark.**

At this extent, it becomes obvious that the Countries layer does not have sufficient spatial detail to be used at this scale.

4. **Turn on the Administrative areas layer.**

The Administrative areas layer of the countries contains moderate spatial detail to be used at a regional scale. The source data for this layer is called Admin_poly and is stored in the ...DesktopAssociate\Chapter06\Oceania.gdb.

5. In the Catalog window, open the Properties of the Admin_poly feature class and click the XY Coordinate System tab.

Decision point: Is the coordinate system of the Admin_poly feature class appropriate for mapping and analysis purposes?

For mapping purposes, you could leave the source data in the WGS 1984 geographic coordinate system and project the layer on the fly. However, if the map is going to be used for analysis purposes, you should project the source data into an appropriate projected coordinate system.

6. Close the Feature Class Properties dialog box.

7. Turn on the Administrative lines layer.

The Administrative lines layer shows the boundaries of the administrative areas with the same spatial detail as the Administrative areas.

8. Open the attribute tables of Administrative lines and Administrative areas.

Decision point: You anticipate that project planners will also need to analyze population density in the potential project areas. Which of the two layers has the appropriate attributes to do that?

The POP_ADMIN field in the Administrative areas layer contains population in the administrative areas. Area is stored in different units in the SQKM and SQMI fields. These fields can be used to calculate population density in a new field.

Note: The Shape_Area field in its current format is not useful for area calculation. Since the source data of the Administrative areas layer is stored in a geographic coordinate system, the Shape_Area field contains the area of the polygons in square decimal degrees. If you wanted to use the Shape_Area field for area calculations, you have to project the Admin_poly feature class into a projected coordinate system that preserves area.

9. Close the table window.

10. Zoom to the full extent.

Decision point: Project planners also would like to see elevation on the map and have the option to derive slope and aspect surfaces of their project areas. What kind of dataset is needed for this task?

Analyzing elevation and deriving slope and aspect require a raster dataset that stores elevation, for example, a digital elevation model.

11. Turn on the Elevation layer.

Elevation represents a 30 arc seconds (+/−1 Km) digital elevation model (DEM). You will quickly explore it using the image analysis window.

12. From the Windows menu, choose Image Analysis.

13. In the Image Analysis window, select the Elevation layer.

Chapter 6: Evaluating data

14. **In the Processing section of the Image Analysis window, click the Shaded Relief button.**

This will create a color-shaded relief layer and add it to the table of contents.

15. **Close the Image Analysis window.**

Question 1: So far, you have evaluated only a few basic properties of the datasets, such as their scale, attributes, and coordinate system. Where would you find the information to evaluate the spatial accuracy and resolution, the data currency, and the reliability of the source of a dataset?

16. **If you are planning to do the next exercise, leave ArcMap open. Otherwise, close ArcMap.**

Challenge

	OBJECTID *	Shape *	CITY_NAME	ADMIN_NAME	CNTRY_NAME	STATUS	POP
	55	Point	Alotau	Milne Bay	Papua New Guinea	Provincial capital	-999
	56	Point	Port Moresby	National Capital	Papua New Guinea	National capital and provincial capi	283733
	57	Point	Funafuti	Tuvalu	Tuvalu	National capital	4749
	58	Point	Honiara	Solomon Is.	Solomon Is.	National capital	56298
	59	Point	Port Vila	Vanuatu	Vanuatu	National capital	35901
	60	Point	Suva	Fiji	Fiji	National capital	77366
	61	Point	Noumea	New Caledonia	New Caledonia	National and provincial capital	93060
	62	Point	Yaren	Nauru	Nauru	National capital	-999
	63	Point	Whangarei	Northland	New Zealand	Provincial capital	50900
	64	Point	Auckland	Auckland	New Zealand	Provincial capital	417910
	65	Point	Tauranga	Bay of Plenty	New Zealand	Provincial capital	110338
	66	Point	Hamilton	Waikato	New Zealand	Provincial capital	152641
	67	Point	Gisborne	East Cape	New Zealand	Provincial capital	34274

Created by the author; data from Esri Data & Maps, 2010, courtesy of ArcWorld.

1. Which attribute queries can you perform in a point dataset of cities in Australia and Oceania as shown in the attribute table above?

    a. Select cities with an area greater than 500 square kilometers
    b. Select cities in New Zealand that have a population greater than 100,000
    c. Select cities that have grown in population by more than 20% in the last 10 years
    d. Select cities that border the ocean

2. You want to find a suitable area for a new vineyard. It must be in an agricultural area with mostly southern exposure that is at least 5 square kilometers in size. Based on these criteria, which of the following datasets can help you find potential sites? (Choose two.)

    a. Vegetation cover
    b. Landuse
    c. Digital elevation model
    d. Geology type

3. You are looking for a feature class to visualize buildings in 3D. Which properties does it need to have?

    a. Exaggeration attribute
    b. M-values
    c. Height attribute
    d. Slope attribute

Data documentation

Information that describes resources in ArcGIS is called metadata. Resources such as datasets, map documents, geoprocessing tools, and toolboxes can be documented with metadata. In the metadata of a dataset, you can find a lot of the information to evaluate the appropriateness of a dataset for a task.

In ArcGIS, metadata can be viewed in different styles. The style filters out different metadata elements and highlights different types of information. The default is the Item Description metadata style, which provides a simple set of metadata for a resource. There are four other metadata styles available in ArcGIS, which allow you to create and view more detailed metadata for a resource.

For more information about metadata styles, refer to the Desktop Help topic "Metadata styles and standards."

Item description

The item description provides a compact, one-page description of a resource, which is indexed and available for searching. The item description metadata style is designed to facilitate metadata creation in ArcGIS and is suitable for anyone who doesn't need to adhere to specific metadata standards. A metadata standard is a document identifying content that should be provided in your metadata. For example, the "ISO 19115 Geographic information—Metadata" is a metadata standard that has been created by ISO (International Organization for Standardization) committees. Most content in the item description is the same for all types of resources (e.g., title, thumbnail, description, and tags), but elements vary slightly. For example, a tool's item description includes information about its usage and parameters.

The elements listed in table 6.2 are included in the item description of a dataset.

Table 6.2 Elements of a dataset's item description	
Element	Description
Title	The name by which the dataset is known.
Thumbnail	A small, static picture of the dataset with default symbology.
Tags	A set of terms that are used by ArcGIS for searching for the dataset (e.g., in the Search window). Terms should be provided as a comma-separated list.
Summary (Purpose)	A summary of the intentions with which the dataset was developed. In metadata standards this information is known as a purpose.
Description (Abstract)	A brief narrative summary of the dataset's content. In metadata standards this information is known as an abstract.
Credits	A recognition of those who created or contributed to the dataset.
Use limitations (ArcGIS 10.1, in ArcGIS 10.0: Access and use limitations)	Describes limitations affecting the appropriate use of the dataset, for example: "Do not redistribute" or "Not for commercial use."
Extent (ArcGIS 10.1)	The bounding coordinates of the dataset, i.e., the westernmost, easternmost, southernmost, and northernmost coordinates.
Scale range (ArcGIS 10.1)	The maximum (zoomed in) and minimum (zoomed out) scale for using the dataset

Exercise 6b: Edit the item description of a feature class

You will continue working with the Australia and Oceania map from the previous exercise. All the datasets in the Oceania.gdb have been extracted from a worldwide dataset in the Data and Maps for ArcGIS media kit. You will use the urban area feature class as an example and update its item description to reflect the new extent of the dataset.

1. **ArcGIS 10.1:** If necessary, start ArcMap and open the ...\DesktopAssociate \Chapter06\OceaniaData.mxd.

    ArcGIS 10.0: If necessary, start ArcMap and open the ...\DesktopAssociate \Chapter06\OceaniaData_100.mxd.

2. Add the Urban_areas feature class to the map.

3. Rename the layer **Urban Areas**.

4. In the Catalog window, right-click the Urban_areas feature class and click Item Description.

Decision point: Based on the current item description, which metadata elements do you think need to be updated?

---

5. In the Item Description window, click Edit.

6. Update the item description with the following information:

    - Thumbnail: Use the ...\DesktopAssociate \Chapter06\UrbanAreas.gif

Hint: Click Update and navigate to the UrbanAreas.gif.

    - Tags: location, Australia, Oceania, boundaries, 2010, polygon, cities, urban areas
    - Summary (Purpose): **Urban Areas represents urban areas in Australia and Oceania. It has been derived from World Urban Areas from Esri Data and Maps 2010. Use for locating urban areas in Australia and Oceania.**
    - Description: **Urban Areas represents urban areas in Australia and Oceania. It has been derived from World Urban Areas from Esri Data and Maps 2010.** World Urban Areas represents the major urban areas of the world.

Hint: You can copy and paste from the Summary.

    - **ArcGIS 10.1 only:** Appropriate scale range Country (left slider) to City (right slider)
    - **ArcGIS 10.1 only:** Extent:
        - West: **-175.214133**
        - East: **178.468740**
        - South: **-53.184814**
        - North: **69.365612**

7. Click Save.

The item description has been updated with the information that you entered.

8. Close the Item Description window.

9. If you are planning to do the next exercise, leave ArcMap open. Otherwise, close ArcMap.

Metadata

To provide more detailed documentation, or if you must create metadata that complies with a metadata standard, choose one of the other metadata styles available in ArcGIS. Besides the Item Description, the following metadata styles are available:

- FGDC CSDGM Metadata
- INSPIRE Metadata Directive
- ISO 19139 Metadata Implementation
- North American Profile of ISO 19115

These metadata styles allow you to create complete metadata for a resource and support different metadata standards. In addition to the information included in the item description, the following pieces of information are included in the metadata of a dataset:

- Contact information of individuals or groups who can answer questions about the dataset
- Names and properties of the attribute fields
- The name and parameters of the dataset's geographic and optionally projected coordinate system
- A dataset's scale (spatial resolution)
- Information about data quality
- Information about a dataset's geoprocessing history
- The dataset's update frequency
- Limitations of use, legal and other constrains
- Contact information and instructions on how to download or obtain the dataset
- Information about when the metadata of the dataset was created

Some metadata elements are updated automatically. For example, the information about a dataset's coordinate system and attribute fields are automatically included and updated in the metadata.

For more information on the ArcGIS metadata format, refer to the ArcGIS Help topic "The ArcGIS metadata format."

Exercise 6c: Examine metadata

1. **ArcGIS 10.1:** If necessary, start ArcMap and open the ...\DesktopAssociate \Chapter06\OceaniaData.mxd.
   ArcGIS 10.0: If necessary, start ArcMap and open the ...\DesktopAssociate \Chapter06\OceaniaData_100.mxd.

2. From the Customize menu, open the ArcMap Options and click the Metadata tab.

3. Change the Metadata Style to the ISO19139 Metadata Implementation Specification.

4. If the option to "Automatically update metadata when it is viewed" is not checked, check it.

5. Click OK to change the metadata style.

6. Add the Pop_places feature class to the map and rename the layer to Populated Places.

7. Open the item description.

8. Scroll down to the bottom of the item description.

9. **ArcGIS 10.0 only:** Click ArcGIS metadata to expand the ArcGIS metadata of the Pop_places feature class.

At the bottom of the item description, the complete metadata of the feature class is now visible. This is because you have changed the metadata style.

10. Scroll down and examine the metadata to answer the following questions:

Question 2: How old is the dataset?

Hint: Press Ctrl+F on your keyboard to bring up the Find tool. Search for occurrences of the word "date."

Question 3: What is the largest scale for displaying the Pop_places feature class?

Hint: Look under Resource Details.

Question 4: Whom do you contact for questions about the Pop_places feature class?

Question 5: What is the coordinate system of the Pop_places feature class?

Question 6: What do the values in the Rank attribute field represent, and what do low numbers mean?

Hint: Using Ctrl+F, search for occurrences of the word "rank."

11. In the Item Description window, click Edit.

With a metadata style other than the Item Description, the editor contains a table of contents on the left side, with several sections that let you provide complete metadata content for the dataset.

12. Click a few sections that interest you and notice that the page in the editor updates.

You are not going to edit any metadata in this exercise, so you will exit the editor without saving.

13. Click Exit.

14. Close the Urban Areas Item Description window.

15. Close ArcMap.

> *Metadata templates*
>
> When you create metadata for multiple datasets in a database, it is very likely that much of the information that you enter in the metadata editor is the same. For example, use constraints, distribution information, or the various kinds of contact information may be the same for all the datasets you are documenting. To save time, you could type in this information once, and then export the metadata to an .xml file. For each new dataset for which you create metadata, you will import the template .xml file first, before you type in only the unique information. The tools for exporting and importing metadata are available from the Item Description window or in the Conversion tools, Metadata toolset.

Answers to chapter 6 questions

Question 1: So far, you have evaluated only a few basic properties of the datasets, such as their scale, attributes, and coordinate system. Where would you find the information to evaluate the spatial accuracy and resolution, the data currency, and the reliability of the source of a dataset?
Answer: You would find this information in the metadata.

Question 2: How old is the dataset?
Answer: Under Citation, the Publication Date states June 30, 2010.

Question 3: What is the largest scale for displaying the Pop_places feature class?
Answer: Under Resource Details, it says Largest scale when displaying the data: 1:250,000

Question 4: Whom do you contact for questions about the Pop_places feature class?
Answer: You contact Esri.

Question 5: What is the coordinate system of the Pop_places feature class?
Answer: The Pop_places feature class is in the WGS 1984 geographic coordinate system.

Question 6: What do the values in the Rank attribute field represent, and what do low numbers mean?
Answer: The values in the Rank field rank populated places based on their level of importance. The lower the number in the Rank field, the more important the feature is.

Answers to challenge questions

Correct answers shown in bold.

	OBJECTID *	Shape *	CITY_NAME	ADMIN_NAME	CNTRY_NAME	STATUS	POP
	55	Point	Alotau	Milne Bay	Papua New Guinea	Provincial capital	-999
	56	Point	Port Moresby	National Capital	Papua New Guinea	National capital and provincial capi	283733
	57	Point	Funafuti	Tuvalu	Tuvalu	National capital	4749
	58	Point	Honiara	Solomon Is.	Solomon Is.	National capital	56298
	59	Point	Port Vila	Vanuatu	Vanuatu	National capital	35901
	60	Point	Suva	Fiji	Fiji	National capital	77366
	61	Point	Noumea	New Caledonia	New Caledonia	National and provincial capital	93060
	62	Point	Yaren	Nauru	Nauru	National capital	-999
	63	Point	Whangarei	Northland	New Zealand	Provincial capital	50900
	64	Point	Auckland	Auckland	New Zealand	Provincial capital	417910
	65	Point	Tauranga	Bay of Plenty	New Zealand	Provincial capital	110338
	66	Point	Hamilton	Waikato	New Zealand	Provincial capital	152641
	67	Point	Gisborne	East Cape	New Zealand	Provincial capital	34274

Created by the author; data from Esri Data & Maps, 2010, courtesy of ArcWorld.

1. Which attribute queries can you perform in a point dataset of cities in Australia and Oceania as shown in the attribute table above?

    a. Select cities with an area greater than 500 square kilometers
    b. Select cities in New Zealand that have a population greater than 100,000
    c. Select cities that have grown in population by more than 20% in the last 10 years
    d. Select cities that border the ocean

2. You want to find a suitable area for a new vineyard. It must be in an agricultural area with mostly southern exposure that is at least 5 square kilometers in size. Based on these criteria, which of the following datasets can help you find potential sites? (Choose two.)

   a. Vegetation cover
   b. Landuse
   c. Digital elevation model
   d. Geology type

3. You are looking for a feature class to visualize buildings in 3D. Which properties does it need to have?

   a. Exaggeration attribute
   b. M-values
   c. Height attribute
   d. Slope attribute

Key terms

Accuracy: The degree to which a measured value conforms to true or accepted values. Accuracy is a measure of correctness. It is distinguished from precision, which measures exactness.

Resolution: In ArcGIS, the smallest allowable separation between two coordinate values in a feature class. A spatial reference can include x-, y-, z-, and m-resolution values. The inverse of a resolution value was called a precision or scale value prior to ArcGIS 9.2.

Metadata: Information that describes the content, quality, condition, origin, and other characteristics of data or other pieces of information. Metadata for spatial data may describe and document its subject matter; how, when, where, and by whom the data was collected; availability and distribution information; its projection, scale, resolution, and accuracy; and its reliability with regard to some standard. Metadata consists of properties and documentation. Properties are derived from the data source (for example, the coordinate system and projection of the data), while documentation is entered by a person (for example, keywords used to describe the data).

Metadata standard: A document identifying content that should be provided to describe geospatial resources such as maps, map services, vector data, imagery, and relevant nonspatial resources. A metadata standard may also describe the format in which the content should be stored.

Resources

- ArcGIS Help 10.1 > Geodata > Geodatabases > Defining the properties of a geodatabase > Spatial references in geodatabases
  - The properties of a spatial reference
- ArcGIS Help 10.1 > Geodata > Geodata > Data types > Metadata
  - What is metadata?
  - Metadata styles and standards
  - Importing and exporting metadata
- ArcGIS Help 10.1 > Geodata > Data types > Metadata > Editing metadata
  - A quick tour of creating and editing metadata

- Instructional Podcast: **http://www.esri.com/news/podcasts/instructional-series.html**
  - QA/QC for GIS Data: Initial Quality Control Checks
  - QA/QC for GIS Data: Visual Inspection and Quality Control
- ArcUser articles: **http://www.esri.com/news/arcuser/**
  - "Maintaining Currency, Minimizing Disruptions"
  - "Speed Metadata Creation with Templates"

chapter 7
Associating tables

Types of table relationships   102
    One-to-one
    Many-to-one
    One-to-many
    Many-to-many cardinalities

Table associations   104
    Table joins
    Exercise 7a: Join tables
    Table relates
    Exercise 7b: Relate tables
    Challenge 1

Relationship classes   109
    Types of relationship classes
    Exercise 7c: Create a simple relationship class
    Exercise 7d: Use the relationship class
    Challenge 2

Answers to chapter 7 questions   114

Answers to challenge questions   114

Key terms   116

Resources   116

Chapter 7: Associating tables

Table associations link the records (rows) in one table with records in another table. Why are they important? Suppose you have a features class of roads with lots of road attributes: road names, address ranges, number of lanes, speed limits, pavement, road owners, their contact information, a road maintenance schedule, and so on. Having all these attributes in the feature attribute table could duplicate a lot of information that may be stored in other tables and make the roads feature class perform very poorly, for example, when geoprocessing is applied to it. Therefore, most database design guidelines promote organizing your database into separate tables, each table being focused on a specific related set of attributes. When you need to access the information that is in another table—for example, for symbolizing or querying your features in a map—you create a table association that links the tables together.

Skills measured
- Explain the circumstances under which it is appropriate to create a new relate in ArcMap.
- Given a scenario, determine appropriate specifications for creating a new join (nonspatial).
- Given a scenario, define the appropriate parameters for a relationship class.

Prerequisites
- Knowledge of the relationship between feature geometry and attributes
- Knowledge of field data properties

Types of table relationships

Once you have decided to associate two tables, you must explore how the records in both tables will be matched. In other words, you must determine the relationship or cardinality between the two tables, that is, how many records in the first table match how many records in the second table. When you evaluate the relationship or cardinality between two tables, you start with the "origin" table, that is, the table to which you want to associate another table. The table containing the attributes you want to associate is the destination table. To determine the table relationship, you must find out how many records in the origin table match to how many records in the destination table, and vice versa. Records are matched based on values in the common or key fields.

One-to-one

The simplest type of relationship between two tables is one-to-one (1:1): each record in the origin table matches exactly one record in the destination table. Figure 7.1 shows a one-to-one relationship. Suppose you have an origin table with country attributes and a destination table listing the status of each country. Each record in the Countries table (left) matches exactly one record in the Country_info table (right). Therefore the table relationship is one-to-one (1:1).

Question 1: In figure 7.1, which is the common key field in the Countries and the Country_Info tables?

Countries (origin)

OBJECTID *	Shape *	CNTRY_NAME *
1	Polygon	American Samoa
2	Polygon	Australia
3	Polygon	Baker I.
4	Polygon	Brunei
5	Polygon	Cambodia
6	Polygon	Christmas I.

One-to-One →

Country_Info (destination)

OBJECTID *	CNTRY_NAME *	STATUS
2	American Samoa	Non-Self-Governing Territory of the USA
24	Australia	UN Member State
11	Baker I.	US Territory
16	Brunei	UN Member State
39	Cambodia	UN Member State
25	Christmas I.	Australian Territory

Figure 7.1 In a one-to-one relationship, each record in the origin table corresponds to exactly one record in the destination table. Created by the author, from Esri Data & Maps, 2010; data courtesy of ArcWorld and ArcWorld Supplement.

Many-to-one

If more than one record in the origin table matches a single record in the destination table, the tables have a many-to-one (M:1) cardinality. Figure 7.2 shows such a many-to-one relationship. Suppose you have an origin table of zoning areas (left) and a destination table of zoning descriptions (right). Since there are multiple zoning areas with the same code, multiple records in the Zoning table match with exactly one record in the Zoning Description table. The table cardinality is many-to-one.

Question 2: In figure 7.2, which is the common key field in the Zoning and the Zoning Description tables?

Zoning (origin)

OBJECTID *	SHAPE *	Zone_Code
1	Polygon	MDR
2	Polygon	VAC
3	Polygon	LDR
4	Polygon	TNS
5	Polygon	MDR
6	Polygon	MDR

Many-to-One

Zoning Description (Destination)

OBJECTID *	Zone_Code	Description
1	HDR	High-Density Residential
2	MDR	Medium-Density Residential
3	LDR	Low-Density Residential
4	VAC	Vacant
5	TNS	Transitional
6	SDP	Special Development Plan

Figure 7.2 In this example, zoning codes are described in more detail in the Zoning Description table (right). Multiple records in the Zoning attribute table match the same record in the Zoning Descriptions table, which makes for a many-to-one (M:1) table cardinality. Created by the author.

One-to-many

This is the reverse of the previous example. Here the Stores table is the origin table, to which you want to associate the Employees table (destination table). When one record in the origin table matches more than one (many) records in the destination table, you have a one-to-many (1:M) cardinality between the two tables (figure 7.3).

Question 3: In figure 7.3, which is the common key field in the Stores and the Employees tables?

Stores (origin)

OBJECTID *	SHAPE *	Store_ID	Revenue
1	Point	25	1,400,000
2	Point	26	354,000
3	Point	27	640,000
4	Point	28	358,600
5	Point	29	680,000
6	Point	30	1,305,000

One-to-Many

Employees (destination)

OBJECTID *	Store_ID	Name	Employee_Number
1	28	Jane Smith	3470
2	28	John Miller	3471
3	29	Theresa Evans	3472
4	30	Richard Tailor	3473
5	28	Andrew Mason	3474
6	26	Mary Daniels	3475

Figure 7.3 In this example, each store in the Stores table (left) has multiple employees in the Employees table (right). Thus, one record in the Stores table matches multiple records in the Employees table, which makes for a one-to-many table cardinality. Created by the author.

It is very important to determine the right type of cardinality between the two tables, because it will determine the type of association you will create between them: a join, a relate, or a relationship class.

> ### Many-to-many cardinalities
>
> A more complex type of cardinality between two tables is many-to-many (M:M). In a many-to-many cardinality, one record in the origin table matches multiple records in the destination table and, conversely, one record in the destination table matches multiple records in the origin table. For example, a many-to-many cardinality could exist between a table of country information and a table of political organizations. Each county can belong to multiple political organizations (e.g., United Nations, UNESCO, World Health Organization, etc.), and each political organization can include multiple countries.

Table associations

In order to link two tables together, the tables must have a common field. This field, also known as the key field, does not need to have the same name in both tables, but it must have the same field data type; and most importantly, the values in the key field must match in both tables. When you establish a table association between two tables, the records (rows) are matched based on their corresponding values in the key field.

Table joins

A table join is a temporary connection between two tables, where the fields of one (destination) table are appended to those of another (origin) table. **You can associate tables through a join when the tables have a one-to-one or a many-to-one cardinality.**

A table join is applied to a layer (not to the layer's source data on disk). On disk, the joined tables remain separate. Since it only "lives" in a layer, a join is saved in the map document or in a layer file. The tables used for a join can be feature attribute tables or any stand-alone tables and can be stored either in the geodatabase or in other file formats, such as shapefile, dBase, or Excel.

Most commonly, you use a join to append additional fields to the attribute table of a feature class. Once you have established the join, you can use the joined fields in the same way as any other fields in the attribute table. For example, you can use the joined fields to calculate fields in the attribute table, or query, symbolize, or label features. However, you cannot edit the joined values from within the feature attribute table. To edit joined attribute values you would first add the destination table to ArcMap and then edit the values directly.

You may also ask yourself, what happens to a join when you change the source data of a layer that participates in a join? If you change the source data for a layer, the key fields and their values might not match any more. The join may break, and you may have to re-create it.

For more information on table joins, refer to the ArcGIS Help topic "Essentials of joining tables."

Exercise 7a: Join tables

For the exercises in this chapter, you will continue to work on the map of Australia and Oceania. You will now explore the cardinality between a Countries attribute table and a stand-alone table of country attributes. Then, you will create a join between the tables, so that you can access the attributes in the stand-alone table from the Countries attribute table.

1. **ArcGIS 10.1:** Start ArcMap and open the ...\DesktopAssociate\Chapter07\OceaniaTables.mxd.

    ArcGIS 10.0: Start ArcMap and open the ...\DesktopAssociate \Chapter07\OceaniaTables_100.mxd.

2. **Open the Countries attribute table.**

Country name (CNTRY_NAME), Shape_Length, and Shape_Area are the only attribute fields in the Countries attribute table describing the counties. You will join a table with additional country attributes.

3. **In the table of contents, open the Country_info table.**

The Country_info table contains additional attributes about the countries.

Hint: You may have to list the layers by source to see the Country_info table. To see the two tables side by side in the table window, you can drag the Countries tab and dock it next to the Country_info table.

Decision point: What is the common field in the two tables?

The CNTRY_NAME field is common to both tables.

Table associations **105**

Decision point: What is the cardinality between the tables?

The cardinality is one-to-one: each record in the Countries table matches to one record in the Country info table.

4. Close the table window.

Decision point: Why is a join appropriate between the Countries table and the Country_info tables?

Because each record in the Countries table matches a record in the Country_info table, you can link these tables using a join. If you had many matching records in the Country_info table, you would use a relate.

5. Right-click the Countries layer, point to Joins and Relates, and choose Join.

6. Run the Join Data tool using the specifications in the following figure:

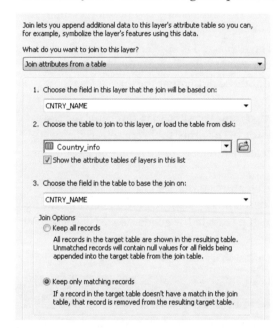

7. Open the Countries attribute table.

The fields from the Country_info table have been appended to the Countries table.

Question 4: What are other ways to access the table join functionality in ArcGIS?

Next you will take a look at the source data of the two tables to confirm that the source tables remain separate.

8. In the Catalog window, open the Item Description of the Countries feature class.

Hint: In the Chapter07\Oceania.gdb, right-click Countries and choose Item Description.

9. In the Item Description window, preview the table.

The Countries table on disk contains only the OBJECTID, Shape, Country name (CNTRY_NAME), Shape_Length, and Shape_Area attribute fields.

10. Preview the Country_info attribute table in the same way.

Country_info contains the same attributes that you have seen before.

11. Close the Item Description window.

To be able to use the table with the joined attributes outside of the current map document, you will export the Countries layer and create a feature class.

12. In the table of contents, right-click Countries layer, point to Data, and choose Export Data.

13. Export the layer to the …Chapter07\Oceania geodatabase and name it **Countries_joined**. Click Yes to add the layer to the map.

Question 5: Joining tables using the layer context menu or the Add Join geoprocessing tool will append all the attributes from the destination table to the origin table. Which tool allows you to select the fields from the destination table that you want to join to the origin table?

Hint: You can use the search terms *join* and *fields* in the ArcMap Search window to answer this question.

Now that you have exported the table you are able to use the joined attributes to symbolize, label, or query the layer also in other map documents. Since you no longer need it, you will remove the Join from the Countries layer now.

14. Right-click the Countries layer, point to Joins and Relates > Remove Join(s), and then choose Countries_info.

15. If you are continuing to the next exercise, leave ArcMap open. Otherwise, close ArcMap.

Table relates

A relate is another temporary association between two tables or layers. However, unlike a join, the fields of the destination table are not appended to the origin table, but you can access the related records when you select the layer's attributes. You can associate tables through a relate when they have a one-to-one, one-to-many, or a many-to-many cardinality.

Similar to a join, a relate is saved in the map document or in a layer file and is applied only to a layer (not to the layer's source data on disk). On disk, the tables remain separate. Also, the source tables for a relate can be attribute tables or stand-alone tables, stored either in the geodatabase or in other file formats, such as shapefile, dBase, or Excel.

You use a relate to view or query the related attributes. When two tables are associated through a relate and you select features or records in one table, all the related records in the other table are selected as well. Similarly, related attributes are displayed in the Identify tool if a feature attribute table is related to another table.

Unlike a join, you cannot calculate fields or symbolize and label features based on attributes from the related table. Also, you cannot edit the related records through the relate. If you anticipate a need for editing related records, you would associate the tables through a relationship class in the geodatabase (see below) or edit the tables directly.

For more information about relates, refer to the Desktop Help topic "Essentials of relating tables."

Exercise 7b: Relate tables

Suppose you wanted to explore the different geographic landmarks that are located in Australia and Oceania. The data that you have available is a feature class of administrative areas and a stand-alone gazetteer table. You will examine the cardinality between the two tables and create a relate between them. This will allow you to explore the geographic landmarks in the stand-alone gazetteer table by administrative area.

1. **ArcGIS 10.1:** If necessary, start ArcMap and open the ...\DesktopAssociate \Chapter07\OceaniaTables.mxd.

   ArcGIS 10.0: If necessary, start ArcMap and open the ...\DesktopAssociate\Chapter07\OceaniaTables_100.mxd.

2. Turn on the Administrative areas layer.

This layer represents administrative areas in the different countries.

3. Open the Gazetteer table and scroll to the right.

Hint: Switch the table of contents view to List By Source, if necessary.

The Gazetteer table contains names and descriptions for geographic features in the map area.

4. Open the attribute table of Administrative areas.

Decision point: What is the common field in the Administrative areas table and the Gazetteer table?

The ADMIN_NAME field is common to both tables.

5. **In both tables, sort the values in the ADMIN_NAME field alphabetically (in ascending order).**

Hint: Right-click the ADMIN_NAME field heading and choose Sort Ascending.

Decision point: What is the cardinality between the Administrative areas table and the Gazetteer table?

The cardinality is one-to-many: the records in the Administrative areas table match to more than one record in the Gazetteer table.

Decision point: Why is a relate appropriate to link the Administrative areas and the Gazetteer tables?

Because each record in the Countries table has more than one matching record in the Gazetteer table, you link these tables using a relate.

6. **Make sure that the Administrative areas table is displayed in the table window.**

7. **Click the Table Options button, point to Joins and Relates, and choose Relate.**

8. **Run the Relate tool using the following specifications:**

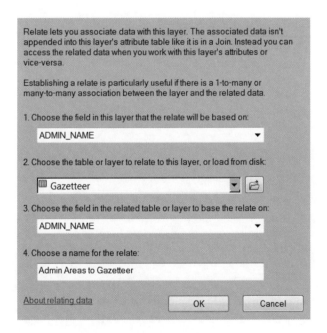

Decision point: Do any of the attributes from the Gazetteer table appear in the Administrative areas table? If not, why not?

With a relate, the two tables are kept separate both on disk and in the layers.

9. In the Administrative areas table, select the row of the Central administrative district in Papua, New Guinea (OBJECTID 131).

Hint: Click the row button next to object ID 131.

10. At the top of the Table window, click the Related Tables button.

11. Click Admin Areas to Gazetteer: Gazetteer.

In the Gazetteer table, the records that match the selected administrative area are selected as well. You can now view the selected geographic landmarks from the attribute table or explore them using the Identify tool.

12. Close the table window.

13. Use the Identify tool to click any administrative area on the map (e.g., Queensland in northeastern Australia).

14. At the top of the Identify window, expand the identified feature and the Gazetteer table.

15. Click a few of the records listed under Gazetteer to view their attributes.

16. If you'd like, identify the related landmarks of a few other administrative areas.

17. Close the Identify window.

Since you have finished exploring the geographic landmarks, you will remove the relate from the Administrative areas layer.

18. Right-click the Administrative areas layer, point to Joins and Relates > Remove Relates, and then choose Admin Areas to Gazetteer.

19. If you are continuing to the next exercise, leave ArcMap open. Otherwise, close ArcMap.

Challenge 1

1. You need to associate a feature class of apartment buildings to a table of tenant information. Each apartment building is occupied by multiple tenants. Which table association should be used?

   a. Join
   b. Relate

2. You want to calculate the values of a population density field in a city layer based on population numbers stored in a stand-alone city demographics table.

   a. Join
   b. Relate

3. You have a feature class of crimes, including attributes such as crime type, date, and neighborhood; and a polygon feature class of neighborhoods, including attributes such as population and median income. Which table association should you use so that you can symbolize crimes by median income?

   a. Join
   b. Relate

4. You have a table of department stores in a mall and a table with customer names and address information. Each department store has been visited by many customers, and each customer may have visited more than one department store. Which table association should you use?

   a. Join
   b. Relate

Relationship classes

Other than joins and relates, relationship classes are elements in the geodatabase that create a permanent association between two tables. You can associate tables through a relationship class when they have a one-to-one, one-to-many, or a many-to-many cardinality.

You can create a relationship class between geodatabase feature classes, between geodatabase feature classes and stand-alone tables, and between geodatabase tables. However, the feature classes or tables need to be stored in the same geodatabase.

Relationship classes help you to enforce the "referential integrity" between related feature classes or tables. This means you can set up a relationship class in a way that, if you modify a feature or value in a table, the related feature or table record is updated as well. For example, you could set up a relationship class between a feature class of utility poles and a feature class of transformers in such a way that, if you move a utility pole, attached transformers move with it.

Relationship classes also allow you to edit related records through the relationship class. In the example above, you could navigate to a transformer attribute from within the utility poles attribute table and edit the transformer attribute. You may also refine the relationship between utility poles and transformers by setting up rules: a pole may support a maximum of three transformers. Or, a steel pole may support class A transformers, but not class B transformers.

Types of relationship classes

There are two types of relationship classes: simple (peer-to-peer) and composite (parent-child). In a simple relationship class, the tables exist independently of each other. For example, to link a parcels feature class with a buildings feature class, a simple relationship class is appropriate. If you edited the parcel (e.g., deleted a parcel and created two new ones) the corresponding buildings on the parcel would not be affected. If you delete the parcel, the values in the key field in the buildings table are set to Null. If you deleted a building, the corresponding parcel is not affected.

Parcels (origin)

Parcel_ID	Zone	Block
...	...	...
123	...	...
...	...	...
...	...	...

Buildings (destination)

Building_ID	Parcel_ID	Floors
...	...	...
...	Null	...
...	Null	...
...	...	...

Figure 7.4 In a simple relationship class, records in origin and destination tables exist independently of each other. For example, if you delete a parcel in the origin table, the key field values in the related Buildings table are set to Null. Esri.

In a composite relationship class, the records in the origin table control the lifetime of the records in the destination table. Destination records cannot exist independently of origin records. For example, suppose you created a composite relationship class that associates a feature class of signal poles to a feature class of traffic lights (that are mounted on the signal poles). If you deleted a signal pole, the related traffic lights would also be deleted in a process called a cascade delete. A composite relationship class can also help you maintain features spatially; moving or rotating an origin feature causes the related destination features to move or rotate with it (when messaging in the composite relationship class is set to Forward). A composite relationship class requires a one-to-one or a one-to-many cardinality.

Relationship classes can be created and edited in ArcGIS for Desktop Advanced and ArcGIS for Desktop Standard; they are read-only in ArcGIS for Desktop Basic. For more information on simple and composite relationship classes, refer to the ArcGIS Help topic "Relationship class properties."

Poles (origin)

Pole_ID	Material	Height
...	...	...
~~123~~	...	...
...	...	...
...	...	...

Traffic lights (destination)

Light_ID	Pole_ID	Type
...	...	...
...	~~123~~	...
...	~~123~~	...
...	...	...

Figure 7.5 In a composite relationship class, the records in the origin table control the lifetime of the records in the destination table. For example, if you delete a pole in the origin table, the related records in the Traffic lights table are deleted as well. Created by the author.

Exercise 7c: Create a simple relationship class

Suppose you wanted to create a permanent association between the Administrative areas and the Gazetteer tables. Once the relationship class is in place, project planners can add Administrative areas to any map document and access—or even edit—the Gazetteer attributes from within the Administrative areas attribute table.

1. **ArcGIS 10.1:** If necessary, start ArcMap and open the ...\DesktopAssociate \Chapter07\OceaniaTables.mxd.

   ArcGIS 10.0: If necessary, start ArcMap and open the ...\DesktopAssociate \Chapter07\OceaniaTables_100.mxd.

2. **Open the Gazetteer table.**

The Gazetteer table contains names and descriptions for geographic features in the map area.

3. **Open the attribute table of Administrative areas.**

Decision point: What is the common field in the Administrative areas table and the Gazetteer table?

___

Note: If you have completed the previous exercise, you may have answered this question already. The ADMIN_NAME field is common to both tables.

4. **In both tables, sort the values in the ADMIN_NAME field alphabetically.**

Decision point: What is the relationship or cardinality between the Administrative areas table and the Gazetteer table?

___

Note: If you have completed the previous exercise, you may have answered this question already. The cardinality is one-to-many: the records in the Administrative areas table match to more than one record in the Gazetteer table.

Decision point: What type of relationship class is appropriate to connect the Administrative areas and the Gazetteer tables?

___

Since administrative areas and geographic features listed in the Gazetteer table can exist independently from each other, a simple relationship class is appropriate.

5. **In the Catalog window, right-click the ... Chapter07\Oceania geodatabase, point to New, and choose Relationship class.**

6. **Name the relationship class AdminAreas_To_Gazetteer.**

7. For the origin table/feature class, select Admin_poly. For the destination table/feature class, select Gazetteer, as shown in the following graphic.

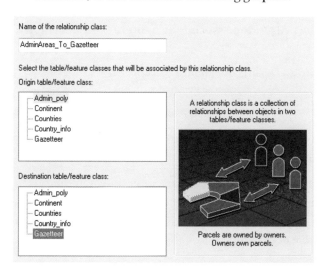

8. **Click Next.**

The second panel of the wizard lets you choose the type of relationship class. Because you want to ensure that administrative areas and gazetteer records exist independently of each other, you will create a simple relationship class.

9. **Choose Simple (peer-to-peer) relationship, and click Next.**

In the third panel, you can add messaging to any simple or composite relationship class, which allows for origin and destination records to notify one another when they are changed, allowing related records to update appropriately. You will not add any messaging here because it will not have any effect on simple relationship classes.

10. **Accept the default of None (no messages propagated), and click Next.**

The fourth panel of the wizard sets the cardinality between the origin and destination tables.

11. **Check 1:M (one-to-many), and then click Next.**

The fifth panel of the wizard allows you to add attributes to the relationship. This is normally only done for many-to-many relationships, where the attributes refine the relationship class. For example, in a parcel database, you may have a relationship class between parcels and owners, where owners can own many parcels and parcels can be owned by many owners. An intermediate table would store which owner owns which parcel. An attribute of the relationship could be the percentage that each owner owns of each parcel. For more information on relationship attributes, see the Relationship attributes section of the ArcGIS Help topic "Relationship class properties."

12. **Check "No, I do not want to add attributes."**

13. **Click Next.**

The sixth panel of the wizard identifies the key fields that link the origin and destination tables. The key field in the origin table is referred to as the *primary key field*. The key field in the destination table is referred to as the *foreign key field*.

14. **For both the primary key field and the foreign key field, select ADMIN_NAME. Then click Next.**

The seventh panel of the wizard summarizes your settings. It should look like this:

15. **Click Finish to create the new relationship class.**

16. **If you are continuing to the next exercise, leave ArcMap open. Otherwise, close ArcMap.**

```
Name: AdminAreas_To_Gazetteer
Origin object class: admin_poly
Destination object class: Gazetteer
Type: Simple
Forward Path Label: Gazetteer
Backward Path Label: admin_poly
Message propagation: None (no messages propagated)
Cardinality: 1 - M
Has attributes: No
Origin Primary Key: ADMIN_NAME
Origin Foreign Key: ADMIN_NAME
```

Exercise 7d: Use the relationship class

Now that you have created the relationship class, you can access and edit the values in the Gazetteer table from within the Administrative areas table.

Chapter 7: Associating tables

Dependency: You need to complete exercise 7b before running this exercise. If you have not completed exercise 7b, please do so before starting this exercise.

1. **ArcGIS 10.1:** If necessary, start ArcMap and open the ...\DesktopAssociate \Chapter07\OceaniaTables.mxd.

    ArcGIS 10.0: If necessary, start ArcMap and open the ...\DesktopAssociate \Chapter07\OceaniaTables_100.mxd.

2. Zoom to the Palau bookmark.

3. Display the Editor toolbar. From the Editor menu, click Start Editing and start an edit session for Administrative Areas. Click Continue in the warning message.

4. Using the Edit tool from the Editor toolbar ▸, select the Palau administrative area shown in the current map extent.

Note: Make sure you select the Administrative area of Palau, not the Country Palau. If you are not sure which feature you have selected, click List By Selection in the table of contents.

5. On the Editor toolbar, click the Attributes button.

6. In the Attributes window, expand the Palau and Gazetteer, as shown in the following graphic.

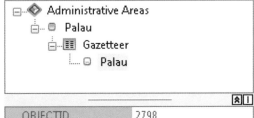

Currently, the Gazetteer table has only one record for Palau. You will add another record for the airport of Palau.

7. In the Attributes window, right-click Gazetteer and choose Add New.

8. Click the new entry and type in the attributes as follows:

NAME	Babelthuap
CATEGORY	Airport
DESCRIPT	International
WITHIN	Oceania
ADMIN_NAME	Palau

9. Stop editing and save your edits.

10. Clear the selected features.

Now that you have added a record to the Gazetteer table, you will identify an Administrative area feature and verify that the record is available.

11. Use the Identify tool ⓘ to identify the administrative area on the map.

12. At the top of the Identify window, expand the identified record and the Gazetteer entry.

13. Verify that both the Babelthuap and the Palau record are listed under the Gazetteer entry.

14. Close the Identify window.

15. Close ArcMap.

Challenge 2

1. A scientist investigating the spread of a plant disease has a feature class of study areas and another feature class of 1 meter ×1 meter quadrants used to collect plant samples. Each study area contains 5 sample quadrants. The scientist needs to associate the two attribute tables in a way that if a study area is moved or rotated, the sample quadrants in it are moved or rotated as well. If a study area is deleted, the corresponding sample quadrants must be deleted as well. Which type of table association would you use?

   a. Simple relationship class
   b. Simple relationship class with backwards messaging
   c. Composite relationship class
   d. Composite relationship class with forward messaging

2. The GIS department of a city maintains a parcels table, and the assessor's department maintains an Excel table of tax information for each parcel. Each parcel record corresponds to exactly one record of tax information. Once a year the GIS department needs to add the updated tax information to the parcel feature class attribute table. Which type of table association will be most appropriate?

   a. Relate
   b. Simple relationship class
   c. Join
   d. Composite relationship class

3. You are working with an ArcGIS for Desktop Basic license and want to associate a layer of shopping centers with a table of stores. Each shopping center contains many stores. You added both the Shopping Centers layer and the stores table to a map document and want to be able to query both the Shopping Center layer and the Stores table and view the corresponding features or records in the other table. Which type of table association is most appropriate?

   a. Relate
   b. Simple relationship class
   c. Append
   d. Join

4. You want to associate a layer of countries and a table of country information. You want to be able to edit the country information table from within the countries layer table, but you do not want to make the country information table dependent on the countries layer attribute table. Which type of table association is most appropriate?

   a. Relate
   b. Simple relationship class
   c. Join
   d. Composite relationship class

Chapter 7: Associating tables

Answers to chapter 7 questions

Question 1: In figure 7.1, which is the common key field in the Countries and the Country_Info tables?
Answer: The CNTRY_NAME field serves as a key field between the two tables.

Question 2: In figure 7.2, which is the common key field in the Zoning and the Zoning Description tables?
Answer: The Zone_Code field serves as a key field between the two tables.

Question 3: In figure 7.3, which is the common key field in the Stores and the Employees tables?
Answer: The Store_ID field serves as a key field between the two tables.

Question 4: What are other ways to access the table join functionality in ArcGIS?
Answer: You can also access the Join tables functionality from the following:
1. Table window (Table Options button)
2. Layer Properties (Joins & Relates tab)
3. Add Join geoprocessing tool (Data Management toolbox, Joins toolset)
4. Join Field geoprocessing tool (Data Management toolbox, Joins toolset)

Question 5: Joining tables using the layer context menu or the Add Join geoprocessing tool will append all the attributes from the destination table to the origin table. Which tool allows you to select the fields from the destination table that you want to join to the origin table?
Answer: The Join Field tool allows you choose which fields from the join table will be appended to the input table.

Answers to challenge questions

Correct answers shown in bold.

Challenge 1

1. You need to associate a feature class of apartment buildings to a table of tenant information. Each apartment building is occupied by multiple tenants. Which table association should be used?

    a. Join
    b. Relate (Since there is a one-to-many relationship between the two tables, a relate is appropriate.)

2. You want to calculate the values of a population density field in a city layer based on population numbers stored in a stand-alone city demographics table.

    a. Join (Since each record in the demographics table corresponds to one city, a join is appropriate. Once the join is established, the user can calculate field values based on the joined fields.)
    b. Relate

3. You have a feature class of crimes, including attributes such as crime type, date, and neighborhood; and a polygon feature class of neighborhoods, including attributes such as population and median income. Which table association should you use so that you can symbolize crimes by median income?

   a. Join (Since there is a many-to-one relationship between the crime and the neighborhood feature classes, a join can be established, and the user can symbolize crimes based on a neighborhoods attribute.)
   b. Relate

4. You have a table of department stores in a mall and a table with customer names and address information. Each department store has been visited by many customers, and each customer may have visited more than one department store. Which table association should you use?

   a. Join
   b. Relate (Since there is a many-to-many relationship between the stores and the customers tables, a relate is appropriate to use.)

Challenge 2

1. A scientist investigating the spread of a plant disease has a feature class of study areas and another feature class of 1 meter × 1 meter quadrants used to collect plant samples. Each study area contains 5 sample quadrants. The scientist needs to associate the two attribute tables in a way that if a study area is moved or rotated, the sample quadrants in it are moved or rotated as well. If a study area is deleted, the corresponding sample quadrants must be deleted as well. Which type of table association would you use?

   a. Simple relationship class
   b. Simple relationship class with backwards messaging
   c. Composite relationship class
   d. Composite relationship class with forward messaging (A composite relationship will allow for the records of the study area table control the existence of the sample quadrants. Forward messaging enables study area features to notify sample patches when they are moved or rotated so they can move or rotate as well.)

2. The GIS department of a city maintains a parcels table, and the assessor's department maintains an Excel table of tax information for each parcel. Each parcel record corresponds to exactly one record of tax information. Once a year the GIS department needs to add the updated tax information to the parcel feature class attribute table. Which type of table association will be most appropriate?

   a. Relate
   b. Simple relationship class
   c. Join (The relationship between the parcels and the tax information table is one-to-one. Since the table connection is only needed once a year, a join is appropriate.)
   d. Composite relationship class

3. You are working with an ArcGIS for Desktop Basic license and want to associate a layer of shopping centers with a table of stores. Each shopping center contains many stores. You added both the Shopping Centers layer and the stores table to a map document and want to be able to query both the Shopping Center layer and the Stores table and view the corresponding features or records in the other table. Which type of table association is most appropriate?

   a. Relate (Since the relationship between the Shopping Centers layer and the Stores table is one-to-many, and you want to be able to query both tables in the map document and view the corresponding records in the other table, a relate is appropriate. The scenario does not ask for a permanent connection between the layer and the table, and also, to create a relationship class, you would need at least an ArcGIS for Desktop Standard license.)
   b. Simple relationship class
   c. Append
   d. Join

4. You want to associate a layer of countries and a table of country information. You want to be able to edit the country information table from within the countries layer table, but you do not want to make the country information table dependent on the countries layer attribute table. Which type of table association is most appropriate?

   a. Relate
   b. Simple relationship class (The relationship between the countries layer and the country information table is one-to-one. Since you want to be able edit the country information within the countries table, a relationship class is required. A simple relationship class will keep the two tables independent.)
   c. Join
   d. Composite relationship class

Key terms

Cardinality: The correspondence or equivalency between sets; how sets relate to each other. For example, if one row in a table is related to three rows in another table, the cardinality is one-to-many.

Destination table: The secondary table in a relationship.

Resources

- ArcGIS Help 10.1 > Geodata > Data types > Tables
  - About joining and relating tables
  - Essentials of joining tables
  - Essentials of relating tables
- ArcGIS Help 10.1 > Geodata > Data types > Relationships and related objects
  - Deciding between relationship classes, joins, and relates
  - Relationship class properties

chapter 8

Georeferencing and spatial adjustment

Georeferencing   118
　Georeferencing raster or CAD datasets
　Exercise 8a: Prepare for georeferencing
　Exercise 8b: Georeference an aerial photo (for ArcGIS 10.1)
　Exercise 8b: Georeference an aerial photo (for ArcGIS 10.0)
　What the root mean square error means

Spatial adjustment   126
　Spatial adjustment methods
　Exercise 8c: Spatially adjust a roads layer
　Challenge

Answers to chapter 8 questions   131

Answers to challenge questions   131

Key terms   132

Resources   132

Chapter 8: Georeferencing and spatial adjustment

Why are we talking about these two topics in one chapter? Because they're both about moving data into its correct place.

Suppose you scanned a paper map. The scanned image is simply a collection of pixels without any information as to where the map is located. It is missing its spatial reference, and its origin (0,0 point) is not in real-world coordinates. To use the map with your other GIS data, you have to georeference the image by assigning a real-world spatial reference to it.

Or, suppose you have a point, line, or polygon layer that does not align properly with the rest of your data. You checked it all: the geographic and projected coordinate systems are defined correctly, you used the right geographic transformation, but there is still a slight offset between this layer and the rest of your GIS data. You can repair this misalignment by matching the layer to a spatially accurate reference dataset. This process is called spatial adjustment.

Skills measured

- Describe circumstances under which it is appropriate to perform georeferencing or spatial adjustment.
- Given a scenario, evaluate whether results of georeferencing or spatial adjustment are acceptable.

Prerequisites

- Knowledge of coordinate systems and spatial references
- Hands-on experience in GIS data editing
- Knowledge of editable data formats in ArcGIS
- Knowledge of the raster data model
- Hands-on experience with CAD data

Georeferencing

Georeferencing provides a correct, real-world spatial reference to raster or CAD data, which is either missing a real-world spatial reference or is in an unknown spatial reference that you cannot identify and define. Georeferencing is the process of aligning geographic data to a reference dataset in a known coordinate system, so it can be viewed, queried, and analyzed with other geographic data.

The reference dataset can be a raster dataset or a vector feature class. You identify distinctive locations that are visible in both datasets as control points, and then establish links from the control points in the original dataset (the raster or CAD dataset that you want to align with your projected data) to the corresponding control points in the reference dataset. Based on these links, the original dataset is aligned with the reference layer (figure 8.1).

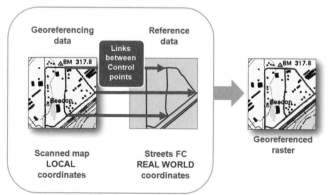

Figure 8.1 The graphic shows a scanned map in local coordinates being georeferenced to a Streets feature class in real-world coordinates. The process of georeferencing is based on links between common control points in the original data and a reference dataset. Esri.

To align a dataset to another dataset, you apply a transformation to it. The transformation will shift, scale, rotate, or, if necessary, skew the original dataset to align it with the reference data. Based on the links that you establish, ArcMap calculates the parameters that define how the original layer will be transformed. In other words, the links that you establish define the shift in x and in y direction that will be applied to the control points in the original layer so they match the control points in the reference layer. The overall transformation will be calculated from the parameters defined by all the links (figure 8.2).

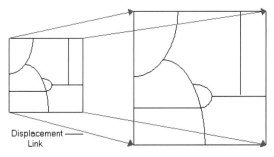

Figure 8.2 To georeference a dataset, a transformation is applied to shift, scale, rotate, or, if necessary, skew the dataset based on the established links. Esri.

Georeferencing raster or CAD datasets

You use georeferencing with any raster or CAD dataset that does not align properly with other GIS data. The misalignment could be either a small shift or a more drastic displacement, where the data displays in a completely different location than the other GIS data that it is supposed to align with.

To georeference a raster dataset, you can use either the update georeferencing or the rectify method. The update georeferencing method will save the transformation information to an external auxiliary (.aux) or world file that will be associated with the raster dataset. The rectify method will create a new raster dataset that is georeferenced using the spatial reference of the data frame. You should use the rectify method if you plan to perform analysis with the georeferenced raster dataset or want to use it with another software package that doesn't recognize the external georeferencing information created with the update georeferencing method. For more information about the update georeferencing and rectify methods, refer to the ArcGIS Help topic "Fundamentals of georeferencing a raster dataset."

When georeferencing a CAD dataset, the update georeferencing method is your only option: you save the transformation information and retain the data in the original CAD format. This will create a world file that defines its transformation. For more information about georeferencing CAD datasets, refer to the ArcGIS Help topic "About georeferencing CAD datasets."

Exercise 8a: Prepare for georeferencing

Suppose you obtained a historical aerial photo to analyze the change over time in an area. The historical aerial photo has been scanned from microfilm and is missing a real-world spatial reference. In order to use the historical aerial photo with your current GIS data, you have to georeference it. But first, you'll gather some coordinate system information.

1. **ArcGIS 10.1:** Start ArcMap and open the ...\DesktopAssociate\Chapter8\Georeferencing.mxd. Maximize your ArcMap window, if necessary.

   ArcGIS 10.0: Start ArcMap and open the ...\DesktopAssociate\Chapter8\Georeferencing_100.mxd. Maximize your ArcMap window, if necessary.

The map document shows an aerial photo of a forested area and highways for reference. It also contains a second aerial photo.

2. Examine the coordinate system information of the Woodside_1991.tif layer and the Layers data frame.

Hint: You find the coordinate system information in the Woodside_1991.tif layer in its layer properties and the coordinate system information of the data frame in the data frame properties.

Decision point: What is the spatial reference of the Woodside_1991.tif image and the data frame?

Both the Woodside_1991.tif image and the data frame have the NAD 1983 UTM Zone 10N projected coordinate system.

3. Using the Identify tool, click the lower left corner of the Woodside_1991 layer and take a note of the approximate location coordinates.

Location: _____ _____ meters

The coordinates of the lower left corner of the Woodside_1991 area are around 550,000 and 4,000,000 meters.

4. Turn on and zoom to the extent of the Woodside1948 layer.

Hint: Right-click Woodside Historical and choose Zoom to Layer.

The Woodside1948 image shows a historical view of the same area that is shown in the Woodside_1991 image. However, it does not display on top of the other image. Soon you will discover why that is.

5. Examine the spatial reference information of the Woodside1948 .tif layer.

The spatial reference information shows as <Undefined>. This means that ArcMap cannot read any spatial reference information for the image.

6. Using the Identify tool, click the lower left corner of the Woodside1948 layer and take a note of the location coordinates.

Location: _____ _____ meters

The coordinates of the lower left corner of the Woodside1948 image are 0,0. This means that the Woodside1948 image is not stored in a real-world coordinate system. Since the image is missing a real-world spatial reference, ArcMap draws it starting at coordinates of 0,0.

Note: If the coordinates of the Woodside1948.tif image were in a real-world coordinate system, you would try to research which coordinate system it is and define it using the Define Projection tool. However, since the image is missing a real-world spatial reference, you must first align it to a known real-world coordinate system (georeference it) before you can define the coordinate system.

7. Zoom back to the extent of the Woodside_1991 image.

8. If you are continuing with the next exercise, leave ArcMap open. Otherwise, close ArcMap.

Exercise 8b: Georeference an aerial photo (for ArcGIS 10.1)

Now you will georeference the Woodside1948 image using the Woodside_1991 image as a reference layer.

Note: These instructions apply to ArcGIS 10.1 only. If you are working in ArcGIS 10.0, follow the ArcGIS 10.0 instructions for exercise 8b later in this chapter.

1. If necessary, start ArcMap and open the ...\DesktopAssociate\Chapter8\ Georeferencing.mxd. Maximize your ArcMap window, if necessary.

2. If necessary, zoom to the extent of the Woodside_1991 layer.

3. Display the Georeferencing toolbar, if necessary. Choose Woodside1948 as the Georeferencing layer, as shown in the following graphic.

All georeferencing operations will be applied to the layer that is set as the Georeferencing layer.

4. On the Georeferencing toolbar, click Georeferencing and clear the Auto Adjust option.

You will add the Woodside1948 image to a Viewer window so you can see the two Woodside images side by side.

5. On the Georeferencing toolbar, click the Viewer button to add this image to a Viewer window.

6. Resize the Viewer window so that it covers about one quarter of your screen.

Now you're ready to georeference the Woodside1948 image. You'll establish links between control points in both images.

Question 1: Which landmarks make good control points?

7. Zoom to the Control Point 1 bookmark. In the Viewer window, zoom to the same location, at about the same scale.

In the bay, in the center of the display, you may notice a white spot in the water. This is a rock outcrop that you will use to establish the first control point.

8. On the Georeferencing toolbar, click the Add Control Points button .

9. Click the left edge of the white spot in the Viewer window. In the main map, click the same location. Use the following graphic as a reference.

You will establish three more links in the same way, but you will choose your own control points from a layer of suggested control points.

10. Turn on and zoom to the extent of the Suggested Control Points layer.

From all of these suggested control points, you will pick three and establish three more links between the Woodside1948.tif and the Woodside_1991.tif images.

11. Examine the points in the Suggested Control Points layer and decide which three control points you want to use.

Hint: Make sure to distribute the control points evenly throughout the images.

12. Using the previously described workflow, establish three more links between control points in the two images.

Hints:
- Feel free to snap to the points in the Suggested Control Points layer in the main map. (On the Snapping toolbar, make sure that Point snapping is enabled.)
- Make sure to create the links from a location in the Image Viewer window to the corresponding control point in the main map.

Note: The control points in the corners are marked as cross-hairs in both the Woodside1948 and Woodside_1991 images. The cross-hairs were added during processing to facilitate image registration.

The graphic on the right shows an example of how you could have established the four links.

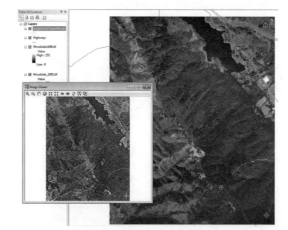

13. Once you have established a total of four links, close the Image Viewer window and open the link table .

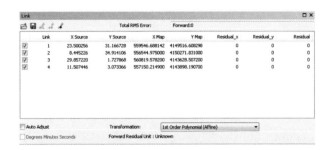

Hint: To open the link table, click the View Link table button on the Georeferencing toolbar.

Every row in the link table represents a link between the coordinates of the control points in the original (georeferencing) layer and the reference layer.

Decision point: Which transformation method is appropriate to use for georeferencing the Woodside1948.tif image?

In order to align the Woodside1948 layer with the Woodside_1991 layer, it needs to be shifted and scaled. The 1st Order Polynomial (Affine) Transformation method is appropriate because it allows for shifting, scaling, rotating, and skewing the georeferencing layer.

Note: For more information on transformation methods, refer to the ArcGIS Help topic "Fundamentals of georeferencing a raster dataset."

14. In the link table, set the 1st Order Polynomial (Affine) for Transformation. Check the Auto Adjust option.

Each link has now an associated residual value. The residual indicates how much each link contributes to the total root mean square (RMS) error, which is shown in the upper toolbar of the link table. The RMS error is a measure of the consistency between the links (see "What the root mean square error means" later in this chapter). You should try to keep your RMS error as close to zero as possible.

A commonly used technique to reduce the total RMS error is to first establish more links than you need for the transformation and then delete the link(s) with the highest residual value(s).

15. In the link table, select the link with the highest residual value (probably the first one you established). Note the current total RMS error and click the Delete button .

The total RMS error should now be reduced. Now you are ready to apply the transformation.

16. From the Georeferencing toolbar, choose Update Georeferencing. Close the link table.

17. Confirm that the two images now align with each other. Zoom in to different locations in the map and visually examine the alignment between the two images. You may want to use the Highways layer for reference, as well.

Hint: The Swipe tool on the Effects toolbar may be helpful to visually compare the alignment between the two layers.

The Woodside1948 image is now georeferenced in the coordinate system of the data frame (NAD 1983 UTM zone 10N). However, although the images now align with each other, ArcMap still does not have the coordinate system information of the Woodside1948.tif layer. If you were continuing to use this data, you would define the coordinate system of the Woodside1948.tif dataset using the Define Projection tool.

18. If you are continuing with the next exercise, leave ArcMap open. Otherwise, close ArcMap.

Exercise 8b: Georeference an aerial photo (for ArcGIS 10.0)

Now you will georeference the Woodside1948 image using the Woodside_1991 image as a reference layer.

Note: These instructions apply to ArcGIS 10.0 only. If you are working in ArcGIS 10.1, follow the ArcGIS 10.1 instructions for exercise 8b.

1. If necessary, start ArcMap and open the …\DesktopAssociate\Chapter8\ Georeferencing_100.mxd. Maximize your ArcMap window, if necessary.

2. If necessary, zoom to the extent of the Woodside_1991 layer.

3. Display the Georeferencing toolbar, if necessary. Choose Woodside1948 as the Georeferencing layer, as shown in the following graphic.

All georeferencing operations will be applied to the layer that is set as the Georeferencing layer.

4. On the Georeferencing toolbar, click Georeferencing and clear the Auto Adjust option.

You will first change the symbology of the Woodside1948.tif layer so you can distinguish it from the Woodside_1991.tif layer.

5. If necessary, turn on and zoom to the extent of the Woodside1948.tif layer. Open its layer properties. Click the Symbology tab and choose a red light-to-dark color ramp. In the Display tab, specify 40% transparency. Then click OK to close the layer properties.

6. Zoom to the extent of the Woodside_1991.tif layer. On the Georeferencing toolbar, click Georeferencing and choose Fit to Display.

The Woodside1948.tif layer now displays in the same area as the Woodside_1991.tif layer. You will now fine-tune the size and position of the Woodside1948.tif layer so it will be easier to georeference it.

7. Click the drop-down arrow next to the Rotate tool, and choose Scale, as shown in the following graphic.

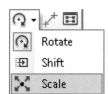

8. Use the Scale tool to resize the Woodside1948.tif layer to approximately the same size as the Woodside_1991.tif layer. If necessary, activate the Shift tool and move the Woodside1948.tif layer over the Woodside_1991.tif layer

Now you're ready to georeference the Woodside1948 image. You'll establish links between control points in both images.

Question 1: Which landmarks make good control points?

9. Zoom to the Control Point 1 bookmark.

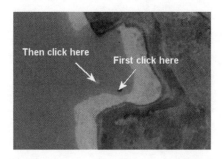

In the bay, in the center of the display, you may notice a spot in the water (displayed in white in the Woodside_1991.tif layer and red in the Woodside1948.tif layer). This is a rock outcrop that you will use to establish the first control point.

10. On the Georeferencing toolbar, click the Add Control Points button. Click the left edge of the red spot in the Woodside1948.tif layer. Click the same location in the Woodside_1991.tif layer. Use the graphic to the left as a reference.

Note: The location of the red and the white spot may vary, depending on how you scaled and shifted the Woodside1948.tif layer.

You will establish four more links in the same way, but you will choose your own control points from a layer of suggested control points.

11. **Turn on and zoom to the extent of the Suggested Control Points layer. Examine the points in the Suggested Control Points layer and decide which three control points you want to use.**

From all of these suggested control points, you will pick four and establish four more links between the Woodside1948.tif and the Woodside_1991.tif images.

Hint: Make sure to distribute the control points evenly throughout the image.

12. **Using the previously described workflow, establish four more links between the Woodside1948 image and the corresponding control points.**

Hints:
- Feel free to snap to the points in the Suggested Control Points layer. (On the Snapping toolbar, make sure that Point snapping is enabled).
- Make sure to create the links from a location in the Wooodside1948.tif layer to the corresponding control point.

Note: The control points in the corners are marked as cross hairs in both the Woodside1948 and Woodside_1991 image. The cross hairs were added during processing to facilitate image registration.

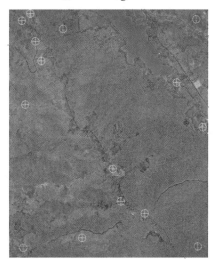

The graphic to the left shows an example of how you could have established the four links.

13. **Once you have established a total of five links, open the link table.**

Hint: To open the link table, click the View Link table button on the Georeferencing toolbar.

Every row in the link table represents a link between the coordinates of the control points in the original (georeferencing) layer and the reference layer.

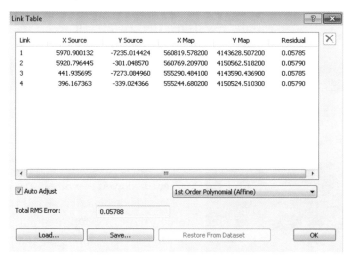

Decision point: Which transformation method is appropriate to use for georeferencing the Woodside1948.tif image?

In order to align the Woodside1948 layer with the Woodside_1991 layer, it needs to be shifted and scaled. The 1st Order Polynomial (Affine) Transformation method is appropriate because it allows for shifting, scaling, rotating, and skewing the georeferencing layer.

Note: For more information on transformation methods, refer to the ArcGIS Help topic "Fundamentals of georeferencing a raster dataset."

14. **In the link table, set the 1st Order Polynomial (Affine) for Transformation. Check the Auto Adjust option.**

Each link has now an associated residual value. The residual indicates how much each link contributes to the total root mean square (RMS) error, which is shown in the upper toolbar of the link table. The RMS error is a measure of the consistency between the links (see "What the root mean square error means" later in this chapter). You should try to keep your RMS error as close to zero as possible.

A commonly used technique to reduce the total RMS error is to first establish more links than you need for the transformation and then delete the link(s) with the highest residual value(s).

15. **In the link table, select the link with the highest residual value (probably the first one you established). Note the current total RMS error and click the Delete button ⊠.**

The total RMS error should now be reduced.

Now you are ready to apply the transformation.

16. **From the Georeferencing toolbar, choose Update Georeferencing. Close the link table.**

17. **Confirm that the two images now align with each other. Zoom in to different locations in the map and visually examine the alignment between the two images. You may want to use the Highways layer for reference, as well.**

Hint: The Swipe tool on the Effects toolbar may be helpful to visually compare the alignment between the two layers.

The Woodside1948 image is now georeferenced in the coordinate system of the data frame (NAD 1983 UTM zone 10N). However, although the images now align with each other, ArcMap still does not have the coordinate system information of the Woodside1948.tif layer. If you were continuing to use this data, you would define the coordinate system of the Woodside1948.tif dataset using the Define Projection tool.

18. **If you are continuing with the next exercise, leave ArcMap open. Otherwise, close ArcMap.**

What the root mean square error means

Each link between the control points in the source and reference layers defines an equation in a system of equations. The solution to this system of equations is a set of numbers, the transformation parameters, which define the transformation. In other words, the transformation parameters determine how the features in the georeferencing layer are shifted, rotated, scaled, and skewed during the transformation.

Since the transformation is applied globally to the entire dataset and the transformation parameters are defined collectively by all links, each individual link only approximates the transformation parameters. After the transformation, each control point will be placed in a slightly different location than defined by the link. The distance between the location of the control point defined by the link and the location where the control point was placed is called the residual error.

(Continued on next page)

> ### *What the root mean square error means (continued)*
> The error of the entire transformation process is calculated by averaging the residual errors of all links that contributed to the transformation. This is called the root mean square (RMS) error. The method that is used to average the RMS error from all the residual values (taking the root mean square sum of the residuals of all links) minimizes the overall error of the transformation. While the total RMS error can also be used as an assessment of the transformation's overall accuracy, it is primarily a measure of consistency between the different control points (links). What is acceptable as a total RMS error depends on the accuracy of both the adjusted and the reference data, as well as on standards set by your organization.
>
>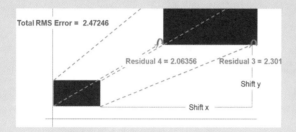
>
> Figure 8.3 The total root mean square (RMS) error is calculated by taking the root mean square sum of the residuals of all links used for the transformation. (Residual distances defined by the orange arrows are exaggerated for clarity.) Created by the author.

Spatial adjustment

Similar to georeferencing, spatial adjustment aligns an input dataset to a reference dataset, based on links between common control points. The major differences between georeferencing and spatial adjustment are in the usage and valid input data: while georeferencing is used to (re)create a missing or unknown spatial reference for raster and CAD data, spatial adjustment is used to correct the alignment of editable vector data.

- Spatial adjustment methods are used in an edit session.
- Only editable vector data formats (geodatabase feature classes and shapefiles) can be spatially adjusted.

Table 8.1 compares georeferencing with spatial adjustment.

Table 8.1 Georeferencing versus spatial adjustment		
	Georeferencing	Spatial adjustment
What it is	The process of aligning data with a missing or unknown spatial reference to reference data in a known coordinate system	ArcMap editing functionality for aligning data with a spatially accurate reference dataset
Input data	CAD or raster data	Vector data in editable formats (shapefile and geodatabase)
Methods	Transformation	Transformation Edgematching Rubber sheeting
Results	Georeference information written to an external world or auxiliary file (update georeferencing method) or create a new dataset with the georeferencing information (rectify method—raster data only)	Coordinates of the adjusted vector data modified to match the reference dataset

Spatial adjustment methods

There are three methods for performing spatial adjustment: transformation, edgematching, and rubber sheeting.

Similar to the transformations used in georeferencing, spatial adjustment transformations will also shift, scale, rotate, and, if necessary, skew a dataset to convert the coordinates from one location to another.

Edgematching is a spatial adjustment method that is typically used for connecting the endpoints of features with each other. For example, for roads layers that have been digitized from adjacent map sheets, you set displacement links to connect the roads along the edge of the map sheet.

In some cases, you may have some features in a layer that align perfectly with the corresponding features in the reference layer, while others need minor geometric adjustments. In these cases, you can apply rubber sheeting, which stretches, shrinks, and reorients features to match the reference layer while maintaining the connectivity between them. It uses two types of links: Identity links act as "pushpins" to keep some features in place, while displacement links define how to adjust other features in the layer.

Table 8.2 compares the three spatial adjustment methods.

Table 8.2 Comparing spatial adjustment methods

	Transformation	Edgematching	Rubber sheeting
What it is	Aligns data to a reference dataset by shifting, rotating, scaling, or skewing a dataset within a coordinate system.	Aligns features along the edge of one layer to features of an adjacent layer.	Aligns data to a reference dataset by stretching, shrinking and reorienting some features in a layer, while maintaining the connectivity among them.
Features adjusted	All features or selected features in the input layer will be adjusted.	Only features that have displacement links will be adjusted.	Displacement link locations will be adjusted; Identity link locations will stay in place.
When to use	Use for aligning data to a reference dataset.	Use for aligning the edges of layers.	Use for fine-tuning after transformation.
Example			

Table 8.2 illustrations from Esri.

Exercise 8c: Spatially adjust a roads layer

Suppose you received a new roads layer that does not align well with your existing GIS data. You confirmed that the misalignment is not due to a spatial reference issue. To be able to use the layer with your other GIS data, you will spatially adjust the new roads layer.

1. **ArcGIS 10.1:** Start ArcMap and open the …\DesktopAssociate\Chapter8\SpatialAdjustment.mxd. Maximize your ArcMap window, if necessary.

   ArcGIS 10.0: Start ArcMap and open the …\DesktopAssociate\Chapter8\ SpatialAdjustment_100.mxd. Maximize your ArcMap window, if necessary.

2. **Zoom to the Compare Layers bookmark and examine the alignment between the layers.**

The Highways layer aligns well with the aerial photo underneath, but there is an offset between the Roads layer and the Highways layer.

Decision point: Which layer should be the input layer for spatial adjustment and which one should be the reference layer?

Since the Highways layer aligns well with the aerial photo underneath, you can assume that it is more accurate than the Roads layer. You will use Highways as a reference layer to spatially adjust (transform) the Roads layer.

3. **Turn off the Woodside_1991.tif layer.**

4. **Display the Editor, Snapping, and Spatial Adjustment toolbars, if necessary.**

5. **From the Editor toolbar, click Editor, Start Editing. Start an edit session for the Roads shapefile.**

6. **On the Snapping toolbar, make sure that End snapping is enabled. For the snapping tolerance, make sure to use at least 10 pixels.**

Hint: To access the snapping tolerance, click the Snapping option on the Snapping toolbar.

First you will choose the input data for the spatial adjustment. You want to apply the spatial adjustment uniformly to all features in the Roads layer.

7. **On the Spatial Adjustment toolbar, choose Spatial Adjustment, Set Adjust Data.**

8. **Check the option to adjust All features in the Roads layer. Then close the Set Adjust Data dialog box.**

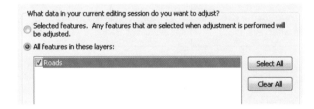

Next, you will create the first displacement link to define the first control points for the adjustment.

9. **On the Spatial Adjustment toolbar, click the New Displacement link tool .**

10. Zoom to the Control Point 1, Control Point 2, Control Point 3, and Control Point 4 bookmarks one after the other, and establish four displacement links as shown in the following graphic.

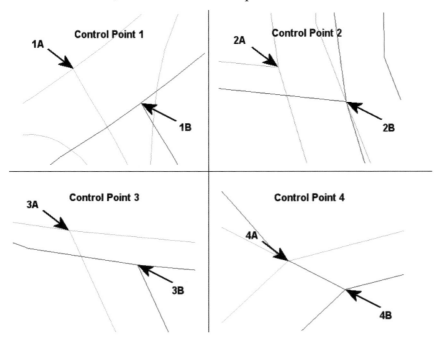

Hints:
- Make sure to establish the links from the orange Roads layer to the red Highways layer (Click point 1A first, then click 1B. Click 2A, and then click the point 2B, etc.).
- When setting the control points, make sure to turn on end snapping on the Snapping toolbar to be able to snap to the ends of the line features.
- The displacement links are added as graphics. If you made a mistake, select the link with the Select Elements tool, and then press the Delete key on your keyboard.

11. Once you have established a total of four links, click the View Link table button 🔲 on the Spatial Adjustment toolbar to open the link table.

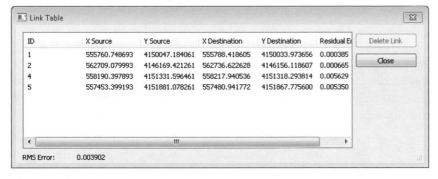

Every row in the link table shows the coordinates of the control points in the source layer and in the destination layer. Each link has an associated residual value, which indicates how much each link contributes to the root mean square (RMS) error, which is shown in the lower left of the link table.

Chapter 8: Georeferencing and spatial adjustment

Decision point: Is the RMS error of the four links acceptable?

If you snapped to the ends of the existing line features, you should see an RMS error of about 0.003902. Although a perfect transformation would produce an RMS error of zero, 0.0039 is a very low and acceptable value. Acceptable values will vary depending on the accuracy of the source and reference data.

12. Close the link table.

13. From the Spatial Adjustment toolbar, click Spatial Adjustment, click Adjustment Methods and ensure that Transformation - Affine method is checked.

An Affine transformation is based on a minimum of three links and can move, scale, rotate, and skew the data. For more information about the different spatial adjustment transformation methods, refer to the ArcGIS Help topic "About spatial adjustment transformations."

14. Zoom to the Compare Layers bookmark and turn on the Woodside_1991.tif aerial photo.

15. Click Spatial Adjustment and choose Adjust.

The Roads layer is moved to align with the Highways layer.

16. Zoom to different areas and visually assess the alignment of the Roads and Highways layers.

Question 2: Could you have used the tools from the Georeferencing toolbar to transform the Roads layer?

17. On the Editor toolbar, click Editor, Save Edits. Then click Editor, Stop Editing.

18. Close ArcMap.

Challenge

1. Your organization is in the process of converting its maps to digital format by scanning them. The images that you receive from the scanner display in a completely different location than the GIS data that covers the same areas. Which technique would you choose to resolve the misalignment?

    a. Georeferencing
    b. Spatial adjustment
    c. Projecting the data
    d. Define projection

2. Your contractor provided you with a layer of building footprints for a new subdivision that does not align properly with your existing building footprints layer and your aerial imagery. After you checked the properties of the new building footprints layer, you confirmed that the data has been projected correctly and that the coordinate system of the layer is defined correctly. Which technique should you apply to resolve the misalignment?

    a. Geocoding
    b. Georeferencing
    c. Edgematching
    d. Spatial adjustment

3. You display a major roads layer on top of a layer of detailed street centerlines. In some areas the layers align perfectly, but in some other areas there is a slight shift between the major roads and the street centerlines. Which technique should you choose to resolve the misalignment?

   a. Transformation
   b. Rubber sheeting
   c. Georeferencing
   d. Edgematching

4. You georeferenced a CAD layer of parcel boundaries using your existing parcel layer as a reference. The total RMS error of the transformation is fairly low. What other method can you use to evaluate the quality of the georeferencing?

   a. Evaluate the difference between the highest and the lowest residual errors
   b. Visually examine the alignment between the two layers
   c. Identify and remove the link with the highest residual value
   d. Use the Feature Compare tool to compare the two layers

Answers to chapter 8 questions

Question 1: Which landmarks make good control points?
Answer: Control points must be clearly identifiable locations in both the georeferencing and in the reference layer. Many different types of features can be used as controls, such as road or stream intersections, the mouth of a stream, the end of a jetty of land, the corner of an established field or building, street corners, or the intersection of two hedgerows.

Question 2: Could you have used the tools from the Georeferencing toolbar to transform the Roads layer?
Answer: No, the tools on the Georeferencing toolbar work only on raster and CAD data. Since Roads is a vector layer, you need to use the tools from the Spatial Adjustment toolbar to transform and align it with the reference layer.

Answers to challenge questions

Correct answers shown in bold.

1. Your organization is in the process of converting its maps to digital format by scanning them. The images that you receive from the scanner display in a completely different location than the GIS data that covers the same areas. Which technique would you choose to resolve the misalignment?

   a. Georeferencing
   b. Spatial adjustment
   c. Projecting the data
   d. Define projection

2. Your contractor provided you with a layer of building footprints for a new subdivision that does not align properly with your existing building footprints layer and your aerial imagery. After you checked the properties of the new building footprints layer, you confirmed that the data has been projected correctly and that the coordinate system of the layer is defined correctly. Which technique should you apply to resolve the misalignment?

   a. Geocoding
   b. Georeferencing
   c. Edgematching
   d. Spatial adjustment

3. You display a major roads layer on top of a layer of detailed street centerlines. In some areas the layers align perfectly, but in some other areas there is a slight shift between the major roads and the street centerlines. Which technique should you choose to resolve the misalignment?

   a. Transformation
   b. Rubber sheeting
   c. Georeferencing
   d. Edgematching

4. You georeferenced a CAD layer of parcel boundaries using your existing parcel layer as a reference. The total RMS error of the transformation is fairly low. What other method can you use to evaluate the quality of the georeferencing?

   a. Evaluate the difference between the highest and the lowest residual errors
   b. Visually examine the alignment between the two layers
   c. Identify and remove the link with the highest residual value
   d. Use the Feature Compare tool to compare the two layers

Key terms

Control point: One of various locations on a paper or digital map that has known coordinates and is used to transform another dataset—spatially coincident but in a different coordinate system—into the coordinate system of the control point. Control points are used in digitizing data from paper maps, in georeferencing both raster and vector data, and in performing spatial adjustment operations such as rubber sheeting.

Georeference: The process of aligning geographic data to a known coordinate system so it can be viewed, queried, and analyzed with other geographic data. Georeferencing may involve shifting, rotating, scaling, skewing, and in some cases warping, rubber sheeting, or orthorectifying the data.

Transformation: The process of converting the coordinates of a map or an image from one system to another, typically by shifting, rotating, scaling, skewing, or projecting them.

Reference data: In georeferencing and spatial adjustment, a layer with a known spatial reference that displays in the correct coordinate space.

Resources

- ArcGIS Help 10.1 > Geodata > Data Types > Rasters and Images > Processing and Analyzing Raster Data > Georeferencing >
  - Fundamentals of georeferencing a raster dataset
  - Georeferencing a raster to another raster
- ArcGIS Help 10.1 > Geodata > Data Types > CAD > Integrating CAD data > Georeferencing CAD datasets
  - About georeferencing CAD datasets
- ArcGIS Help 10.1 > Desktop > Editing > Editing existing features > Performing spatial adjustments
  - About spatial adjustment
  - About spatial adjustment transformations
  - About spatial adjustment edgematching
  - About spatial adjustment rubber sheeting
- ArcGIS Help 10.1 > Desktop > Editing > Fundamentals of editing > Editing tutorial
  - Exercise 5: Using spatial adjustment
- Georeferencing Rasters in ArcGIS, Esri web course

chapter 9
Geocoding

Geocoding components   134
　Address tables
　Reference data
　Address locators
　Address locator styles
　Exercise 9a: Create an address locator for street addresses
　Composite address locators

Address matching   139
　Exercise 9b: Geocode a table of addresses

The geocoding environment   142
　Challenge
　Geocoding options
　Exercise 9c: Modify geocoding options

Geocoding results   146

Answers to chapter 9 questions   147

Answers to challenge questions   147

Key terms   147

Resources   148

Chapter 9: Geocoding

How often do you have an address and need to locate it on a map? We need to do this all the time. When you use a GIS to find an address on a map, we call that geocoding. ArcGIS for Desktop allows you to do even more: you can take hundreds or thousands of addresses and convert them to point locations on a map. And not only that, you can also map other kinds of location descriptions such as ZIP Codes or place-names. Geocoding, also known as address matching, is the process of transforming any location description, usually a street addresses, into spatial data that can be displayed as features on a map.

Since addresses are the most commonly used location descriptions, this chapter will focus on address geocoding. Why is this important? Geocoding addresses and visualizing them on a map allows you to query them, find spatial patterns, and use them in routing applications.

Skills measured
- Given a scenario, determine appropriate batch geocoding specifications (e.g., locator, output feature class and location, geocoding environment, input parameters).

Prerequisites
- Knowledge of geographic data types and storage formats, especially table formats
- Hands-on experience applying coordinate systems in ArcGIS

Geocoding components

Address geocoding can be as simple as finding a single address on a map or it can be as involved as finding thousands or even millions of addresses. This latter process is called batch geocoding. Typically, large organizations will use batch geocoding to convert an address database to points on a map.

There are three main "ingredients" to batch geocoding in ArcGIS for Desktop:
1. Address table: The address information must be organized in an address table that ArcGIS for Desktop can read.
2. Reference data: A reference dataset is used by the software to compare and match the location descriptions.
3. Address locator: An address locator interprets the location descriptions and matches them to the reference data.

Later in this chapter you will review these different address components in more detail. Figure 9.1 describes the workflow for batch geocoding (i.e., address matching).

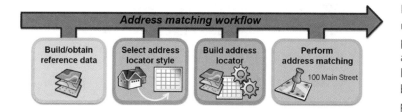

Figure 9.1 After you have built or obtained reference data for the area where you plan to locate addresses, you select an address locator style and build the address locator based on that style. Once you have built the address locator, you are ready to perform the address matching. Esri.

Address tables

What exactly is an address? You can think of an address as a set of instructions to find a location. Addresses are built from descriptive elements, including house numbers, street names, street types, city and state names, and ZIP Codes. For example, for standard United States addresses, these elements are typically arranged as the house number followed by the street name, city, state, and ZIP Code. Some parts of the United States and other countries structure the address elements differently or use different elements.

In order to perform batch geocoding, you need to organize addresses in a table. Most address locator styles require the address table to store the street address (house number, street name, and street direction) in a single field. The address table can be stored in any table format supported by ArcGIS for Desktop, including geodatabase table format, dBase, or INFO format (figure 9.2).

OBJECTID *	NAME	ADDRESS	CITY	STATE	ZIP
9	Camp Service Station	169 HUNNICUTT ST NW	ATLANTA	GA	3031
10	Central Petroleum	1100 CENTER ST NW	ATLANTA	GA	3031
11	Charlie Cota Inc.	400 8TH ST NW	ATLANTA	GA	3031
12	City Food Market	501 ETHEL ST NW	ATLANTA	GA	3031
13	Clamerty's	421 SPRING ST NW	ATLANTA	GA	3030
14	Crossroads Theater	120 MEMORIAL DR SE	ATLANTA	GA	3031
15	Damar Sales	388 7TH ST NE	ATLANTA	GA	3030

Figure 9.2 A typical address table used for geocoding standard US addresses contains the address information in one field, followed by fields for the city name, state, and ZIP Code. Created by the author based on Esri tutorial data (comes with the software).

Reference data

Reference data for geocoding includes point, line, or polygon feature classes that have address information in their attribute table. For geocoding US street addresses, the reference dataset must contain separate fields for some common address elements, such as house number range, street name, and street type. This information is used by the address locator to find the correct street features and locate the addresses (figure 9.3).

OBJECTID *	Shape *	L_F_ADD	L_T_ADD	R_F_ADD	R_T_ADD	PREFIX	PRE_TYPE	NAME	TYPE	SUFFIX
236	Polyline	136	154	137	155			Flat Shoals	Ave	SE
237	Polyline	156	166	157	159			Flat Shoals	Ave	SE
238	Polyline	168	178	161	175			Flat Shoals	Ave	SE
239	Polyline	180	192	177	193			Flat Shoals	Ave	SE
240	Polyline	700	898	701	899			Forrest	St	NW
241	Polyline	601	799	600	798			Forrest	Trl	NW
242	Polyline	700	758	731	759			Forrest	Way	NW
243	Polyline	760	784	761	785			Forrest	Way	NW
244	Polyline	800	860	801	861			Kilgore	St	NW

Figure 9.3 In a typical street centerline reference feature class, the house number range along each street feature is stored in fields called L_F_ADD (Left From Address), L_T_ADD (Left To Address), R_F_ADD (Right From Address), and R_T_ADD (Right To Address). This information is used by the address locator to locate the addresses on the correct side of the street. Other fields include the prefix and suffix direction of the street and the prefix type (e.g., I for Interstate in I-70). Created by the author based on Esri tutorial data (comes with the software).

Reference data should cover the geographic area for which you want to match addresses and can be stored as geodatabase feature classes or shapefiles. The completeness and spatial and attribute accuracy of the data often affects the match results of a geocoding process.

Depending on what kind of location description you want to geocode, you use different kinds of reference data. For example, for geocoding street addresses, you typically use street centerlines or points with address information as reference data. For geocoding ZIP Codes, you would most likely use a ZIP Code polygon layer as reference data.

Address locators

The address locator is the main dataset for geocoding in ArcGIS. When you perform geocoding, the address locator interprets the addresses and matches them to the reference data (figure 9.4).

From within ArcGIS for Desktop, you can use either an existing address locator, for example, the nationwide address locator provided in the StreetMap folder on the Esri Data and Maps media; access an ArcGIS Online geocode service; or build your own address locator based on a predefined address locator style. Building your own address locator has the advantage of letting you use your own reference data and refine all the settings to meet your needs.

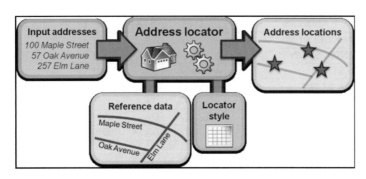

Figure 9.4 Geocoding is performed in ArcGIS with an address locator. An address locator contains information about local conventions for addresses (defined by a locator style) and a snapshot of the reference data. A locator interprets the addresses, matches them against the reference data, and generates points on the map. Esri.

Address locator styles

To geocode different types of street addresses, ZIP Codes, and geographic place-names, you use different types of address locators. ArcGIS provides predefined templates for building different types of address locators. These are called address locator styles.

An address locator style is a template on which an address locator is based. The style determines what type of addresses or location descriptions can be geocoded, the required information and format of reference data, and the output information that is returned. It also contains information on how addresses are divided into their elements, methods used to search for possible matches, and default values for geocoding options (see "Geocoding options" later in this chapter).

When you build an address locator, you select an address locator style appropriate for the type of geometry (point, line, or polygon) in your reference data and the format of the addresses you want to geocode (figure 9.5). For example, the US Address—Dual Ranges locator style is used for the majority of United States street addresses, with street centerlines as reference data.

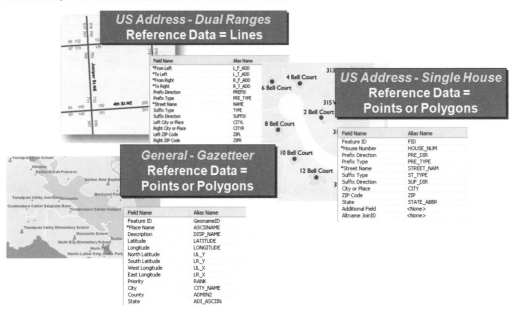

Figure 9.5 Different address locator styles allow you to geocode different types of addresses and location descriptions. Depending on the geometry type of the reference data, US addresses are commonly geocoded using either the US Address—Dual Ranges or the US Address–Single House locator style. The General—Gazetteer style can be used for finding geographic place-names or landmarks in an area of the world with points or polygons as reference data. Created by the author, data courtesy of Hamilton County.

The address locator style also defines the format of the fields required in the reference data. For example, an address locator for street addresses will usually require fields in the reference data for address elements such as street name, street type, from- and to- address ranges for each street segment, postal zones, and city. Table

9.1 lists the address locator styles provided with ArcGIS, the reference data they require, and their use. Address locator styles can be customized to support international addresses.

Table 9.1 Address locator styles provided with ArcGIS			
Styles	Reference data	Use for finding...	Examples
US Address—Dual Ranges	Lines	A street address on a specific side of the street	320 Madison St. N2W1700 County Rd. 105-30 Union St.
US Address—One Range	Lines	A street address where street side is not needed, or is stored as an attribute of each street segment	2 Summit Rd. N5200 County Rd PP 115-19 Post St.
US Address—Single House	Points or polygons	Parcels, buildings, or address points	71 Cherry Ln. W1700 Rock Rd. 38-76 Carson Rd.
US Address—Street Name	Lines	Streets by street name (without house numbers)	Raspberry Lane, San Antonio, TX
US Address—City State	Points or polygons	City and state names in the United States	River Forest, IL
US Address—ZIP 5 Digit	Points or polygons	A specific ZIP Code location	22066
US Address—ZIP+4	Points or polygons	A specific ZIP+4 location	96822-2323
US Address—ZIP+4 Range	Points or polygons	A specific ZIP+4 location	63703-0078
General—City State Country	Points or polygons	A specific city in a state and country	Rice, WA, USA Toronto, Ontario, Canada
General—Gazetteer	Points or polygons	Geographic place-names or landmarks in an area or the world	Leeds Castle, England Sapporo, Japan
General—Single Field	Points or polygons	Any features that are identified by a name or code (in one field)	Cafe Cabrillo N1N115 (office number) 125638452 (Unique ID for borehole, pipeline, watershed, etc.)

For more information on the characteristics of each address locator style, refer to the ArcGIS Help topic "Commonly used address locator styles."

Exercise 9a: Create an address locator for street addresses

In preparation for geocoding some customer addresses later in this chapter, you will now build an address locator.

1. **ArcGIS 10.1:** Start ArcMap and open the …\DesktopAssociate\Chapter9\Geocoding.mxd.

    ArcGIS 10.0: Start ArcMap and open the …\DesktopAssociate\Chapter9\Geocoding_100.mxd.

2. Open the Centerlines Attribute table.

The L_LO_ADD, L_HI_ADD, R_LO_ADD and R_HI_ADD fields store the house number ranges along both sides of each street. You will create an address locator that will enable you to geocode the input addresses at the correct side of the street.

You will use the Centerlines layer as a reference layer for building the address locator.

3. Close the table.

4. In the Catalog window, right-click the …\DesktopAssociate\Chapter9\Hamilton.gdb geodatabase and point to New and choose Address Locator.

5. For the Address Locator Style, click the Browse button and choose US Address—Dual Ranges.

Note: You will learn more about different address locator styles later in this chapter.

An address locator contains the rules and settings for interpreting the addresses and creating the output points and a copy of the reference data. This way, once the address locator is built, it is independent from the reference data: you could move or delete the reference data without affecting the capability of the address locator. If there are any changes or updates to the original reference data, you can use ArcGIS tools to update the address locator.

6. For Reference Data, choose Centerlines. Click the Role field for Centerlines, and then click the drop-down arrow and choose Primary Table.

Now you will choose the appropriate fields in the reference data for each address locator field. Fields with an asterisk (*) next to their names are required by the address locator style. These fields must be matched to valid fields in the reference data. Optional fields can be left as <None> if the fields do not apply.

7. Choose the following fields under Alias Name for each of the locator Field Names:
   - From Left: L_LO_ADD
   - To Left: L_HI_ADD
   - From Right: R_LO_ADD
   - To Right: R_HI_ADD

8. Save the Output Address Locator in the …\DesktopAssociate\Chapter9\Hamilton.gdb and name it **Hamilton_DualRanges**.

You can store an address locator either at the root level of a geodatabase or in a folder as a .loc file. You will learn more about the available address locator styles later in this chapter.

9. Click OK to build the address locator and close the progress window when the tool is finished.

Next, you will test the new locator by finding a single address.

10. Display the Geocoding toolbar.

11. On the Geocoding toolbar, select the Hamilton_DualRanges locator as the active locator as shown in the following graphic.

12. **Click the <Type an address...> text box, type 511 E Main St, Westfield, IN 46074, and then press Enter.**

The point location flashes on the map. You will create a callout graphic at the address location.

13. **Right-click the address on the Geocoding toolbar and click Add Callout.**

14. **If you are continuing with the next exercise, leave ArcMap open. Otherwise, close ArcMap.**

> ### *Composite address locators*
>
> Sometimes it is beneficial to use a combination of different address locators. For example, suppose you are geocoding customer addresses with a locator and reference data for your city. For most addresses this works well, but every once in a while you have a customer address from another city that remains unmatched with your current locator. This is when a composite address locator can help increase the number of matched addresses. A composite locator consists of two or more individual address locators and/or geocode services. If the first address locator fails to geocode the address, the second one takes over. When you build the composite locator, you specify the order in which the participating locators will be used to match the addresses. When you geocode a single address with a composite locator, the address will be matched against all the participating locators, and all the candidates will be displayed based on the order of the participating locators. When you geocode a table of addresses, the participating locators are used in the order you specified. If the first participating locator finds candidates above its minimum match score, addresses are matched automatically to the best of these candidates. Otherwise, the address is passed to the next participating locator.
>
> A composite address locator simply references the participating address locators and geocode services; it does not contain any address information or other data to geocode the addresses. The addresses are matched based on settings and geocoding options of the locator that was used.
>
> A composite address locator is especially useful for geocoding spatially disjointed address data and for building a fallback solution when addresses cannot be geocoded by just one address locator. This situation may arise when geocoding addresses or location descriptions from multiple data sources or when you have different types of addresses and/or location descriptions in one dataset. For example, you might have a table of customer information, where some customers provided their full addresses, while other customers only disclosed their ZIP Code.
>
> For more information on composite address locators, refer to the ArcGIS Help topic "Creating a composite address locator."

Address matching

When an address locator interprets the input addresses and matches them to the reference data, it finds several candidates (addresses in the reference data) for matching the address. Depending on how closely the candidate matches the input address, the address locator assigns a score between 0 and 100 to each candidate and ranks them by score. The candidate with the highest score will be the match. The result of geocoding is a point feature class, where each point represents a matched address. The attribute table of that feature class contains a copy of the address fields from the input address table, with records for all input addresses, whether or not they could be matched. The Status field in the output attribute table shows whether an address has been matched (M), matched with candidates tied (T), or remains unmatched (U).

The Score field in the output attribute table indicates the quality of the matched addresses: The score can be in a range of 0 to 100, where 100 indicates that the candidate is a perfect match—the higher the score, the greater the confidence in a correct match. If the input address can be found without any ambiguity on the map, the match score will be 100. A match score between 85 and 99 can generally be considered a good match. If some of your address elements have minor spelling errors (such as Valdes instead of Valdez) or an address element such as "St" is missing, then the address locator will probably find a match, but the match score will be about 70 or 80.

Optionally, you can specify additional fields to be added to the output feature class. For more information about the fields in the output feature class, refer to the ArcGIS Help topics "About geocoding a table of addresses" and "Locator output fields."

The output point feature class can be stored either in a geodatabase or as a shapefile (figure 9.6). The coordinate system of the output feature class will be the same as the coordinate system of the reference data.

ObjectID *	Shape *	Status	Score	Match_addr	LocAddress
69	Point	T	88.03	1010 S RANGELINE RD, CAR, IN, 46032	1010 RANGELINE RD
70	Point	M	97.06	1024 PRINCETON GATE, CAR, IN, 46032	1024 PRINCETON GATE
71	Point	M	97.06	709 VILLAGE DR E, CAR, IN, 46032	709 VILLAGE DR E
72	Point	M	97.06	411 JENNY LN, CAR, IN, 46032	411 JENNY LN
73	Point	M	97.06	10425 FERGUS AVE, CAR, IN, 46032	10425 FERGUS AVE
74	Point	M	97.06	3709 W 98TH ST, CAR, IN, 46032	3709 W 98TH ST
75	Point	U	0		100 WOODLAND DR

Figure 9.6 The graphic shows some of the fields in the attribute table of a geocoded feature class. Besides the Status and Score fields, it shows the Match_addr and the original address field (LocAddress) side by side. Notice that addresses that are tied to more than one candidate have a lower score than matched addresses, while unmatched addresses have a score of 0.

Exercise 9b: Geocode a table of addresses

You will now geocode a table of customer addresses against the Hamilton_DualRanges address locator you created in the previous exercise.

1. **ArcGIS 10.1:** If necessary, start ArcMap and open the ...\DesktopAssociate\Chapter9\Geocoding.mxd.

    ArcGIS 10.0: If necessary, start ArcMap and open the ...\DesktopAssociate\Chapter9\Geocoding_100.mxd.

2. **Open the CustomerInfo table.**

Hint: If the CustomerInfo table is not listed in the table of contents, make sure that the List By Source view is active at the top of the table of contents.

The LocAddress field in the CutomerInfo table contains the addresses that you will geocode in the next exercise. The house number, street names, and street types are stored in a single field, followed by fields for the city, state, and ZIP Code. Although there could be additional fields in this table, for example, for customer names, their purchasing habits, and so on, this is a typical address table. (**Note:** The field containing the customer names has been deleted from this table.)

3. **Close the table window.**

4. **In the table of contents, right-click the CustomerInfo table and choose Geocode Addresses.**

5. **If you completed exercise 9a, click OK and move on to step 6.**

If you have not completed exercise 9a:
- **ArcGIS 10.1:** In the Choose Address Locator to use window, click Add and add the Hamilton_DualRanges from your ...\DesktopAssociate\Chapter9\Locators folder, and then click OK.
- **ArcGIS 10.0:** In the Choose Address Locator to use window, click Add and add the Hamilton_DualRanges_100 from your ...\DesktopAssociate\Chapter9\Locators folder, and then click OK.

6. Make sure that CustomerInfo is selected for Address table.

7. In the Address Input Fields section, select the following fields for the required information:

Street or Intersection	LocAddress
City or Place-name	LocCity
State	LocState
ZIP Code	LocZip

8. In the Output section, browse to save the output feature class as **CustomerPoints** in your ...\DesktopAssociate\Chapter9\Hamilton.gdb. Your Geocode Addresses dialog box should now match the one shown in the following graphic.

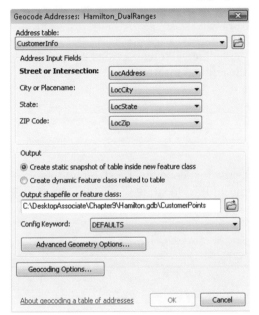

9. **Click OK to start batch geocoding.**

When the batch geocoding process is finished, the Geocoding Addresses dialog box shows the final statistics of the addresses that were matched or not matched.

10. **Review the statistics listed in the dialog box.**

Most of the addresses were matched. A few addresses were tied to more than one candidate, and a few remain unmatched. The Rematch button in the dialog box would give you the option to revisit the tied and unmatched addresses and rematch them.

11. **Close the dialog box.**

A point layer containing the geocoding results is added to the map. There is one point feature for each of the matched addresses.

12. **Open the attribute table of Geocoding Result: CustomerPoints. Scroll to the right and examine the fields in the table.**

The Status field shows whether the address record has been matched and the Score field shows the score of that match. The Match_addr field contains the address where the matched location actually resides. Depending on the score of the match, this address may or may not be the same as the input address.

13. **Close the table window and save the map document.**

14. **If you are continuing with the next exercise, leave ArcMap open. Otherwise, close ArcMap.**

The geocoding environment

The ArcGIS geocoding environment gives you several options for controlling and fine-tuning the geocoding process.

In the properties of an address locator, you find many settings that define how input addresses will be interpreted and matched to the reference data, and what the output data will look like. After performing geocoding, you may wonder to what extent you can trust that the points have been placed in their correct locations. To increase the level of confidence in the geocoded feature class, you can increase the minimum match score, a setting in the address locator properties, that controls the extent to which the address must match a location before it is considered a match. For example, does the address have to match exactly or can it be close? Let's say there are a lot of spelling errors in the address table. To control how much spelling variation the address locator allows when matching addresses to the reference data, you can set or modify the spelling sensitivity.

Challenge

For each of the following scenarios, choose the correct address locator style.

1. You have a feature class of parcel polygons to use as reference data for geocoding US street addresses. Which address locator style would you choose when building the address locator?

   a. US Address—Street Name
   b. US Address—Dual Ranges
   c. US Address—Single House
   d. US Address—One Range

2. You have a feature class of street centerlines to use as reference data for geocoding US street addresses. Which address locator style would you choose when building the address locator?

   a. US Address—Street Name
   b. US Address—Dual Ranges
   c. US Address—Single House
   d. US Address—One Range

3. Which address locator style allows you to geocode the telephone number 909 793 2853?

   a. General—Gazetteer
   b. General—Single Field
   c. General—City State Country
   d. US Address—ZIP+4

4. Which address locator style allows you to geocode the ZIP Code 92373?

   a. US Address—ZIP+4
   b. US Address—ZIP+4 Range
   c. US Address—ZIP Code 5
   d. US Address—ZIP 5 Digit

Geocoding options

The geocoding options in the address locator properties define how the input addresses are matched to the reference data. Table 9.2 lists some geocoding options that control how the locator matches addresses to the candidates and some consideration of when to modify them.

Table 9.2 Geocoding options

Option	Description	Default	Considerations
Minimum match score	The minimum score that a candidate needs in order to be considered for a match in batch geocoding	85	• Increase if you need a high level of confidence in the matched addresses • Decrease to maximize the number of matched addresses, accepting that some addresses are potentially matched incorrectly
Minimum candidate score	The minimum score that a potential match location requires to be considered a candidate	10	Decrease if the address locator is unable to find any likely candidates for an address that you want to geocode
Match if best candidates tie	A setting that matches an address to the first location found if there are multiple candidates with the same highest score	Yes	Turn off if you wish to review the tied candidates and match the address manually to one of the tied candidates
Spelling sensitivity	The degree (in values of 0 to 100) to which spelling variation is allowed during the search for likely candidates in the reference data	80	• Increase to restrict candidates to exact matches • Decrease if your addresses contain many spelling errors to allow more candidates to be evaluated
Intersection connectors	A list of characters, symbols, or words that can be used to connect the two streets at an intersection address	& @ | and at	• Use for geocoding street intersections, e.g., Hollywood Blvd. & Vine Ave • Add connector symbols to this list that are used in your address table
Match without house number	A setting that allows for searching and matching input addresses containing no house numbers	No	Use for geocoding street names without house numbers, e.g., Pennsylvania Avenue
Match with no zones	A setting that allows for searching and matching input addresses containing no zones	No	Use for geocoding addresses without a city or state name or a ZIP Code, e.g., 380 New York Street

Chapter 9: Geocoding

Exercise 9c: Modify geocoding options

Suppose you want to use the Hamilton_Dual Ranges address locator for an emergency response application that requires high confidence in the spatial accuracy of the geocoded feature class. To increase the confidence in the geocoding results, you will modify the geocoding options in the address locator properties.

Dependency: You need to complete exercise 9b before running this exercise. If you have not completed exercise 9b, please do so before starting this exercise.

1. **ArcGIS 10.1:** If necessary, start ArcMap and open the ...\DesktopAssociate\Chapter9\Geocoding.mxd.

   ArcGIS 10.0: If necessary, start ArcMap and open the ...\DesktopAssociate\Chapter9\Geocoding_100.mxd.

2. Display the Geocoding toolbar, if necessary.

3. Open the Attribute table of the Geocoding Result: CustomerPoints layer that you created in the previous exercise.

4. Right-click the Status field and choose Advanced Sorting. Sort by Status, and then by Score (both Ascending).

Decision point: What is the range of match scores for the matched addresses?

You will increase the minimum match score to 95 and disable the option to match if the best candidates tie.

5. Close the table window.

6. On the Geocoding toolbar, make sure the Hamilton_DualRanges locator is selected as the active locator.

7. In the Catalog window, open the properties of the Hamilton_DualRanges locator.

Note: The Hamilton_DualRanges (for ArcGIS 10.1) and Hamilton_DualRanges (for ArcGIS 10.0) locators in the Locators folder are the same as the one that you built in exercise 9a. If you have completed exercise 9a, open the properties of the Hamilton_DualRanges locator in the Hamilton.gdb. If you have not completed exercise 9a, open the properties of the Hamilton_DualRanges locator in the Locators folder.

8. In the Address Locator Properties window, expand the Geocoding options.

The Minimum match score determines the minimum score that a candidate needs in order to be considered for a match.

9. **ArcGIS 10.1:** Set the Minimum match score to 95. Set the Match if best candidates tie option to No (see the following graphic). Click OK to close the Address Locator Properties dialog box.

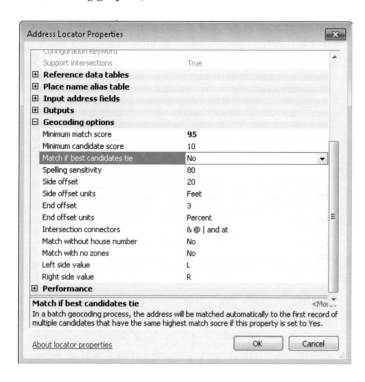

ArcGIS 10.0: Set the Minimum match score to 90. Clear the box to Match if candidates tie (see the following graphic). Click OK to close the Address Locator Properties dialog box.

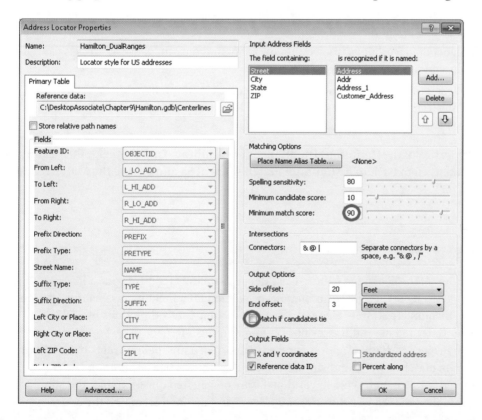

Question 1: Would increasing the spelling sensitivity increase spatial accuracy of the geocoded addresses?

You will now geocode the addresses in the CustomerInfo table using the options you have just set.

10. Geocode the CustomerInfo table using the Hamilton_Dual Ranges address locator.

Hint: Use the address locator that you just modified. Use the following table to set the correct Address Input Fields:

Street or Intersection	LocAddress
City or Place-name	LocCity
State	LocState
ZIP Code	LocZip

11. In the Geocode Addresses dialog box, click Geocoding Options and examine the settings.

The Minimum match score and the disabled Match if best candidates tie option are reflected here.

Note: If you change any settings in the current Geocoding Options dialog box, these changes would apply only to the current geocoding operation and would be set back to their defaults for the next one. The settings that you changed in the Address Locator Properties dialog box are applied permanently to the address locator.

12. Save the output feature class as CustomerPoints2 in your ...\DesktopAssociate\ Chapter9\Hamilton.gdb.

13. Click OK to start geocoding.

Now that you increased the minimum match score and disabled the Match if best candidates tie option, the number of matched addresses decreased. Since these addresses are now matched with a higher minimum match score, you can have a higher level of confidence in these results. There are no addresses tied to more than one candidate, and the number of unmatched addresses increased.

If you were continuing the geocoding workflow, you would now try to find out why these addresses cannot be matched, fix them if possible, and manually rematch them.

For more information about rematching, refer to the ArcGIS Help topics "About rematching a geocoded feature class" and "Rematching a geocoded feature class using the Interactive Rematch dialog box."

14. Close the dialog box.

15. If you modified the properties of the Hamilton_DualRanges or Hamilton_DualRanges_100 locators in the Locators folder, change them back to a Minimum match score of 85 and enable the Match if best candidates tie option.

16. Close ArcMap.

Geocoding results

The completeness and accuracy of the geocoded feature class depends on the quality of your address and reference data. Some reasons why a match may be ambiguous or fail to match include misspellings in input addresses or reference data, incomplete address elements (such as missing prefix directions like "West"), changes in street names over time, and new addresses not yet found in the reference data. For addresses that cannot be matched, you can inspect each input address and either correct a misspelling, relax the spelling sensitivity, lower the minimum match score, or individually match the address to a point on the map.

Addresses that are matched with a match score lower than 80 are considered questionable. You should try to identify and correct the problem that causes the low match score and manually rematch the address.

Answers to chapter 9 questions

Question 1: Would increasing the spelling sensitivity increase spatial accuracy of the geocoded addresses?
Answer: No, increasing the spelling sensitivity will not affect the accuracy of the geocoded addresses; it only controls how many candidates the address locator considers.

Answers to challenge questions

Correct answers shown in bold.

1. You have a feature class of parcel polygons to use as reference data for geocoding US street addresses. Which address locator style do you choose when building the address locator?

    a. US Address—Street Name
    b. US Address—Dual Ranges
    c. US Address—Single House
    d. US Address—One Range

2. You have a feature class of street centerlines to use as reference data for geocoding US street addresses. Which address locator style do you choose when building the address locator?

    a. US Address—Street Name
    b. US Address—Dual Ranges
    c. US Address—Single House
    d. US Address—One Range

3. Which address locator style allows you to geocode telephone number 909 793 2853?

    a. General—Gazetteer
    b. General—Single Field
    c. General—City State Country
    d. US Address—ZIP+4

4. Which address locator style allows you to geocode the ZIP Code 92373?

    a. US Address—ZIP+4
    b. US Address—ZIP+4 Range
    c. US Address—ZIP Code 5
    d. US Address—ZIP 5 Digit

Key terms

Geocoding: A GIS operation to transform a description of a location—such as a pair of coordinates, an address, or a name of a place—to a location on a map.

Address matching: A process that compares an address or a table of addresses to the address attributes of a reference dataset to determine whether a particular address falls within an address range associated with a feature in the reference dataset. If an address falls within a feature's address range, it is considered a match, and a location can be returned.

Resources

- ArcGIS Help 10.1 > Guide Books > Geocoding
  - What is geocoding?
  - Geocoding tutorial
- ArcGIS Help 10.1 > Guide Books > Geocoding > Preparing for geocoding
  - Commonly used address locator styles
- ArcGIS Help 10.1 > Guide Books > Geocoding > Building an address locator
  - Creating a composite address locator
  - Outputs
  - Geocoding options
- ArcGIS Help 10.1 > Guide Books > Geocoding > Locating addresses
  - About geocoding a table of addresses
- ArcGIS Help 10.1 > Guide Books > Geocoding > Understanding geocoding
  - What is an address?
  - The geocoding workflow
  - The geocoding process

chapter 10

Creating feature geometry

Feature templates 150
    Exercise 10a: Create feature templates

Feature construction tools 152
    Using snapping
    Exercise 10b: Digitize features using construction tools

Segment construction methods 155
    Exercise 10c: Create features using construction methods
    Challenge 1

Creating features from existing features 159
    Exercise 10d: Create features from existing features
    Challenge 2

Answers to challenge questions 162

Key terms 163

Resources 164

Chapter 10: Creating feature geometry

When you build your GIS database, you may discover that some data you need does not exist or is missing. If features do not exist, you may have to create them from scratch. You can create new features in three different ways: you can visually identify and trace them on a base layer, for example, an aerial photograph; you can construct features using geometric relationships, for example, at right angles or parallel to other features; or you can create them from existing geometry.

You can use the editing workflow in figure 10.1 to create new features from scratch.

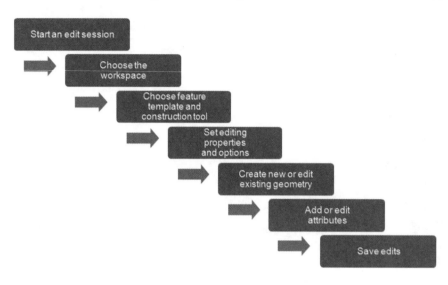

Figure 10.1 The editing workflow shown is the most efficient one for creating new features or editing existing ones. Esri.

Skills measured
Given a scenario, determine specifications for creating a new simple feature.

Prerequisites
Knowledge of ArcGIS editing, including the following:
- Elements of the edit sketch: vertices and segments
- The editing workspace
- Editable data types

Hands-on experience with the ArcGIS editing environment, including the following:
- Feature templates
- Snapping
- Tools and methods on the Editor toolbar

Feature templates

Feature templates contain much of the information required for creating new features, considerably streamlining the editing process. Feature templates define the name of the feature class that will store the new features, the default tool used to create the new features, and the default attributes associated with the new features. Feature templates also have a name, description, and tags that are used when searching for feature templates in the ArcMap Create Features Window. Feature templates are saved in the map document (.mxd), in a layer file (.lyr), layer package (.lpk), or map package (.mpk).

Feature templates are automatically created for editable layers in the table of contents at the beginning of an edit session. If you add a layer from the same edit workspace after you start the edit session, you will have to manually create a feature template for it or stop the edit session and start a new one. To create feature templates for layers that reference source data in a geodatabase or folder other than the edit workspace, you have to start an edit session for that workspace.

A layer can have multiple templates associated with it. For example, if a layer is symbolized based on unique attribute values for road type (e.g., highways, major roads, and local roads), feature templates will be created for each road type. You can specify different default tools and default attributes for each of these templates.

Exercise 10a: Create feature templates

During quality control, you discover that features in some feature classes in your geodatabase are missing. Before adding these missing features, you will create the necessary feature templates in the corresponding layers.

1. **ArcGIS 10.1: Open the ...\DesktopAssociate\Chapter10\FeatureTemplates.mxd**

   ArcGIS 10.0: Open the ...\DesktopAssociate\Chapter10\FeatureTemplates_100.mxd.

The map document shows an area in Zilker Park in the city of Austin, Texas. Later in this chapter, you will digitize a missing street feature in the park.

2. **Display the Editor toolbar and start an edit session. Open the Create features window if necessary.**

When you started the edit session, feature templates were automatically created for the Streets layer. Since it has been symbolized using the Unique Values method, feature templates have been created for each streets category in the layer.

3. **Open the properties of the Local Streets feature template.**

Hint: In the Create Features window, right-click the Local Streets feature template and choose Properties.

4. **Enter your name and today's date as the default attribute for the MODIFIED_B and MODIFIED_D fields. Then close the template properties.**

Next, you will add a few more layers to the map.

5. **From the ...\DesktopAssociate\Chapter10\Austin.gdb, add the ZilkerPark feature class as a layer.**

6. **From the ...\DesktopAssociate\Chapter10 folder, add the Historical_Landmarks shapefile as a layer.**

Decision point: No feature template has been created in the Create Features window for the ZilkerPark and Historical_Landmarks layers. Why not?

The feature template for the Zilker Park layer was not automatically created because you added the layer after starting the edit session for the Austin.gdb geodatabase. The feature template for Historical_Landmarks is not created because its source data is not stored in the Austin.gdb, and therefore, the layer is currently not editable.

Before you create the feature templates for these layers, you will first spend some time symbolizing them and setting other layer properties.

7. **Open the layer properties of the ZilkerPark layer and symbolize it with a green color and 60% transparency.**

Hint: To set the transparency, click the Display tab.

8. **Open the layer properties of Historical Landmarks and symbolize it with a color that contrasts well with the background image and the park polygon.**

9. **In the Fields tab, turn off all fields except for the BUILDING_N, DATE_BUILT, and ADDRESS fields. Then, click OK to close the layer properties.**

Hint: Hold down the Ctrl key and uncheck any field name to turn all fields off. Then turn on the fields listed above.

Now you are ready to create the feature templates for the new layers.

10. In the Create Features window, click the Organize Templates button 🖽 to open the Organize Feature Templates dialog box.

11. Click New Template and select the ZilkerPark layer to create a template for it. Then click Finish and close the Organize Feature Templates dialog box.

To create a feature template for the Historical_Landmarks layer, you will start an edit session for this layer.

12. Right-click the Historical Landmarks layer, point to Edit Features, and choose Start Editing. Click Yes if prompted to save your changes.

When you start an edit session for the Historical Landmarks layer, a feature template is created for it.

In the next exercise, you will use the feature templates you created to digitize some missing features.

13. Stop editing and save your edits. Save the map document. If you are continuing with the next exercise, leave ArcMap open. Otherwise, close ArcMap.

Feature construction tools

After you select a feature template, you can use any of the construction tools listed at the bottom of the Create Features window to create new features. Construction tools define how new features will be created. For example, construction tools define whether a line or polygon feature will be created click-by-click or freehand, by dragging the mouse. The Auto Complete Polygon construction tool automatically completes polygons using the boundaries of existing, adjacent polygons. ArcGIS provides different construction tools for creating points, lines, and polygons. Any of the construction tools available for a particular geometry type can be set as default tools in a feature template.

Tables 10.1, 10.2, and 10.3 list the different construction tools available for points, lines, and polygons, and when to use them.

Table 10.1 Construction tools for points

Task	Construction tool	When to use
Create a point feature at the location you click	Point	The location of the point can be visually identified on a base layer.
Construct a point feature at a defined direction and distance from a known location	Point At End of Line	The location of the point is described by a direction and/or distance from another point.

Table 10.2 Construction tools for lines

Task	Construction tool	When to use
Create a line feature with straight segments	Line	For tracing line features on a base layer
Create a line feature with rectangular shape	Rectangle	For creating lines with rectangular, circular, or elliptical shapes, e.g., a walking path around an artificial (rectangular) pond, a roundabout, a racetrack
Create a line feature with circular shape	Circle	
Create a line feature with elliptical shape	Ellipse	

Continued on next page

Table 10.2 Construction tools for lines (continued)

Task	Construction tool	When to use
Create a line feature quickly	Freehand	For creating free-form designs, e.g., redlining boundaries

Table 10.3 Construction tools for polygons

Task	Construction tool	When to use
Create a polygon	Polygon	For tracing polygons on a base layer
Create a rectangle	Rectangle tool	For polygons that have rectangular, circular, or elliptical shapes, e.g., a building, a water tower, or a gas plume
Create a circle	Circle tool	
Create an ellipse	Ellipse tool	
Create adjoining polygons	Auto Complete Polygon	For tracing adjoining polygons on a base layer, e.g., vegetation types
	Auto Complete Freehand	For quickly sketching adjoining polygon outlines without exact boundaries, e.g., progress of a wildfire on consecutive days

Using snapping

When you create new features, you use snapping to connect the elements of the edit sketch to each other or to other features. Snapping is enabled by default when you open ArcMap (unless you or somebody else turned it off) and applies to all visible layers in the map. You can snap to point features, endpoints of line features, and vertices and segments (edges) of line or polygon features. In addition, when snapping to line or polygon features, you can snap to the intersection, midpoint, or tangent point of features.

When you create new features, the snapping tolerance determines the distance within which the mouse pointer snaps to another feature or the distance within which a feature is moved to another existing feature. The default snapping tolerance is 10 pixels. You specify the snapping tolerance depending on the density of the features you snap to.

- Increase the snapping tolerance when your features are widely spread and you want to snap more easily to the intended location.

- Decrease the snapping tolerance when your features are very dense and your mouse pointer inadvertently snaps to locations that you do not intend. **Note:** In these situations consider pressing the spacebar on your keyboard to temporarily suspend snapping.

For more information on snapping, refer to the ArcGIS Help topic "About snapping."

Exercise 10b: Digitize features using construction tools

During quality control, a few features were found to be missing from some of the feature classes in your geodatabase. You will now add these missing features.

   1. **ArcGIS 10.1: Open the …\DesktopAssociate\Chapter10\ConstructionTools.mxd.**

      ArcGIS 10.0: Open the …\DesktopAssociate\Chapter10\ConstructionTools_100.mxd.

The map document shows the Zilker Park area in the city of Austin, Texas, that you may already be familiar with from the previous exercise.

   2. **Display the Editor toolbar, if necessary. Start an edit session for the Austin.gdb file geodatabase and open the Create Features window 🗒 if necessary.**

   3. **Zoom to the Lou Neff Drive bookmark.**

You will digitize the missing street feature between City Collector and Local Streets.

   4. **Display the Snapping toolbar and verify that end snapping is enabled.**

   5. **In the Create Features window, select the Local Streets feature template.**

Decision point: Which construction tool is the default tool in the Local Streets feature template?

You will use the Line construction tool to digitize the new street feature.

   6. **Refer to the following graphic:** Snap to the Streets endpoint on the left (marked A in the graphic), and digitize the missing street feature. Finish the sketch by snapping to the end of the existing street feature (marked B in the graphic).

   7. **On the Editor toolbar, open the Attributes window 📄 and enter Lou Neff Road for the FULL_STREET attribute.**

Next, you will create an additional park polygon next to Zilker Park.

   8. **Zoom to the Zilker Preserve bookmark.**

   9. **On the Snapping toolbar, make sure that vertex snapping is enabled.**

   10. **Select the Zilker Park feature template in the Create Features window. Under Construction Tools, select Auto Complete Polygon.**

   11. **Digitize the new park polygon counterclockwise, as shown in the graphic on the right. Start by snapping to the vertex marked A in the graphic. Digitize along the bank of the river until you reach point B. Then draw a straight line to point C and snap to the streets vertex. Draw a short straight segment and snap to the vertex of the existing park polygon marked D. Double-click to finish the sketch.**

Since you used the Auto-Complete Polygon construction tool, the boundary with the adjacent Zilker Park polygon was automatically created.

   12. **In the Attributes window enter Zilker Nature Preserve for the PARK_NAME attribute of the new polygon.**

Next, you will create a new historical landmark point. Since the source data of the Historical Landmarks layer is stored as a shapefile, you will switch the edit session to the folder containing the shapefile.

13. Right-click Historical Landmarks, point to Edit Features and choose Start Editing. When you are prompted to save edits, click Yes.

14. Zoom to the Lou Neff Drive bookmark and pan the map slightly to the right until you see the complete white building footprint shown in the aerial photo.

15. Select the Historical Landmarks feature template and create a point feature inside the white structure across the parking lot from the new street, as shown in the following graphic.

16. In the Attributes window, enter the values as follows:

BUILDING_N	Old Zephyr Station
DATE_BUILT	1906
ADDRESS	2201 Lou Neff Rd

17. Stop editing and save your edits. If you are continuing with the next exercise, leave ArcMap open. Otherwise, close ArcMap.

Segment construction methods

By default, the Line and Polygon construction tools create straight line segments between the vertices you click. To construct vertices and segments differently, these tools have additional construction methods that allow you to create curved segments, trace existing features, construct vertices at a distance and direction from a known point, and more.

Table 10.4 lists the common construction tasks, construction methods used to accomplish those tasks, and when to use each.

Table 10.4 Construction methods

Task	Construction method(s)	When to use
Create straight segments between vertices	Straight	For tracing line or polygon features on a base layer, e.g., power lines
Create straight segments perpendicular to the previously sketched segment	Right Angle	For creating features with exact 90-degree angles, e.g., rectangular building footprints
Create curved segments	• Arc • End-Point Arc (constructs the same kind of curve, but operates differently)	For creating true (parametric) curves, e.g., the curb line of a cul-de-sac with a 180-degree radius

Continued on next page

Table 10.4 Construction methods (continued)

Task	Construction method(s)	When to use
Create curved segments that are tangential to the previously sketched segment	Tangent	For creating tangent curves, e.g., railroads or highway off ramps
Create Bezier curved segments	Bezier	For creating smooth, potentially s-shaped curves, e.g., road casements
Create segments parallel to or on top of other lines or polygons	Trace	For creating features that are coincident with or parallel to other features, e.g., a street casement that is 15 feet from a parcel boundary
Create midpoints or vertices of centerlines	Midpoint	For creating features that are located exactly in the middle, between two other features or locations
Construct vertices at a specified distance and direction from a known location	Direction-Distance	For creating features at a specified direction and distance from a known location, e.g., located 30 feet from one corner of a building and 45 degrees from another corner
Construct vertices at specified distances from two known locations	Distance-Distance	For creating features at specified directions from two other locations, e.g., a light located 125 feet from one building corner and 50 feet from a survey marker
Construct vertices at the intersection of two lines	Intersection	For creating features located at the intersection of two temporary construction lines (pointing in two directions), e.g., construct a light pole by snapping a construction line to the front side of a building and another construction line to a street centerline that runs perpendicular to the building.

For more information on construction methods, refer to the ArcGIS Help topic "Segment Construction methods."

Exercise 10c: Create features using construction methods

You will now use the Line tool and different construction methods to digitize a few missing features.

1. **ArcGIS 10.1:** Open the …\DesktopAssociate\Chapter10\ConstructionMethods.mxd.

    ArcGIS 10.0: Open the …\DesktopAssociate\Chapter10\ConstructionMethods_100.mxd.

The map document shows an area in the northwestern part of the city of Austin, Texas.

2. If necessary, display the Editor toolbar. Start an edit session.
3. Display the Snapping toolbar, if necessary, and enable vertex, edge, and end snapping.
4. Zoom to the Crown Ridge Street bookmark.

5. In the Create Features window, select the Local Streets feature template. Snap to the endpoint of the existing City Collector street and click to start the edit sketch, as shown in the following graphic on the left.

You will digitize the missing street by tracing the lot boundary on the left side of the street with an offset of 25 feet.

6. On the Editor toolbar, click Trace. Then right-click anywhere on the map and choose Trace Options. For the Offset, enter 25 and click OK to close the Trace Options.

7. Refer to the following graphic. To start the tracing, click the lot boundary at the left side of the street, as shown in the following graphic. Drag your mouse pointer up until you reach the end of the street, as shown in the graphic, and click.

The new street feature is created at a 25-foot distance, parallel to the lot boundary you traced.

8. On the Editor toolbar, click the Straight Segment method and double-click the end of the existing street to finish the sketch.

Next you will digitize the cul-de-sac using the Midpoint method.

9. With the Straight Segment method still enabled, click the end of the street that you just digitized. Then, on the Editor toolbar, click the drop-down arrow next to the Trace tool and select the Midpoint method from the palette, as shown in the following graphic.

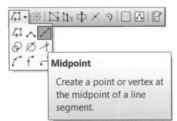

10. With the Midpoint method active, snap to and click the upper street casement, and then click the opposite lower street casement as shown in the following graphic.

Hint: Press the spacebar on your keyboard to temporarily suspend snapping. You want to avoid snapping to the lot boundaries. Alternatively, you can turn off the Lots layer to avoid snapping to it.

A new vertex is created at the midpoint between the two locations you clicked.

11. Continue clicking opposite locations along the street casement of the cul-de-sac until you reach the circle. Then select the Straight Segment method and double-click the end of the light pole in the middle of the circle to finish the sketch, as shown in the following graphic.

Last, you will digitize the missing street on the left using a combination of the Tangent Curve and Straight Segment methods.

12. With the Straight Segment method enabled, click the end of the street that you just digitized. Then, on the Editor toolbar, click the drop-down arrow next to the Trace tool and select the Tangent Curve Segment method from the palette.

13. Experiment with the Tangent Curve Segment method and create the slight curve to the left along the road, as shown in the graphic on the left. Then, switch to the Straight Segment method and finish the sketch at the intersection with the City Collector street.

Hint: If you make a mistake, click the Undo button. Press C on your keyboard to pan the map to the left.

14. Stop editing and save your edits. If you are continuing with the next exercise, leave ArcMap open. Otherwise, close ArcMap.

Challenge 1

For each of the following situations, select the most appropriate feature construction tool or method.

1. You need to create the first vertex of a building footprint 45 feet from one corner of a building and 60 feet from another corner of the same building.

    a. Distance-Direction tool
    b. Midpoint tool
    c. Distance-Distance tool
    d. Intersect tool

2. You need to create adjacent soil polygons without having to click every vertex along the common boundary.

    a. Line
    b. Auto Complete Polygon
    c. Auto Complete Freehand
    d. Polygon tool

3. You need to create the round end of a cul-de-sac.

    a. End Point Arc tool
    b. Circle
    c. Bezier
    d. Tangent

4. You need to create a river centerline.

   a. Circle
   b. Midpoint
   c. Trace
   d. Distance-Direction

5. You need to digitize light poles that are 15 feet from the street centerline.

   a. Point tool
   b. Absolute X,Y tool
   c. Distance-Distance tool
   d. Point at the end of the line

Creating features from existing features

ArcGIS provides a number of editing tools and commands that allow you to create features from the geometry of existing features. To use these tools and commands, you must select features of the appropriate geometry type from editable layers.

Table 10.5 below lists editing tasks and commands used for creating features from existing geometry that are available on the Editor toolbar.

Table 10.5 Editing tasks and commands for creating features from existing geometry

Task	Command	Existing geometry type	When to use
Split a line at a particular length	Split	Line	For splitting a line at a specified distance, measure, or percentage; or into equal parts
Split a line interactively	Split tool (on the Editor toolbar)	Line	For splitting a line by snapping to a vertex, point, or endpoint
Create points along a line	Construct Points	Line	For adding points at regular intervals coincident with a line feature, e.g., create light poles along a street at specified intervals
Create lines parallel to a selected line	Copy Parallel	Line	For creating parallel lines at one or both sides of a line, e.g., create a sewer line parallel to a street centerline
Combine two or more features in the same layer into one	Merge	Line, polygon	For creating a new feature from two or more input features, e.g., create one large parcel from two smaller ones
Combine two or more features from the same or different layers into one	Union	Line, polygon	For creating a new feature from two or more input features, e.g., create a new sales area from two ZIP Code polygons and an existing sales area.

Continued on next page

Table 10.5 Editing tasks and commands for creating features from existing geometry (continued)

Task	Command	Existing geometry type	When to use
Create a feature at a specified distance around another feature	Buffer	Points, lines, polygons	For creating a buffer zone around a feature, e.g., a school, a street, or a ZIP Code area.
Clip an area out of a polygon feature	Clip	Polygon	For creating doughnut polygons, e.g., clipping out the area of a lake from a land parcel; clipping out an area with a buffer, e.g., clipping parcels within 25 feet of a street centerline

Exercise 10d: Create features from existing features

As part of your data quality-control efforts, you will construct and update some features in your Parks layer.

1. **ArcGIS 10.1:** Open the …\DesktopAssociate\Chapter10\ConstructFeatures.mxd and start an edit session.

   ArcGIS 10.0: Open the …\DesktopAssociate\Chapter10\ConstructFeatures_100.mxd and start an edit session.

The map document shows the area in Zilker Park in the City of Austin, Texas, that you are already familiar with from previous exercises.

2. **Turn on the Zilker Park layer.**

The Parks layer is missing the portion of the park south of the major arterial street that runs through it. To fix this, you will combine the Zilker Park polygon from the Parks layer with the southern Zilker Park polygon from the Zilker Park layer.

3. **Using the Edit tool, select the Zilker Park polygon from the Parks layer.**

Hint: Click the drop-down arrow in the Selection Chip to make sure you select the features from the correct layer.

4. **Holding down the Shift key, select the southern Zilker Park polygon from the Zilker Park layer (the one that is NOT covered by a Parks polygon).**

Hint: You may want to check the List By Selection tab in the table of contents to ensure that one polygon is selected from each layer: Parks and Zilker Park.

5. **From the Editor menu, use the Union command to combine the two selected polygons in the Parks layer.**

Hint: In the Union dialog box, choose the template for the Parks layer (green outline).

6. **On the Editor toolbar, open the Attributes window and enter Zilker Park 2 for the PARK_NAME attribute. For the sake of time, leave the rest of the attribute fields blank.**

Decision point: How many polygons in the Parks layer currently represent the northern portion of the park?

Hint: You may use the Selection Chip or the Identify tool to answer this question.

7. In the Parks layer, delete the original Zilker Park polygon and rename the Zilker Park 2 polygon to Zilker Park.

Next, you will create a right-of-way feature around the major arterial street that runs through Zilker Park and remove that area from the park polygon. You will buffer the street and then use the buffer polygon to clip out the right-of-way from the park polygon.

8. Select the six major arterial features that run through Zilker Park, as shown in the following graphic.

Hint: Hold down the Shift key on your keyboard to select the features. Make sure to click exactly on the street features to select them. Ignore the Selection Chip.

9. From the Editor menu, use the Buffer command to create a 60-foot buffer around the selected street features in the Right of Way feature template.

Hint: Make sure that the Right Of Way layer is turned on. In the Buffer dialog box, choose Right of Way for the Template, and enter **60** for Distance.

Next, you will clip out the street casement from the Park polygon. Since the Clip command expects a single selected feature, you need to first merge the buffer polygons you just created.

10. From the Editor menu, use the Merge command to merge the selected buffer polygons into one. It doesn't matter to which polygon you merge the other features.

11. From the Editor menu, use the Clip command with the default buffer distance of 0. Discard the area that intersects.

12. Turn off the Right of Way layer and take a closer look at the right-of-way that you clipped from the park polygon.

The upper end of the right-of-way that has been clipped still needs some reshaping, but you won't do that for now.

13. Stop editing and save your edits. Close ArcMap.

Challenge 2

For each of the following situations select the most appropriate command.

1. You need to divide a selected trail into two equal parts to place a halfway marker.

    a. Construct points
    b. Split
    c. Buffer
    d. Copy Parallel

2. You need to remove parts of a residential zoning feature that intersect a parcel polygon.

    a. Intersect
    b. Union
    c. Clip
    d. Split

3. You need to create a new polygon from the area of overlap of two sales territories.

   a. Intersect
   b. Union
   c. Clip
   d. Split

4. You need to create a polygon in a states layer from all county polygons in a state.

   a. Merge
   b. Copy Parallel
   c. Intersect
   d. Union

Answers to challenge questions

Correct answers shown in bold.

Challenge 1

1. You need to create the first vertex of a building footprint 45 feet from one corner of a building and 60 feet from another corner of the same building.

   a. Distance-Direction
   b. Midpoint
   c. Distance-Distance tool
   d. Intersect

2. You need to create adjacent soil polygons without having to click every vertex along the common boundary.

   a. Line
   b. Auto Complete Polygon
   c. Auto Complete Freehand
   d. Polygon tool

3. You need to create the round end of a cul-de-sac.

   a. End Point Arc tool
   b. Circle
   c. Bezier
   d. Tangent

4. You need to create a river centerline.

   a. Circle
   b. Midpoint
   c. Trace
   d. Distance-Direction

5. You need to digitize light poles that are 15 feet from the street centerline.

   a. Point tool
   b. Absolute X,Y tool
   c. Distance-Distance tool
   d. Point at the end of the line

Challenge 2

1. You need to divide a selected trail into two equal parts to place a halfway marker.

    a. Construct points
    b. Split
    c. Buffer
    d. Copy Parallel

2. You need to remove parts of a residential zoning feature that intersect with a parcel polygon.

    a. Intersect
    b. Union
    c. Clip
    d. Split

3. You need to create a new polygon from the area of overlap of two sales territories.

    a. Intersect
    b. Union
    c. Clip
    d. Split

4. You need to create a polygon in a states layer from all county polygons in a state.

    a. Merge
    b. Copy Parallel
    c. Intersect
    d. Union

Key terms

Geometry: The measures and properties of points, lines, and surfaces. In a GIS, geometry is used to represent the spatial component of geographic features.

Segment: An element of the edit sketch. A segment consists of a start point, an endpoint, and a function that describes a straight line or curve between these two points. Segments may be straight, circular arcs, elliptical arcs, or Bezier curves.

Vertex: An element of the edit sketch. One of a set of ordered x,y coordinate pairs that defines the shape of a line or polygon feature.

Chapter 10: Creating feature geometry

Resources

- ArcGIS Help 10.1 > Desktop > Editing > Introduction >
  - What is editing?
  - A quick tour of editing
- ArcGIS Help 10.1 > Desktop > Editing > Fundamentals of editing
  - Editing tutorial
    - Exercise 1: Getting started with creating features
    - Exercise 2: Creating and editing features
    - Exercise 2a: Defining new types of features to create
    - Exercise 2b: Creating features from existing features
  - Common Editing tasks
    - Common point editing tasks
    - Common line editing tasks
    - Common polygon editing tasks
  - Using Snapping
    - About snapping
- ArcGIS Help 10.1 > Desktop > Editing > Creating new features > Creating features > Using feature templates
  - About feature templates
  - Best practices for using feature templates
  - Setting feature template properties
- ArcGIS Help 10.1 > Desktop > Editing > Creating new features
  - Segment construction methods
- ArcGIS Help 10.1 > Desktop > Editing > Creating features from other features
  - Creating a buffer around a feature
  - Merging features in the same layer
  - Combining features from different layers (Union)
  - Constructing polygons from the shapes of other features

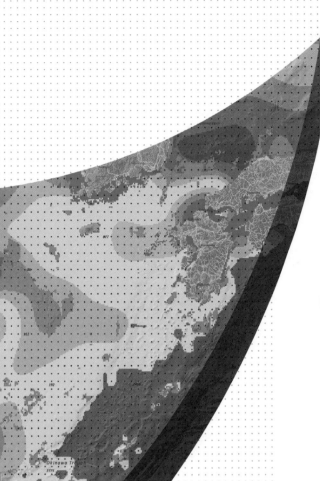

chapter 11

Updating feature geometries

Modifying feature shape 166
    Reshaping features
    Generalizing and smoothing
    Direct editing of feature coordinates
    Exercise 11a: Reshape a polygon feature
    Extending and trimming lines
    Exercise 11b: Extend and trim lines

Dividing features into parts 172
    Exercise 11c: Split lines and cut a polygon
    Challenge
    Versioned editing

Answers to challenge questions 177

Key terms 177

Resources 177

Chapter 11: Updating feature geometries

Have you ever encountered a feature in your database that is out of date, such as a building footprint or a street centerline, or whose geometry doesn't match up with the aerial photo underneath? Editing and updating feature geometries are important tasks of maintaining a database. You may edit geometries in response to real-world features changing over time, to clean up after data conversion, or to repair errors that have been found during quality control.

Out of the many workflows for updating feature geometries, this chapter focuses on two of them: modifying the shapes of features and dividing features into parts.

Skills measured

- Given a scenario, determine how to update the geometry (e.g., edit function type, data, feature snapping) of simple features in ArcGIS Desktop.
- Explain the circumstances under which versioning should be used.
- Explain the difference between reconcile and post operations.

Prerequisites

- Hands-on experience editing features in ArcGIS, including feature templates

Modifying feature shape

Probably the most common editing task is to modify the shape of a line or polygon feature. You can modify the shape of a feature in two different ways: you can either edit the individual sketch elements of the feature, for example, delete vertices, insert new vertices, move vertices, or you can reshape a feature by redigitizing portions of it.

Reshaping features

For minor edits, you would most likely just edit individual vertices and segments. The Edit Vertices toolbar provides all the necessary tools for adding, deleting, and moving individual vertices. You can either work on vertices one by one, or you can select multiple vertices and edit (e.g., move or delete) them as a group.

For more major edits it is more efficient to redigitize portions of the features using the Reshape tool from the Editing toolbar. For example, after a wildfire you may want to redigitize portions of a polygon boundary representing a forested area to exclude the burn scars. You will get some practice reshaping features in the next exercise.

Generalizing and smoothing

Generalizing is used to simplify the geometry of a feature: it decreases the number of vertices representing the feature based on a specified maximum allowable offset, which limits how much the generalized feature can differ from the original one. You generalize features to eliminate unnecessary detail from the data, often to prepare data for display at scales smaller than the scale the data was captured at.

Smoothing is mostly used to prepare data for cartographic purposes to be more visually pleasing. Based on a maximum allowable offset, smoothing fits Bezier curves through the vertices to soften straight edges and angular corners. Generalize and Smooth are available as commands on the Advanced Editing toolbar and as geoprocessing tools.

> ### *Direct editing of feature coordinates*
>
> If you know the exact coordinates of a feature, you can edit them directly in the Edit Sketch Properties window. Once you display the edit sketch of a feature, you can display the x,y coordinates stored for each vertex in the Edit Sketch Properties window, and then move vertices by editing their x,y coordinates. If features contain m- or z-coordinates, you can edit these in the Edit Sketch Properties window as well. For more information about editing feature coordinates directly, refer to the ArcGIS Help topics "Using the Edit Sketch Properties window" and "Editing a vertex's m-value or z-value."

Exercise 11a: Reshape a polygon feature

During quality control, you discovered a water reservoir polygon that doesn't meet your standards. Instead of redigitizing the feature, you will reshape it.

1. **ArcGIS 10.1:** Start ArcMap, open the ...\DesktopAssociate\Chapter11\Reshape.mxd, and start an edit session.

   ArcGIS 10.0: Start ArcMap, open the ...\DesktopAssociate\Chapter11\Reshape_100.mxd, and start an edit session.

The map document shows a water reservoir, Barton Springs Pool, in the City of Austin, Texas. You will reshape the polygon feature to match the outline of the reservoir as it is shown in the aerial photo.

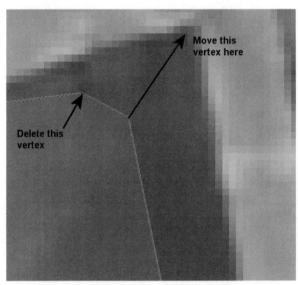

Decision point: What are two ways to display the vertices of a selected feature?

To display the vertices of a selected feature, you can either double-click the feature with the Edit tool, or you can use the Edit Vertices tool.

2. Select the Barton Springs Pool polygon and display its vertices. Zoom to the Fix Number 1 bookmark.

3. From the Edit Vertices toolbar, use the Delete Vertex tool to delete the red vertex, as shown in the previous graphic. Using the Modify Sketch Vertices tool, move the other vertex to the corner of the reservoir, as shown in the previous graphic. Then click the Finish Sketch button to finish the sketch.

4. Save your edits.

You just reshaped a feature by deleting and moving a vertex. Next, you will move multiple vertices at the same time.

5. Zoom to Fix Number 2 bookmark.

6. Display vertices of the Barton Springs Pool feature. Using the Modify Sketch Vertices tool, select the six vertices shown in the following picture. Then move them to the edge of the pool. Finish the sketch.

Hint: To select multiple vertices at the same time, drag a box around them.

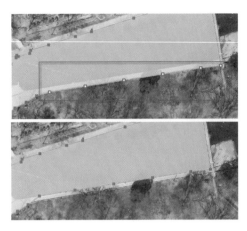

You just reshaped the feature by moving multiple vertices at once. For smaller edits like this one, this workflow is appropriate. However, for major edits, modifying individual vertices will be too time-consuming. Next, you will reshape the feature by redigitizing a portion of the polygon boundary.

7. Zoom to Fix Number 3 bookmark. If necessary, select the Barton Springs Pool feature. On the Snapping toolbar, make sure Vertex and Edge snapping are enabled.

8. Open the Editor Options. In the General tab, uncheck the option to Use symbolized feature during editing. Then click OK.

9. Display the vertices of the Barton Springs Pool feature. Using the Reshape Features tool, snap to the vertex indicated in the graphic to the left.

Note: It is important to remember which vertex to snap to, since you are no longer in vertex editing mode when you activate the Reshape features tool.

10. Reshape the polygon as shown in the following graphic. Finish the sketch by snapping and double-clicking the edge.

A portion of the Barton Springs Pool polygon boundary is replaced with the boundary that you just digitized.

11. **Save your edits and stop editing. Zoom to the full extent.**

12. **If you are continuing with the next exercise, leave ArcMap open. Otherwise, close ArcMap.**

Extending and trimming lines

A very common error with line features are disconnected lines. Features that should be connected with each other are not connected, because one or both of the lines are either too short (undershoot) or too long (overshoot). To connect the lines, you can extend or trim them.

Tables 11.1 and 11.2 list the commonly used editing tools for extending and trimming line features and describe how they work and when to use them.

Table 11.1 Tools for extending lines

Tool	What it does	When to use
Extend tool (Advanced Editing toolbar)	Extends a line to the intersection with another line by adding a straight segment at the end	To interactively extend undershoots to connect with another line
Continue Features (Edit Vertices toolbar)	Allows you to continue a line by digitizing any shape	To extend a line to any shape (e.g., as a curve) and optionally connect to another feature using snapping
Line Intersection (Advanced Editing toolbar)	Extends two line features to their intersection by adding a straight segment at the end	To extend two undershoots to connect with one another

Table 11.1 illustrations created by the author.

Chapter 11: Updating feature geometries

Table 11.2 Tools for trimming lines

Tool	What it does	When to use
Trim tool (Advanced Editing toolbar)	Trims a line to the intersection with another line by removing the portion that extends beyond the intersection	To interactively trim overshoots to connect with another line
Trim to Length (Sketch Context menu)	Trims the line to the distance you enter, measured from the first vertex in the sketch	To interactively trim a line when you know the exact length of it

Table 11.2 illustrations created by the author.

For more information on how to use these tools, refer to the ArcGIS Help topics "Extending a line to an intersection with another line," "Extending a line by sketching," "Splitting lines at intersections," "Trimming a line to an intersection with another line," and "Trimming a line to a specific length."

Exercise 11b: Extend and trim lines

During quality control, you found some errors in your streets data: a few disconnected streets that need to be extended or trimmed. You will use various feature editing tools to connect these line features.

1. **ArcGIS 10.1:** Start ArcMap, open the ...\DesktopAssociate\Chapter11\ Extend_Trim.mxd, and start an edit session.

   ArcGIS 10.0: Start ArcMap, open the ...\DesktopAssociate\Chapter11\ Extend_Trim_100.mxd, and start an edit session.

The map document shows street centerlines in a residential area in the city of Austin, Texas.

2. **Zoom to the Extend/Trim bookmark. On the Snapping toolbar, make sure that end, Vertex, and Edge snapping are turned on. If necessary, open the Advanced Editor toolbar.**

Floradale Dr and Cy Ln are currently disconnected. To connect the two streets, you will extend Cy Lane to meet Floradale Dr. Then, you will trim Floradale Dr to the end of Cy Ln.

3. Select the Floradale Dr feature. Using the Extend tool, extend Cy Ln to Floradale Dr.

4. Select the Cy Ln feature. Use the Trim tool to trim Floradale Dr to Cy Ln, as shown in the following graphic.

The two street features are now properly connected. Next, you will trim a street feature to a particular length.

5. Save your edits, and then zoom to the Trim to Length bookmark.

Decision point: What is the length of Childress Dr from the intersection with Hampshire Dr to the end of the cul-de-sac (in feet)?

Hint: Use the Measure tool to answer this question.

The length of the street feature is approximately 140 feet.

6. Select the Childress Dr feature that extends into the empty lot and expose its vertices.

7. Right-click the red vertex and choose Trim to Length. Trim the feature to a length of 140 feet and finish the sketch.

The Childress Dr feature is now trimmed to a length of 140 feet.

Trimming or extending features works well for connecting features by a straight line. However, there may be situations when you have to add a few more vertices to connect features.

8. Save your edits. Then zoom to the Continue Feature bookmark.

9. **ArcGIS 10.1:** Select Middle Fiskville Rd at the lower end of the map extent and expose its vertices. From the Edit Vertices toolbar, use the Continue Features tool to digitize the missing portion of the street. Double-click the street's endpoint to finish the sketch, as shown in the following graphic.

ArcGIS 10.0:

- Using the Streets feature template, snap to the last (red) vertex at the lower part of Middle Fiskville Rd and digitize the missing piece, as shown in the graphic below. Double-click the street's endpoint to finish the sketch.

- Select both the newly digitized feature and the lower part of Middle Fiskville Rd. From the Editor menu, use the Merge command to merge the two lines into one. (Keep the attributes of the lower part of Middle Fiskville Rd.)

You completed Middle Fiskville Rd to connect properly with the other street features.

10. Save your edits and stop editing. Zoom to the full extent.

11. If you are continuing with the next exercise, leave ArcMap open. Otherwise, close ArcMap.

Dividing features into parts

Another very common editing task is to divide an existing line or polygon feature into two or more new ones. For example, to update for a recent change in parcel ownership, you may have to split a parcel polygon into two parts and assign different owner names to the two new parcels. Or after creating a new street centerline that crosses an existing one, the existing street centerline has to be split at the new intersection.

There are two ways to divide features into parts: you can either click to indicate where to split features, or you can use existing geometry to split them. For example, you can draw a line to split a polygon, or you can use an existing overlapping line feature to split it.

Tables 11.3 and 11.4 list the commonly used editing tools for splitting line and polygon features, describe how they work, and describe when to use them.

Table 11.3 Tools for splitting lines

Tool	What it does	When to use
Split tool (Editor toolbar)	Splits a line feature at the location you click	To manually split lines at any location
Split command (Editor menu on the Editor toolbar)	Splits a line feature into two or more lines of a specified length (as measured from the start or end point of the line)	To split lines • At a distance • At a percentage • Into an equal number of parts • By measure (if it has m-values)
Planarize Lines (Advanced Editing toolbar)	Splits line features at all intersections with other lines	To split multiple selected line features at their intersections (e.g., to split streets at their intersections)
Line Intersection tool (Advanced Editing toolbar)	Splits two line features where they intersect with other lines. Lines are split only at the intersection(s) you click.	To split line features only at specified intersections, e.g., to split streets at regular street intersections, but not at overpasses or underpasses

Table 11.4 Tools for splitting polygons

Tool	How it works	When to use
Cut Polygons tool (Editor toolbar)	Splits a polygon feature along a line that you draw	• To manually split polygons along any linear shape • To create donut polygons (polygons with a hole)
Split Polygons tool (Advanced Editing toolbar)	Splits a polygon feature along an existing line feature	To split polygons along an overlapping line feature

For more information on how to use these tools, refer to the ArcGIS Help topics "Splitting lines manually," "Splitting lines at a specified distance or percentage," "Splitting lines into an equal number of parts," "Splitting lines at intersections," and "Splitting lines at intersections with Planarize Lines."

For more information on how to use these tools, refer to the ArcGIS Help topics "Splitting a polygon" and "Splitting a polygon by an overlapping feature."

Exercise 11c: Split lines and cut a polygon

During quality control, you have found some errors in your streets and parks data. To fix these errors, you will need to split some street features and cut a park polygon into two parts.

1. **ArcGIS 10.1:** Start ArcMap, open the ...\DesktopAssociate\Chapter11\Split_Cut.mxd, and start an edit session.

   ArcGIS 10.0: Start ArcMap, open the ...\DesktopAssociate\Chapter11\Split_Cut_100.mxd, and start an edit session.

The map document shows street centerlines in an area of downtown Austin.

2. Make sure that Vertex Snapping is turned on. If necessary, open the Advanced Editing toolbar.

3. In the center of the map, select Congress Ave between 10th and 6th St.

Congress Ave is one feature that crosses four blocks. Since it is good practice to have separate street features per block, you will split Congress Ave at the intersections with 9th St, 8th St, and 7th St, as shown in the following graphic.

4. From the Editor toolbar, activate the Split tool . Then click the intersection of Congress and 9th St.

5. Select Congress Ave between 9th and 6th St and use the Split tool again at the intersection with 8th St. Repeat these steps to split Congress Ave at the intersection with 7th St.

Splitting lines individually is very systematic and allows you to control exactly where a line is split. However, it can be very time consuming if you have many lines that need to be split. In this situation, the Planarize Lines tool is more efficient.

6. Holding down the Shift key, select Colorado St and Lavaca St.

These two streets have the same problem as Congress Ave. They need to be split at their intersections with 9th, 8th, and 7th streets.

7. Holding down the Shift key, add 9th, 8th, and 7th St to the selection, as shown in the graphic to the right.

8. **ArcGIS 10.1:** From the Advanced Editing toolbar, click the Planarize Lines tool . Accept the default Cluster Tolerance and click OK.

   ArcGIS 10.0: Open the Topology toolbar, if necessary. From the Topology toolbar, click the Planarize Lines tool . Accept the default Cluster Tolerance and click OK.

Colorado St and Lavaca St have been split at their respective intersections with 9th, 8th, and 7th St.

Now that you have successfully split some line features in your Streets layer, you will fix the error in the Parks layer.

9. Zoom to the Zilker Park Area bookmark.

10. Using the Edit tool, select the Zilker Nature Preserve polygon.

11. Zoom to the Zilker Nature Preserve bookmark.

The portion of the Zilker Nature Preserve that is enclosed by the Nature Center Dr (black), Stratfort Dr (orange), and the selected park boundary actually belongs to Zilker Park. To correct this error, you will cut off the portion of the Zilker Nature Preserve polygon and merge it with the Zilker Park polygon.

12. On the Snapping toolbar, click Snapping and turn on Intersection snapping. Turn off all other snap types.

13. From the Snapping menu, open the Snapping Options and turn on SnapTips.

14. On the Editor toolbar, click the Cut Polygons tool.

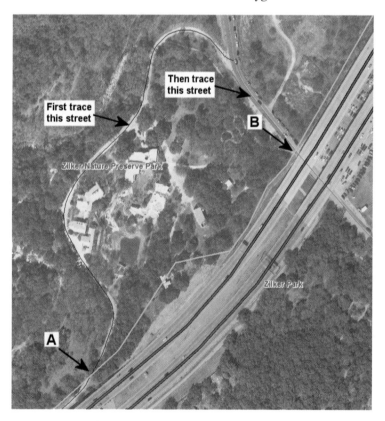

15. **Refer to the previous graphic and do the following:**
    - With the Cut Polygons tool active, click the Intersection marked A in the graphic.
    - On the Editor toolbar, click the Trace tool . Press O on your keyboard to open the Trace Options. Make sure that the Offset is set to 0, and then click OK to close the Trace Options.
    - Click the same Intersection again to start tracing. Trace Nature Center Dr (black) and Stratfort Dr (orange). Snap to and double-click the Intersection marked B to finish the sketch.

Note: Make sure that you properly snap to point B before you finish the sketch. Otherwise, you will get an error message that the CutPolygons task could not be completed.

16. **Clear the selected features. Holding down the Shift key, select both the new Zilker Nature Preserve polygon (the one that you just created) and the Zilker Park polygon. Merge the Zilker Nature Preserve polygon with the Zilker Park polygon.**

You have successfully cut off a portion of the Zilker Nature Preserve polygon and merged it with the Zilker Park polygon.

17. **Save your edits and stop editing. Close ArcMap.**

Challenge

Which editing tool or command is the best choice for each of the following scenarios?

Suppose you have been tasked to do the following:

1. Update a layer of fire roads to match an aerial photo.

   a. Smooth
   b. Stretch proportionally
   c. Reshape
   d. Generalize

2. Connect disconnected street centerline features, such as the one shown in the previous graphic.

   a. Split line
   b. Extend line
   c. Planarize lines
   d. Reshape

3. Reshape a rectangular parking lot so that its coordinates match the ones you received from a surveyor.

   a. Reshape
   b. Move
   c. Continue features
   d. Sketch properties

4. Split a set of 20 selected line features at their intersections (in one step)

   a. Planarize lines
   b. Split
   c. Trim
   d. Modify sketch vertices

Versioned editing

Versioning allows multiple users to edit the same data in an ArcSDE geodatabase without applying locks or duplicating data. The workflow for version editing in an ArcSDE geodatabase differs from the editing workflow in single user geodatabase. To support multiuser editing in an ArcSDE geodatabase, you can perform edits in a version, which is an isolated snapshot of the geodatabase created as a child version from a parent version. For more information on versioning, refer to the ArcGIS Help topics "What is a version?" and "A quick tour of versioning."

There are two common versioned editing workflows:

1. **Multiple users perform edits in separate versions.**

When multiple editors edit the same data in separate versions, they initially save their edits to their respective versions. Conflicts between the edits made in the different versions are resolved when the editors post their changes to the parent or target version. However, after starting an edit session in the versions, the target version may have been changed by other users in a way that conflicts with the edits. Therefore, the editors must first reconcile their versions with the target version, which will update the child version with any changes made in the target version. This process is called an explicit reconcile (between a child and a target version). If the reconcile process reveals any conflicts between the reconcile version and the edited version, editors have to review and resolve them before they can post their changes to the target version. Posting will update the target version with any edits made in the child versions.

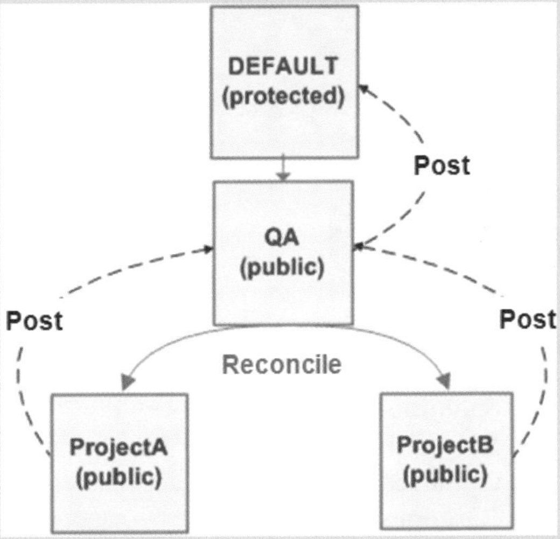

Figure 11.1 In the previous example, a QA version has been created from the DEFAULT (root) version of the geodatabase. Two versions have been created from QA: ProjectA and ProjectB. Before the editors in the ProjectA and ProjectB versions can post their changes to the QA version, they must first reconcile their versions with QA to update their versions with any changes that may have been made in QA. Esri.

2. **Multiple users perform edits in the same version.**

When multiple editors edit in the same version, every editor initially works with their own representation of the version and cannot see any of the changes made by the other editors until they save their edits. When saving the edits, ArcGIS detects any conflicts between edit sessions within that version of the geodatabase and resolves conflicts based on the save preferences you set on the Versioning tab of the Editing Options dialog box. Because this reconcile process takes place based on predetermined settings, it is an implicit one.

For more information on versioned editing, refer to the ArcGIS Help topics "The version editing process," "Saving edits to a version," "Reconciling a version," and "Posting changes."

Answers to challenge questions

Correct answers shown in bold.

Suppose you are tasked to do the following:

1. Update a layer of fire roads to match an aerial photo.

   a. Smooth
   b. Stretch proportionally
   c. Reshape
   d. Generalize

2. Connect disconnected street centerline features, such as the one shown in the previous graphic.

   a. Split line
   b. Extend line
   c. Planarize lines
   d. Reshape

3. Reshape a rectangular parking lot so that its coordinates match the ones you received from a surveyor.

   a. Reshape
   b. Move
   c. Continue features
   d. Sketch properties

4. Split a set of 20 selected line features at their intersections (in one step)

   a. Planarize lines
   b. Split
   c. Trim
   d. Modify sketch vertices

Key terms

Bezier curve: A curved line whose shape is derived mathematically rather than by a series of connected vertices.

Resources

- ArcGIS Help 10.1 > Desktop > Editing > Fundamentals of editing > Editing tutorial
  - Exercise 2: Creating and editing features
    - Exercise 2c: Editing polygon features
    - Exercise 2d: Editing vertices and segments
- ArcGIS Help 10.1 > Desktop > Editing > Fundamentals of editing > Common editing tasks > Common polygon editing tasks
  - Creating and editing multipart polygons
  - Splitting a polygon by an overlapping feature

Chapter 11: Updating feature geometries

- ArcGIS Help 10.1 > Desktop > Editing > Editing existing features >
  - Reshaping a line
  - Reshaping a polygon
  - Splitting a line
    - ... manually
    - ... at a specified distance or percentage
    - ... at intersection with Planarize
  - Extending a line
    - ...by sketching
    - ...to an intersection with another line
    - ...using a geoprocessing tool
  - Trimming a line
    - ...to a specific length
    - ...to an intersection with another line
    - ...using a geoprocessing tool
  - Methods for splitting line features
  - Splitting a polygon
  - Simplifying and smoothing a feature
- ArcGIS Help 10.1 > Desktop > Editing > Editing existing features > Editing vertices and segments
  - The Edit Vertices toolbar
  - Adding a vertex > manually
  - Deleting a vertex
  - Moving a vertex > by dragging it
  - Editing a vertex's m-value or z-value
  - Using the Edit Sketch Properties window
- ArcGIS Help 10.1 > Geodata > Geodatabases > Working with versioned data
  - What is a version?
  - A quick tour of versioning
- ArcGIS Help 10.1 > Geodata > Geodatabases > Working with versioned data > Editing versions
  - The version editing process
  - Saving edits to a version
  - Reconciling and posting edits to a version

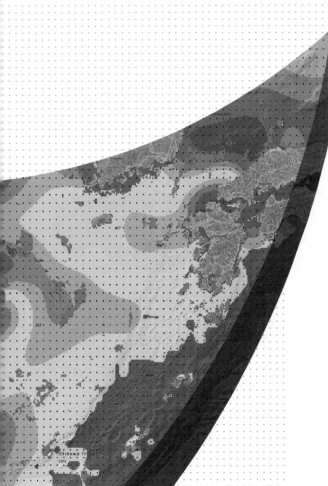

chapter 12

Editing attributes

Attribute editing methods 180
   Editable attribute values
   Exercise 12a: Use different methods for editing feature attributes
   Exercise 12b: Transfer attribute values

Field calculations 183
   Exercise 12c: Perform simple field calculations
   Challenge
   Exercise 12d: Perform advanced field calculations
   Calculating Geometry

Answers to challenge questions 189

Resources 189

Chapter 12: Editing attributes

Editing GIS features includes editing both feature geometries and feature attributes. When you create GIS features, you digitize the geometry (points, lines, or polygons), and then you add attribute values to the empty fields in the attribute table. Later on, when you maintain GIS features, you may need to edit the geometry and attributes to update them for changes in the real world or to correct some errors.

Skills measured

- Given a scenario, determine specifications (e.g., data, attribute update environment) for updating feature attributes.
- Calculate the value of a field using other field values.

Prerequisites
- Knowledge of ArcGIS editing workflow
- Knowledge of field data types

Attribute editing methods

Different editing situations require different methods. To update attributes of a few selected features, you can use the Attribute window to edit individual values. To assign the same attribute value to a group of features, you can calculate attribute values using the Field Calculator or, if the attributes already exist in another feature class, you can transfer them from one feature class to another. For example, you might have a layer with outdated geometry, but accurate and valid attributes. When you replace the outdated geometry layer with a new one, you can transfer the attributes from the outdated layer to the new one.

Editable attribute values

You can edit the attributes of data stored in any editable vector format: attribute values of shapefiles or geodatabase feature classes, or values stored in a stand-alone dBase (.dbf) or geodatabase table. However, you cannot edit the values of any ArcGIS system fields, (i.e., fields that are generated and maintained by the software). For example, values stored in the Shape, OBJECTID (OID), FID, Shape_Length, and Shape_Area fields are not editable.

Any edits you perform on a field value must conform to the data type and other properties of the field. For example, values entered in a Short Integer field must be short integers. Or, the length of a value entered in a text field must comply with the length that has been defined for that field. For more information about field data types, refer to the ArcGIS Help topic "ArcGIS field data types."

Exercise 12a: Use different methods for editing feature attributes

This exercise exposes you to a variety of the manual and automatic methods for assigning attribute values to GIS features.

1. **ArcGIS 10.1: Start ArcMap and open the ...\DesktopAssociate\Chapter12\EditingAttributes.mxd.**
   ArcGIS 10.0: Start ArcMap and open the ...\DesktopAssociate\Chapter12\EditingAttributes_100.mxd.

The map shows a residential area that is just being developed. Street centerlines have been created even though some of the houses have not yet been built.

2. **Zoom to the Pioneer Crossing bookmark. Use the Identify tool** to explore the attributes of the street centerlines in this area.

The geometries of the street centerlines in this area have been created, but they are still missing most of their attribute values.

3. **Start an edit session.**

4. **Using the Edit tool** , select the street centerline labeled number 1 in the following graphic. From the Editor toolbar, open the Attributes window .

5. Type in the street name and street type attributes as follows:

STREET NAME	NOBLEMAN
STREET TYPE	DR

Entering the attribute values one-by-one is necessary when features have unique attribute values. However, when multiple features have the same values, it is more efficient to assign them in one step.

6. Holding down the Shift key, select the street centerlines labeled 2, 3, 4, and 5 in the preceding graphic.

You will assign the same street name and street type to all of the selected features.

7. In the top panel of the Attributes window, click Centerlines (the layer name) as shown in the following graphic. Enter the attribute values as shown in the following table:

STREET NAME	FRIENDSHIP
STREET TYPE	DR

8. Click each of the selected features underneath the layer name, and verify that the street name and street type attribute have been assigned to all of the selected features.

Next, you will use the Field Calculator to assign attribute values to multiple selected features at once.

9. Holding down the Shift key, select the street centerlines labeled by the numbers 6, 7, and 8 in the preceding graphic.

10. Open the attribute table of the Centerlines layer. At the bottom of the table window, click Show Selected records.

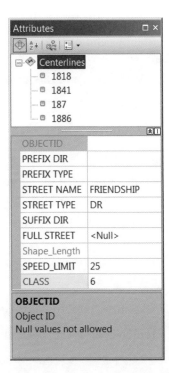

11. Right-click the STREET NAME field and open the Field Calculator.

Note: STREET NAME has been set as a field alias. The name of the field is STREET_NAM.

12. Calculate the values of the STREET_NAM field as "**TIMBER HEIGHTS**".

Hint: Type the following expression in the expression text box: "**TIMBER HEIGHTS**".

For the three selected records, TIMBER HEIGHTS has been added as an attribute value to the STREET NAME field.

13. **Open the field calculator for the STREET TYPE field and modify the expression to calculate the field as "DR". Run the field calculator again with the new expression.**

14. **Clear the selected features and close the table window.**

15. **Save your edits.**

16. **If you are continuing with the next exercise, leave ArcMap open. Otherwise, close ArcMap.**

Exercise 12b: Transfer attribute values

Continuing the scenario from the previous exercise, you will transfer the street name and type attribute values from a layer of address points to a layer of street centerlines.

1. **ArcGIS 10.1: If necessary, start ArcMap and open the ...\DesktopAssociate\Chapter12\ EditingAttributes.mxd.**

   ArcGIS 10.0: If necessary, start ArcMap and open the ...\DesktopAssociate\Chapter12\ EditingAttributes_100.mxd.

2. **If necessary, start an edit session and zoom to the Pioneer Crossing bookmark.**

3. **Turn on the Address Points layer. Open the attribute table of the Address Points layer and examine the fields.**

Among many other attributes, the Address Points table contains the street name and street type for each of the houses.

The tool that you will use to transfer the attributes from the Address Points to the Centerlines layer is located on the Spatial Adjustment toolbar. Since the tool works by clicking on each feature involved in the transfer one at a time, you will use snapping to make the attribute transfer more efficient.

4. **Close the table. Display the Spatial Adjustment toolbar and the Snapping toolbar, if necessary.**

You will set up for attribute transfer by defining the source and the target layer, specifying the matching fields in both layers, and then physically transferring the attributes on a feature-by-feature basis.

5. **On the Spatial Adjustment toolbar, click Spatial Adjustment and choose Attribute Transfer Mapping.**

You will link the matching fields in the source and the target layers to prepare for the attribute transfer.

6. **For the Source, make sure that Address Points is selected. For the Target, select Centerlines.**

7. **Click Auto Match.**

Auto Match matches the common fields between the source and the target layer. To match additional fields, you can manually select them and add them to the list of matched fields.

8. **Select both the PRE_TYPE field in the Address Points layer and the PREFIX_TYP field in the Centerlines layer. Then click Add.**

9. **Uncheck the Transfer Geometry option. Then click OK to close the Attribute Transfer Mapping dialog box.**

10. **On the Snapping toolbar, make sure that Point and Edge snapping are turned on.**

11. **On the Spatial Adjustment toolbar, select the Attribute Transfer tool .**

You will use the attribute transfer tool to transfer attribute values to the centerlines labeled 9, 10, and 11 in the following graphic.

12. Click any of the address points along centerline number 9, and then snap to and click centerline number 9. Repeat this process for address points and centerlines marked 10 and 11.

13. Use the Identify tool to verify that the street name and the street type attributes have been transferred to the centerlines layer.

14. Zoom to the full extent and save your edits.

15. If you are continuing with the next exercise, leave ArcMap open. Otherwise, close ArcMap.

Field calculations

In exercise 12a, you used the Field Calculator to assign attribute values to multiple selected records at one time. The Field Calculator allows you to perform different kinds of field calculations: simple field calculations include copying values from one field to another, concatenating (i.e., combining) values from existing fields into a new field, mathematical operations such as dividing or multiplying field values to create new ones, or simply populating one or more fields with new values. The Field Calculator can be used to assign or update values for all records or a selected set of records, as you discovered in exercise 12a.

The Field Calculator also allows you to perform some advanced calculations, such as entering code blocks that preprocess the data before the calculations are made. For example, you can create a script that uses IF... THEN statements to perform calculations on attribute values based on whether or not a condition is true.

You can use the Field Calculator inside or outside an edit session. However, when a geodatabase feature class participates topology or a network, the Calculates Values command will only be available within an edit session.

You can use either VBScript or Python as a parser language to write expressions in the Field Calculator. However, the syntax for mathematical expressions is the same for VBScript and Python.

Refer to tables 12.1, 12.2, and 12.3 for examples of some common mathematical, text, and date expressions you can write.

Chapter 12: Editing attributes

Table 12.1 Mathematical expressions

Task	Syntax	Example
Add two or more fields	[numericField1] + [numericField2]	[SiteRisk] + [RouteRisk]
Subtract two or more fields	[numericField1] – [numericField2]	[FloodRisk] – [SiteRisk]
Divide two fields	[numericField1] / [numericField2]	[AGE_UNDR21] / [POP2000]
Find percentage of multiple fields	([population_field1] + [population_field2] + [population_field3]) / [total_population_field]	([AGE_UNDER5] + [AGE_5_17] + [AGE_18_21]) / [POP2000]
Convert units, e.g., from meters to feet.	[meters] * 3.2808	30 * 3.2808 returns 98.4
Convert degrees/minutes/seconds to decimal degrees.	[DEG] + ([MIN] / 60) + ([SEC] / 3600)	121 + (8 / 60) + (6 / 3600) returns 121.135

Table 12.2 Text expressions

Task	VB Script syntax	Python syntax
Populate field with new text	"Text"	"Text"
Concatenate two text fields, e.g., Joe and Smith to Joe Smith	[textField1] & [textField2] (without a space) [textField1] & " " & [textField2] (with a space)	!textField1!+!textField2! (without a space) !textField1!+ " " + !textField2! (with a space)
Concatenate new text with a text field, e.g., Current owner: [owner name stored in a text field]	"Text" & " " & [textField]	!Text!+ " " + !textField!
Convert text to lowercase e.g., JOE SMITH to joe smith	LCase([textField])	!textField!.lower ()
Convert text to uppercase e.g., joe smith to JOE SMITH	UCase([textField])	!textField!.upper ()
Convert text to titlecase, e.g., JOE SMITH to Joe Smith	No direct function in VB script	!textField!.title ()
Trim space(s) from beginning or end of text, e.g., __New York__ to New York	Trim([textField])	! textField!.strip ()
Replacing text, e.g., Road with Rd	Replace ([Street], "Road", "Rd")	!Street!.replace ("Road", "Rd")

Table 12.3 Date expressions

Task	VB Script syntax	Python syntax
Enter today's date, e.g., October 1, 2012	Date()	datetime.datetime.now( )
Calculate current date and time	Now()	datetime.datetime.now( )
Calculate the difference between two dates	DateDiff ("d", [dateField1], [dateField2]) (returns days, use "m" for months)	datetime.timedelta(seconds=0)

For more information on the Field Calculator, refer to the ArcGIS Desktop Help topic "Fundamentals of Field Calculations."

Exercise 12c: Perform simple field calculations

You will now use Python expressions in the Field Calculator to concatenate values from different text fields into one and eliminate blank spaces in the middle, beginning, and end of a new text string. Then you will convert the text string from uppercase to lowercase.

1. **ArcGIS 10.1:** If necessary, start ArcMap and open the ...\DesktopAssociate\Chapter12\ EditingAttributes.mxd. Start an edit session, if necessary.

    ArcGIS 10.0: If necessary, start ArcMap and open the ...\DesktopAssociate\Chapter12\ EditingAttributes_100.mxd. Start an edit session, if necessary.

2. Open the attribute table of the Centerlines layer.

You will use the Field Calculator to concatenate the values of the PREFIX_DIR, PREFIX_TYPE, STREET_ NAME, STREET_TYPE and SUFFIX_DIR fields into a single field.

3. Open the Field Calculator for the FULL STREET field.

Hint: Right-click the FULL STREET field heading and choose Field Calculator.

4. Run the Field Calculator with the following expression:

[PREFIX_DIR] + [PREFIX_TYP] + [STREET_NAM] + [STREET_TYP] + [SUFFIX_DIR]

Hint: To add the fields to the expression, double-click the field names in the fields list.

5. Examine the calculated values in the FULL STREET field.

Decision point: What problems do you see with the calculated values?

The values from the concatenated fields have been added to the FULL STREET field. However, since many cells in the PREFIX DIR, PREFIX TYPE, and SUFFIX DIR fields were empty, there are also many blank spaces at the beginning and end of the full street names. On the other hand, there are blank spaces missing between the street name and the street type.

You will rewrite the previous Field Calculator expression and modify it to put spaces between the calculated fields.

To get practice with another parser, you will use Python as the parser language.

6. Open the Field Calculator for the FULL NAME field and set the Parser to Python.

7. Modify the current Field Calculator expression to read:

!PREFIX_DIR! +" "+ !PREFIX_TYP! +" "+ !STREET_NAM! +" "+ !STREET_TYP! +" "+ !SUFFIX_DIR!

Hint: Beware of typos! In Python, field names are enclosed in exclamation marks. After each field name, make sure to enter a plus sign, a double-quote, a blank space, another double-quote, and a plus sign.

8. Run the field calculator with the modified expression.

9. Examine the calculated values in the FULL NAME field. Pay special attention to streets that have a prefix direction (e.g., N for North or S for South).

Since most streets do not have a prefix type value, there are now three blank spaces between the prefix direction and the street name, instead of one blank space as desired. You will fix this by using the replace function in the Field Calculator. Do not worry about the blank spaces at the beginning and end of the text strings in the FULL STREET field; you will eliminate them later.

10. **Open and run the Field Calculator for the FULL STREET field using the following expression: ! FULL_STREE!.replace(" ", " ")**

Hint: Make sure to have three blank spaces between the first set of double-quotes, and one blank space between the second set of double-quotes.

Note: FULL STREET has been set as a field alias, which is not used in a Field Calculator expression. The name of the field is FULL_STREE.

Next you will eliminate the blank spaces at the beginning and end of the text strings in the FULL STREET field.

11. **Open and run the field calculator for the FULL STREET field using the following expression: !FULL_STREE!.strip()**

The strip function eliminates the leading and trailing blanks at the beginning and end of the full street name.

Last you will convert the text in the FULL STREET field to title case.

12. **Open and run the Field Calculator for the FULL STREET field using the following expression: !FULL_STREE!. title()**

13. **Stop editing and save your edits. Close the table window.**

14. **If you are continuing with the next exercise, leave ArcMap open. Otherwise, close ArcMap.**

Challenge

For each of the following scenarios, choose the correct answer:

1. In an attribute table of a feature class that participates in a topology, you right-click a field heading, but the Field Calculator option is unavailable. What may be the reasons? (Choose two.)

    a. The field is a Shape_Area field
    b. The field is a date type
    c. The field is type double
    d. You are outside of an edit session

Hint: For help answering the following question, refer to the ArcGIS Help topic "Fundamentals of field calculations."

2. In an Address field, the first four characters represent the house number. Which of the following expressions will extract the house number into a new field?

    a. Mid([Address], 1,4)
    b. LTrim([Address], 1,4)
    c. Right([Address], 4)
    d. Left ([Address], 4)

Hint: For help answering the following question, refer to the ArcGIS Help topic "Calculate Field examples."

3. Which Field Calculator expression would you use to eliminate extra blank spaces inside of a city name in a text field called City, (e.g., New York has three blank spaces between New and York)?

    a. !City!.replace(" ", " ")
    b. Trim ([City], " ", " ")
    c. Replace ([City], " ", " ")
    d. !City!.strip(" ", " ")

Exercise 12d: Perform advanced field calculations

You will now use a Python code block in the Field Calculator to preprocess the data before the calculations are made.

1. **ArcGIS 10.1:** If necessary, start ArcMap and open the ...\DesktopAssociate\Chapter12\ EditingAttributes.mxd. Start an edit session, if necessary.
   ArcGIS 10.0: If necessary, start ArcMap and open the ...\DesktopAssociate\Chapter12\ EditingAttributes_100.mxd. Start an edit session, if necessary.

2. Open the attribute table of the Address Points layer and locate the INPUT_DATE and STATUS fields.

The INPUT_DATE field contains the date on which the address points were established. The STATUS field is empty.

3. Sort the INPUT_DATE field in Descending order. Then scroll through the records to the bottom of the table.

Near the bottom of the table some of the records have an input date of 1/1/1111 and 1/1/1801, which are placeholder dates for unknown and historic address points. You will load a code sample into the Field Calculator to calculate the STATUS field as UNKNOWN, HISTORIC, or CURRENT based on the values in the INPUT_DATE field.

4. Open the Field Calculator for the STATUS field.

5. Set the Parser to Python.

Field Calculator expressions can be saved to and loaded from a text file with a .cal extension.

6. Click Load and open DateCalc.cal from your ...\DesktopAssociate\Chapter12 folder.

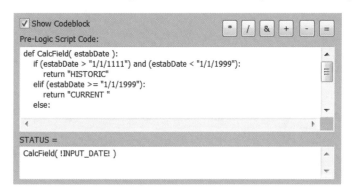

The code block you loaded evaluates the values in the INPUT_DATE field. It calculates the STATUS field values as HISTORIC for input dates later than 1/1/1111 and earlier than 1/1/1999; as CURRENT for input dates later than 1/1/1999. If the input date does not fall into any of these time frames, the STATUS field will be calculated as UNKNOWN.

7. Click OK to run the code.

8. Examine the values in the STATUS field and verify that they have been calculated correctly.

9. Stop editing and save your edits. Close the table window.

10. Close ArcMap.

Calculating Geometry

The Calculate Geometry dialog box is a tool used for common geometric calculations. For example, you can calculate the x,y coordinates of a point, the length of a line, or the area and centroid of a polygon. If features store a z-value as part of their geometry, you can also perform geometric calculations in 3D. For more information on the Calculate Geometry dialog box, refer to the ArcGIS Help topic "Fundamentals of Field Calculations."

Answers to challenge questions

Correct answers shown in bold.

1. In an attribute table of a feature class that participates in a topology, you right-click a field heading, but the Field Calculator option is unavailable. What may be the reasons? (Choose two.)

   a. The field is a Shape_Area field
   b. The field is a date type
   c. The field is type double
   d. You are outside of an edit session

2. In an Address field, the first four characters represent the house number. Which of the following expressions will extract the house number into a new field?

   a. Mid([Address], 1,4)
   b. LTrim([Address], 1,4)
   c. Right([Address], 4)
   d. Left ([Address], 4)

3. Which Field Calculator expression would you use to eliminate extra blank spaces inside of a city name in a text field called City, e.g. New York (e.g., New York has three blank spaces between New and York)?

   a. !City!.replace(" ", " ")
   b. Trim ([City], " ", " ")
   c. Replace ([City], " ", " ")
   d. !City!.strip(" ", " ")

Resources

- ArcGIS Help 10.1 > Desktop > Editing > Editing attributes
  - About editing attributes
  - Applying the same attribute values to multiple features in a layer
- ArcGIS Help 10.1 > Desktop > Editing > Editing attributes >
  - About maintaining attribute integrity while editing
  - Editing attributes with subtypes and attribute domains
- ArcGIS Help 10.1 > Desktop > Editing > Performing spatial adjustment
  - About spatial adjustment attribute transfer
  - Transferring attributes between features
- ArcGIS Help 10.1 > Geodata > Data types > Tables > Calculating field values
  - Fundamentals of field calculations
  - Making simple field calculations
- ArcGIS Help 10.1 Desktop > Geoprocessing > Tool reference > Data Management toolbox > Fields toolset
  - Calculate field examples
- ArcUser article: "The Field Calculator Unleashed," Tom Neer, 2005 (**http://www.esri.com/news/arcuser/0405/files/fieldcalc_1.pdf**)

chapter 13

Maintaining attribute integrity

Attribute editing with default values 192
   Setting default values
   Subtypes and default values
   Exercise 13a: Set default values

Attribute editing with domains 194
   Workflow for creating domains
   Exercise 13b: Create a domain
   Exercise 13c: Convert a table to a domain
   When to use subtypes versus domains
   Challenge

Answers to challenge questions 199

Key terms 199

Resources 200

Chapter 13: Maintaining attribute integrity

Maintaining attribute integrity is an essential goal of attribute editing. Entering incorrect values, typos, or inconsistent values are very common errors that can occur during attribute editing. For example, when multiple users edit the same street dataset, they can easily enter inconsistent values for street type (e.g., Av, AV, Ave, or AVE for Avenue). Entering invalid or inconsistent attribute values in a field not only compromises the integrity of your data and potentially leads to incorrect analysis results, it also takes time and energy to discover and fix the errors. The geodatabase offers different mechanisms for maintaining attribute integrity and making attribute editing more efficient.

Skills measured
- Compare and contrast the different types of geodatabase domains.
- Given a scenario, convert a table of field values to an appropriate domain or domains.
- Given a scenario, determine the appropriate subtype and domains to create.

Prerequisites
- Knowledge of the ArcGIS editing workflow
- Knowledge of field data types

Attribute editing with default values

Setting up default values for attributes makes attribute editing much more efficient and helps to enforce data integrity. If, for example, in a layer of transmission main lines, most of the water lines are 24 inches in diameter, you can set 24 inches as a default value for the DIAMETER field. This way, the DIAMETER attribute value for any new feature you create will automatically be set to 24 inches, and you only have to modify the value if its diameter is different.

Setting default values

You can set default values at two different levels. At the geodatabase feature class level, you can set default values in the field properties. If you then add the feature class as a layer to ArcMap and start an edit session, the default values set in the field properties are automatically set in the feature template properties.

You can also manually set default values in the feature template properties, with or without default values being set in the field properties. If you set different default values in the field properties than in the feature template properties, the values in the feature template properties will override the values in the field properties.

For more information on default values, refer to the ArcGIS Help topic "About maintaining attribute integrity while editing."

Subtypes and default values

Subtypes are subsets of features within a feature class, or records in a table, based on the values in one attribute field. For example, in a Streets feature class, if you have values of Local Streets, Main Streets, and Highways, you could create a subtype based on each of these values. Creating subtypes in a feature class or table requires either a short or a long integer field. In the example above, you would create an integer field in the Streets attribute table to contain code values for each street type—for example, 1 for Local Streets, 2 for Main Streets, and 3 for Highways. This field would then be used to build subtypes for the Streets feature class.

Instead of setting one default attribute value for all the features in a feature class or all records in a table, you can set different default values for each subtype. For example, suppose you added a speed limit attribute. The Local Streets subtype (1) could have a default value of 25 miles per hour, while the Main Streets subtype (2) could have a default value of 35 miles per hour, and the Highways subtype (3) could have a default value of 65 miles per hour.

When you add a feature class with subtypes to ArcMap and start an edit session, ArcMap automatically creates a feature template for each subtype. If you create new features using, for example, the Local Streets feature template, the speed limit value of the new features will automatically be set to 25 miles per hour. When you create a new feature using the Highways feature template, its speed limit value will be 65 miles per hour.

For more information on subtypes, refer to the ArcGIS Help topics "A quick tour of subtypes," "Working with subtypes," and "Creating subtypes."

Exercise 13a: Set default values

Suppose you frequently add new features to a street centerlines layer and want to make attribute entry more efficient and error-proof. Subtypes have been built for the Centerlines feature class, and you find that the speed limit value in one subtype of the street centerlines is almost always the same. Therefore, you will experiment with setting default values for the speed limit of this subtype.

1. If necessary, start ArcMap. Open a new blank map. Add the ...\DesktopAssociate\ Chapter13\Austin. gdb\Centerlines layer to the map.

When you add the Centerlines layer, it automatically displays with Unique Values symbology based on the CLASS field. This is because subtypes have been set up for the Centerlines feature class.

2. Open the attribute table of the Centerlines layer. Scroll to the right and examine the properties of the CLASS field.

The CLASS field contains code values for different road classes.

Decision point: What field data type is the CLASS field?

Since the CLASS field is of type Short Integer, it can be used to group the Centerlines features into subtypes.

3. In the Catalog window, open the properties of the Centerlines feature class and click the Subtypes tab.

The features of the Centerlines layer have been grouped into subtypes based on the CLASS field. Each road class represents a subtype.

4. Under Subtypes, select the CLASS 6 subtype, as shown in the following graphic.

Hint: Click the small, gray box next to it.

5. Under Default Values and Domains, set the Default Value for the SPEED_LIMI field to 35, as shown in the following graphic. Then click OK to close the Feature Class Properties.

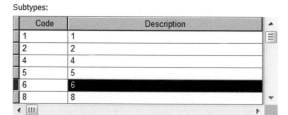

6. Start an edit session and display the Create Features window, if necessary.

Hint: To display the Create Features window, click the Create Features button on the Editor toolbar.

7. **Right-click the CLASS 6 feature template and open its properties.**

Since you have set a speed limit of 35 as a default value for the CLASS 6 subtype, the corresponding feature template inherited that value. However, if you would like to change the default value while in an edit session, you can override the default value set in the feature class properties with one that you set in the feature template properties.

8. **In the Feature Template Properties, set the default value for the SPEED_LIMI field to 25, as shown in the following graphic. Then click OK to close the Feature Template Properties window.**

PREFIX DIRECTION	<Null>
PREFIX TYPE	<Null>
STREET NAME	<Null>
STREET TYPE	<Null>
SUFFIX DIRECTION	<Null>
FULL_STREE	<Null>
SPEED_LIMI	25
CLASS	6

To verify that the default values have been correctly set for the subtypes, you will create a new centerline in the CLASS 6 subtype.

9. **Using the CLASS 6 feature template, draw a line of any shape, finish the sketch, and open the Attributes window.**

The new feature has been created with a default speed limit value of 25.

10. **Stop editing. If you are continuing with the next exercise, leave ArcMap open.**

Attribute editing with domains

Attribute domains are another mechanism in the geodatabase to make attribute editing more efficient and to enforce data integrity. Attribute domains are rules that define which values are allowed in a field. When you edit the values in a field that has an attribute domain applied to it, the field will only accept values contained in that domain. For example, if in a Streets feature class, the valid values for the Surface Material field are Asphalt, Concrete, Aggregate, or Gravel, you can apply a domain to it to limit its values to this list. When you edit the Surface Material attribute values, you will be presented with a drop-down list that allows you to enter only one of these four values.

There are two types of attributes domains: coded-value domains and range domains. Coded-value domains specify a list of valid values that can be entered in a field. This can be a list of text values that can be entered in a text field, such as the street surface materials listed in the example above, or it can be a list of numbers that can be entered in a numeric field. For example, in a feature class of distribution mains, pipes can have a diameter of only 10, 24, or 30 inches. You can set up a coded value domain containing values of 10, 24, and 30 and apply it to the Diameter field.

Range domains specify a range of valid values for a numeric attribute. All values within the range domain—all numbers between a minimum and a maximum value—are valid values for the attribute. For example, Streets features may have between 1 and 8 lanes specified in the Lanes attribute field. A range domain with a minimum value of 1 and a maximum value of 8 would limit the valid attribute values in the Lanes field to between 1 and 8. Range domains can only be applied to numeric and date fields (short integer, long integer, float, double, or date), while coded-value domains can be applied to any field type, including text fields.

For more information on attribute domains, refer to the ArcGIS Help topics "A quick tour of attribute domains" and "About maintaining attribute integrity while editing."

Workflow for creating domains

You create an attribute domain in the properties of a geodatabase. After you created an attribute domain, you can then apply it to one or more attribute fields in any feature class or table in that geodatabase. For example, a coded value domain of street types can be applied to a Street Type attribute in a Streets feature class and in an Address table.

A typical workflow for creating attribute domains and using them during editing includes the following steps:

1. **Select the field:** Select the field to which you will apply an attribute domain. The data type of the field can be short integer, long integer, float, double, text, or date.
2. **Choose a domain type:** Choose whether to create a coded-value domain or a range domain.
3. **Create the domain:** You can create domains manually using the Domains tab in the geodatabase properties, or you can create domains using geoprocessing tools such as the Create Domain or Table to Domain tools.
4. **Apply the domain to an attribute field:** You can apply attribute domains either to an entire field (all records) or only to the records within a subtype. Once attribute domains are applied to a field or a subtype, you can validate them using the Validate Features command on the Editor toolbar to ensure that the existing values comply with the domain.
5. **Edit attribute values:** To edit the values in a field that has a coded-value domain applied to it, pick the correct value from the provided drop-down list. To edit values in a field with a range domain, type in the value manually and use the Validate Features command on the Editor toolbar to confirm that the value complies with the range domain.

For more information on editing attribute values with domains, refer to the ArcGIS Help topic "Editing attributes with subtypes and attribute domains."

Exercise 13b: Create a domain

Continuing the scenario from the previous exercise, you are still working on making the attribute entry for a street centerlines layer more efficient and error-proof. You will now create a domain for the prefix directions of the streets, apply it the appropriate attribute field, and use it for attribute editing.

1. If necessary, start ArcMap with a new blank map and add the …\DesktopAssociate\ Chapter13\Austin. gdb\Centerlines layer to the map.

2. Open the attribute table of the Centerlines layer and examine the values in the PREFIX DIR field.

Decision point: What kind of domain is appropriate for the PREFIX DIR field?

You will create a coded-value domain for the different street directions and apply it to the PREFIX DIR field.

Decision point: What field data type is the PREFIX DIR field?

---

3. Close the table. In the Search window, locate the Create Domain geoprocessing tool and open it.

4. Set following parameters:
   - Input Workspace: …\DesktopAssociate\Chapter13\Austin.gdb
   - Domain Name: **StreetDirections**
   - Domain Descriptions: **Prefix directions for streets**

The data type of the domain must match the data type of the field that it will be applied to.

5. For the Field Type, choose TEXT. Make sure that CODED is selected for the Domain type. Then click OK to run the tool.

6. In the Catalog window, right-click the …\DesktopAssociate\Chapter13\Austin.gdb and choose Properties. Click the Domains tab.

Chapter 13: Maintaining attribute integrity

The StreetDirections domain has been created in the geodatabase properties, but the coded values and their descriptions still need to be filled in.

7. **In the Coded Values section, enter the following Code and Description values.**

Code	Description
N	North
S	South
E	East
W	West

8. **Click Apply, and then click OK to close the Database Properties window.**

Next, you will apply the domain to the PREFIX DIR field in the Centerlines feature class.

9. **In the Catalog window, open the properties of the Centerlines feature class and click the Fields tab.**

10. **Click the PREFIX_DIR field to select it. In the Field Properties section, click the empty cell next to Domain and choose the StreetDirections domain, as shown in the following graphic. Click Apply, and then click OK to close the Feature Class Properties window.**

Field Name	Data Type
OBJECTID	Object ID
Shape	Geometry
PREFIX_DIR	Text
PREFIX_TYP	Text
STREET_NAM	Text
STREET_TYP	Text
SUFFIX_DIR	Text
FULL_STREE	Text
SPEED_LIMI	Double
Shape_Length	Double
CLASS	Short Integer

Click any field to see its properties.

Field Properties

Alias	PREFIX DIRECTION
Allow NULL values	Yes
Default Value	
Domain	StreetDirections
Length	2

To verify that the domain has been correctly applied to the PREFIX_DIR field, you will create a new centerline feature and assign a prefix direction attribute value.

11. **Start an edit session and display the Create Features window if necessary.**

Hint: To display the Create Features window, click the Create Features button on the Editor toolbar.

12. **Using any of the Centerlines feature templates, draw a line of any shape, finish the sketch, and open the Attributes window.**

Since you have successfully set up a coded-value domain, you can now pick a value for the PREFIX_DIR field from a drop-down list.

13. **Click the PREFIX_DIR field and choose North from the drop-down list**

14. **Stop editing and save your edits. If you are continuing with the next exercise, leave ArcMap open.**

Exercise 13c: Convert a table to a domain

You want to do more to make the attribute entry for a street centerlines layer efficient and error-proof. Entering values into the STREET TYPE field has been cumbersome because there are many abbreviations for the different street types. To use a consistent abbreviation scheme, you used a look-up table containing valid abbreviations for the street types and then entered the values manually. To streamline this workflow, you will create a coded-value domain from this table.

1. If necessary, start ArcMap with a new blank map and add the …\DesktopAssociate\ Chapter13\Austin.gdb\Centerlines layer to the map.

2. Open the attribute table of the Centerlines layer and examine the values in the STREET TYPE field.

3. Add the StreetTypes table to the map and open it.

The StreetTypes table contains a list of valid street types that can be used in the STREET TYPE field in the Centerlines attribute table. In the Austin geodatabase, you will convert the StreetTypes table to a coded value domain.

4. Close the table window. In the Search window, locate the Table to Domain geoprocessing tool and open it.

5. Run the Table to Domain tool with the following parameters:

    - Input table: StreetTypes
    - Code Field: STREET_TYPE
    - Description Field: STREET_TYPE
    - Input Workspace: …\DesktopAssociate\ Chapter13\Austin.gdb
    - Domain Name: STREET_TYPE

    Click OK.

6. In the Catalog window, open the properties of the …\DesktopAssociate\Chapter13\Austin.gdb. Click the Domains tab and verify that the STREET_TYPE domain has been created. Then close the Database Properties windows.

7. In the Catalog window, open the Properties of the Centerlines feature class and apply the STREET_TYPE domain to the STREET_TYP field.

Note: The field name is STREET_TYP; for readability, STREET_TYPE has been set as an alias for the field.

Hint: If you need more detailed instructions to perform this step, refer to step 10 of the previous exercise.

To verify that the domain has been correctly applied, you will create a new centerline feature and assign a street type attribute value.

8. Start an edit session and display the Create Features window, if necessary.

9. Using any of the Centerlines feature templates, draw a line of any shape, finish the sketch, and open the Attributes window.

Since you have successfully converted the StreetType table to a coded-value domain in the Austin.gdb, you can now pick street type from a drop-down list.

10. Click the STREET_TYPE field and choose any value from the drop-down list.

11. Stop editing and save your edits. Close ArcMap.

When to use subtypes versus domains

Recall that subtypes are classifications of features or records within a feature class or table based on common values in one attribute field. Attribute domains are lists or ranges of valid values that can be entered in a field. Knowing when to create subtypes or domains (or both) will help you make editing more efficient and maintain the integrity of your database.

Use subtypes for the following:
- To categorize features within a feature class without creating multiple feature classes
- To automatically symbolize features by subtype when you add the feature class as a layer to ArcMap
- To set different default values or attribute domains by subtype

Use domains for the following:
- To create a consistent attribution scheme for a field (limit input of values to a specific numeric range or a specific list)
- To apply attribute validation to the field
- When the values in one field are limited but do not affect the values in other fields—that is, they are not suited to group features into subtypes, but should be restricted to a list or numeric range of values

Challenge

For each of the following scenarios, choose the correct answer.

1. What kind of geodatabase functionality should be implemented in a Highways feature class to constrain the Speed Limit field to values between 55 and 75 miles per hour?

    a. Subtypes
    b. Range domain
    c. Spatial domain
    d. Coded-value domain

2. What kind of geodatabase functionality should be implemented to set different default values for the surface material of high-elevation, medium-elevation, and low-elevation trails?

    a. Subtypes
    b. Range domain
    c. Spatial domain
    d. Coded-value domain

3. Suppose you have a table of 55 land use descriptions that can be assigned to a Landuse_Description field. Which type of geodatabase functionality should be implemented to ensure that land use descriptions are spelled consistently and that typos are avoided?

    a. Subtypes
    b. Range domain
    c. Spatial domain
    d. Coded-value domain

4. In a feature class of water distribution mains, the pipes have either a diameter of 3 inches or 6 inches. Three-inch pipes are designed to have a pressure between 100 and 250 psi, while 6-inch pipes are designed for 100 to 450 psi. What is the best way to enforce this kind of attribution?

   a. Subtypes based on diameter and two coded-value domains for the diameter
   b. Subtypes based on pressure and two coded-value domains for the diameter
   c. Subtypes based on pressure and two range domains for the diameter
   d. Subtypes based on diameter and two range domains for the pressure

Answers to challenge questions

Correct answers shown in bold.

1. What kind of geodatabase functionality should be implemented in a Highways feature class to constrain the Speed Limit field to values between 55 and 75 miles per hour?

   a. Subtypes
   b. Range domain
   c. Spatial domain
   d. Coded-value domain

2. What kind of geodatabase functionality should be implemented to set different default values for the surface material of high-elevation, medium-elevation, and low-elevation trails?

   a. Subtypes
   b. Range domain
   c. Spatial domain
   d. Coded-value domain

3. Suppose you have a table of 55 land use descriptions that can be assigned to a Landuse_Description field. Which type of geodatabase functionality should be implemented to ensure that land use descriptions are spelled consistently and that typos are avoided?

   a. Subtypes
   b. Range domain
   c. Spatial domain
   d. Coded-value domain

4. In a feature class of water distribution mains, the pipes have either a diameter of 3 inches or 6 inches. Three-inch pipes are designed to have a pressure between 100 and 250 psi, while 6-inch pipes are designed for 100 to 450 psi. What is the best way to enforce this kind of attribution?

   a. Subtypes based on diameter and two coded-value domains for the diameter
   b. Subtypes based on pressure and two coded-value domains for the diameter
   c. Subtypes based on pressure and two range domains for the diameter
   d. Subtypes based on diameter and two range domains for the pressure

Key terms

Attribute integrity: The degree to which attribute data is correct and free of errors according to the database design.

Data integrity: The degree to which the data in a database is accurate and consistent according to data model and data type.

Resources

- ArcGIS Help 10.1 > Geodata > Data types > Domains
  - A quick tour of attribute domains
  - Creating a new attribute range domain
  - Creating a new attribute coded value domain
  - Associating default values and domains with tables and feature classes
- ArcGIS Help 10.1 > Geodata > Data types > Subtypes
  - A quick tour of subtypes
  - Working with subtypes
  - Creating subtypes
- ArcGIS Help 10.1 > Geodatabases > Defining the properties of data in a geodatabase > Geodatabase table properties
  - ArcGIS field data types
- ArcGIS Help 10.1 > Geodata > Geodatabases > Building a geodatabase tutorial
  - Exercise 3: Creating subtypes and attribute domains

Editing with topology

Topology 202
    Topology in ArcGIS
    Choosing a topology
    Topology elements
    Exercise 14a: Edit with map topology (ArcGIS 10.1)
    Exercise 14a: Edit with map topology (ArcGIS 10.0)

Geodatabase topology 208
    Geodatabase topology rules
    Challenge
    Designing and building a geodatabase topology
    Geodatabase topology errors
    Exercise 14b: Edit with geodatabase topology (ArcGIS 10.1)
    Exercise 14b: Edit with geodatabase topology (ArcGIS 10.0)
    Using predefined fixes
    Exercise 14c: Edit with geodatabase topology and predefined fixes

Answers to challenge questions 220

Key terms 220

Resources 220

Chapter 14: Editing with topology

When you edit and maintain features in your database, you want to preserve the basic spatial relationships that features have in the real world: streets should remain connected, ZIP Codes should remain adjacent, and manholes should be on top of sewer lines, even after you may have reshaped the sewer line.

Almost every database contains datasets in which features share at least portions of their geometry, as in the following example (figure 14.1). Features can share geometry with features in the same feature class or features in a different feature class. For example, when street centerlines are connected, they share their endpoints with one another, but they may also share geometry with bus routes that run along them.

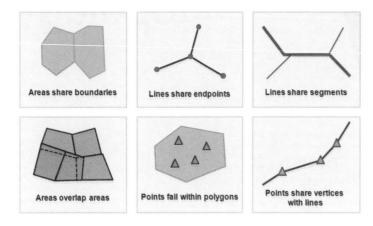

Figure 14.1 These graphics show some examples of how features share geometry. Esri.

Skills measured

- Identify required components for a topological edit.
- Explain the significance of topological relationships and the impact of edits on those relationships.

Prerequisites

- Knowledge of spatial relationships such as connectivity, adjacency, intersection, or containment
- Knowledge of the ArcGIS editing workflow
- Hands-on experience with common editing operations such as moving vertices or reshaping features

Topology

Topology is a mechanism for maintaining the spatial integrity of your data and for preserving shared geometries. Topology provides you with specific editing tools. These tools allow you to edit shared portions of the features' geometry simultaneously with one edit. This edit cascades into all of the affected features. For example, you can use these tools to reshape the boundary between two adjacent city area polygons, which updates both affected features at the same time, without introducing any gaps or overlaps.

There are two strategies to maintain your database with topology:
1. **Find – fix – prevent:** If you know your database already contains spatial errors, you may use topology to find and fix these errors. In addition, you can use topology editing tools for making edits (e.g., updates that reflect changes in the real world) and preventing new spatial errors.
2. **Prevent:** If your database is already free of errors, you can simply use topology editing tools for updating features and preventing new spatial errors.

Topology in ArcGIS

There are two kinds of topologies in ArcGIS: map topology and geodatabase topology.

A map topology is based on the layers in a map, and is set up for the duration of an edit session and discovered coincident geometry on-the-fly. You create a map topology by specifying the following:

- Which layers to include in the map topology
- A cluster tolerance (the minimum distance between vertices before they are considered coincident)

Once you have created the map topology, the software evaluates the participating layers for coincident geometry. For example, if two polygons are adjacent to each other (i.e., closer to each other than the cluster tolerance), the common boundary of the polygons is treated as coincident geometry. If you wanted to edit this common boundary, special topology editing tools allow you to perform edits on both polygons at once, maintaining coincidence between them. Map topology is available with all license levels and can be used with geodatabase feature classes and shapefiles.

Geodatabase topology uses rules to model more complex spatial relationships such as containment, adjacency, connectivity, and coincidence. Geodatabase topology not only allows you to maintain coincidence during editing, but also allows you to find and fix spatial errors in your database. Geodatabase topology contains a set of rules that define how features should share geometry. For example, a geodatabase topology may include a rule that specifies that ZIP Codes must not overlap and another rule that they must not have gaps. If you validate your ZIP Code polygons against these rules, you will find the ones that violate them.

Choosing a topology

Since both types of topologies allow you to simultaneously update the coincident geometry of features, the topology type to use depends entirely on whether or not you want to model spatial relationships. To move, modify, or reshape features that are partly or entirely coincident, it is easiest to create a map topology in an edit session and perform the edits using the topology editing tools. For example, suppose after a hurricane, you have to reshape the coincident coastline in land use, parks, and a forested areas layer. Map topology provides the tools you need to perform this edit.

When you want to ensure that feature geometry is consistent with real-world spatial relationships, and you want the software to find spatial errors in your data, a geodatabase topology is the best choice. For example, you can use a geodatabase topology to check if streams are properly connected with each other and if they are properly contained within the watershed boundaries. For future edits, you can use a geodatabase topology to simultaneously update coincident geometry without introducing new errors.

Topology elements

When you edit coincident geometry with topology tools, you do not edit individual features, but perform your edits on the topology elements: edges and nodes. Once you have selected a topology in a map document, line features and the outlines of polygon features become topological edges, and point features, the endpoints of lines, and the places where edges intersect become nodes. The edits that you perform on edges and nodes update all features that participate in them. For example, when you reshape an edge, the edit updates all features that participate in that edge (figure 14.2).

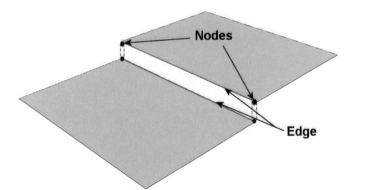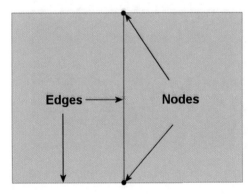

Figure 14.2 The graphic shows the edge and nodes between two polygons. In the left example, polygons have been offset in height to show the coincident boundaries of both polygons participating in the edge. The coincident corner vertices of both polygons participate in the nodes. The right example shows the same polygons as they would appear in ArcMap. The arrows in the right example point to two edges: one selected edge (shown in magenta) and an unselected one. If you reshaped the selected edge or moved the nodes, both polygons would be updated. Created by the author.

Chapter 14: Editing with topology

Exercise 14a: Edit with map topology (ArcGIS 10.1)

Suppose you contracted with a land trust to maintain their GIS database. You start working on the land trust boundary. Since the boundary is coincident with some layers maintained by the city (annexation and precinct boundaries), you obtained these layers and extracted the relevant features that are adjacent to the land trust.

Note: These instructions apply to ArcGIS 10.1 only. If you are working in ArcGIS 10.0, follow the ArcGIS 10.0 instructions for exercise 14a below.

1. Start ArcMap and open the …\DesktopAssociate\Chapter14\TopoEditing.mxd.

2. Display the Editor, Snapping, and Topology toolbars, if necessary.

The map shows land trust, annexation, and precinct boundaries north of the City of Noblesville, Indiana.

3. Zoom to the Coincident Boundaries bookmark. Turn the Land Trust and Annexation layers off and on.

This extent shows the southeastern corner of the land trust. In this extent, most of the land trust boundary is coincident with the annexation and precinct boundaries.

4. Zoom to the Modify Edge bookmark.

In this extent, land trust, annexation, and precinct boundaries run through private backyards. After checking with the city, you learn that this is not correct. You will use a map topology to reshape the coincident boundary to run along the forest edge.

5. Start an edit session. From the Editor menu, open the Editor Options. In the Topology tab, check the box to display Unselected Nodes. Then click OK to close the Editor Options.

6. On the Topology toolbar, click the Select Topology button . Select Map Topology and check all three layers to include them in the map topology. Then click OK.

7. On the Snapping toolbar, make sure that Vertex and Edge snapping are turned on.

8. From the Topology toolbar, click the Topology Edit tool .

9. Click the horizontal boundary to build the topology cache. Then click the horizontal edge again to select it. Then, holding down the Shift key, select the edge to the right, as shown in the following graphic.

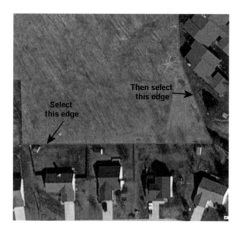

10. From the Topology toolbar, click Show Shared Features .

The Show Shared features window lists the features that share geometry in the selected edges.

11. From the Topology toolbar, click the Reshape Edge tool .

12. Reshape the edge as shown in the following graphic. Start by clicking the vertex in the lower left corner of the selected edge. Then click the forest corner inside the land trust area. Finish the sketch by snapping to and double-clicking the edge near the corner of the house.

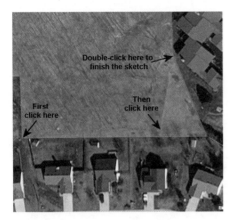

The coincident boundary has been reshaped and all participating features have been updated, as shown in the following graphic.

13. Using the Topology Edit tool, right-click the selected edge and choose Clear Selected Topology Elements.

14. Turn the Land Trust and Annexation layers off and on to confirm that all participating features have been updated.

15. Save your edits and stop editing. If you are continuing with the next exercise, leave ArcMap open. Otherwise, close ArcMap.

Exercise 14a: Edit with map topology (ArcGIS 10.0)

Suppose you contracted with a land trust to maintain their GIS database. You start working on the land trust boundary. Since the boundary is coincident with some layers maintained by the city (annexation and precinct boundaries), you obtained these layers and extracted the relevant features that are adjacent to the land trust.

Note: These instructions apply to ArcGIS 10.0 only. If you are working in ArcGIS 10.1, follow the ArcGIS 10.1 instructions for exercise 14a above.

1. Start ArcMap and open the …\DesktopAssociate\Chapter14\TopoEditing_100.mxd.

The map shows land trust, annexation, and precinct boundaries north of the City of Noblesville, Indiana.

2. Display the Editor, Snapping, and Topology toolbars, if necessary.

3. Zoom to the Coincident Boundaries bookmark. Turn the Land Trust and Annexation layers off and on.

This extent shows the southeastern corner of the land trust. In this extent, most of the land trust boundary is coincident with the annexation and precinct boundaries.

  4. **Zoom to the Modify Edge bookmark.**

In this extent, land trust, annexation, and precinct boundaries run through private backyards. After checking with the city, you learn that this is not correct. You will use a map topology to modify and reshape the coincident boundary to run along the forest edge.

  5. **Start an edit session. From the Editor menu, open the Editor Options. In the Topology tab, check the box to display Unselected Nodes. Then click OK to close the Editor Options.**

  6. **On the Topology toolbar, click the Map Topology button . Check all three layers to include them in the map topology. Then click OK.**

  7. **On the Snapping toolbar, make sure that Vertex and Edge Snapping are turned on.**

  8. **From the Topology toolbar, click the Topology Edit tool . Double-click the horizontal boundary to select the edge and expose its vertices, as shown in the following graphic.**

  9. **From the Edit Vertices toolbar, click the Add Vertex tool , using the Add Vertex as shown in the following graphic.**

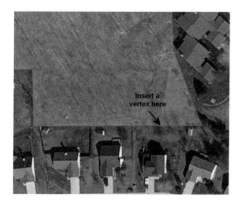

  10. **Using the Modify Sketch Vertices tool , move the new vertex to the forest corner inside the land trust area, as shown in the following graphic. Then finish the sketch.**

Hint: To finish the sketch, either click the Finish Sketch button or press F2 on your keyboard.

To adjust the remaining boundary, you will move the node at the right end of the selected edge and reshape the edge above it.

11. Using the Topology Edit tool, 🔩, select the node at the right edge of the map display. Move the selected node to the forest edge, as shown in the following graphic.

Hint: To select the node, you can press the N key on your keyboard and draw a box around it.

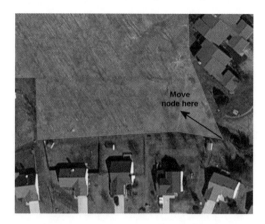

12. Using the Topology Edit tool 🔩, select the right edge of the map display. From the Topology toolbar, click the Reshape Edge tool 🖉.

13. Reshape the edge as shown in the graphic below. Start by snapping to the vertex the vertex near the corner of the house. Then finish the sketch by double-clicking the node that you just moved.

The coincident boundary has been reshaped and all participating features have been updated, as shown in the following graphic.

14. Using the Topology Edit tool 🔩, right-click the selected edge and choose Clear Selected Topology Elements.

15. Turn the Land Trust and Annexation layers off and on to confirm that all participating features have been updated.

16. Save your edits and stop editing. If you are continuing with the next exercise, leave ArcMap open. Otherwise, close ArcMap.

Geodatabase topology

Geodatabase topology is implemented as an item in a feature dataset. All feature classes that participate in a geodatabase topology must be stored in the same feature dataset as the topology. A given feature class can participate only in one geodatabase topology. Creating or modifying a geodatabase topology requires an ArcGIS for Desktop Standard or ArcGIS for Desktop Advanced license.

Geodatabase topology rules

Before you build a geodatabase topology, you have to design it. You have to identify which spatial relationships in your data are important to preserve and determine which rules are suitable to model these relationships. You can specify rules that control relationships between features in the same feature class and between features in different feature classes or subtypes. Often times there are multiple rules for enforcing the same spatial relationship or slight variations of a spatial relationship. The ArcGIS Geodatabase Topology rules poster provides assistance in choosing the right topology rule(s) for a given spatial relationship. It contains 32 topology rules and descriptions of spatial relationships they apply to, the types of spatial relationships that cause rule violations, and examples of real-world applications for each topology rule. To access the topology rules poster, refer to the ArcGIS Help topic "Geodatabase topology rules and topology error fixes" and click the ArcGIS *Geodatabase Topology Rules* link.

In the ArcGIS Geodatabase Topology Rules poster, topology rules are categorized by the geometry type of the feature classes they model (point, line, and polygons rules). For each rule, icons indicate whether the topology rule occurs within a single or between two feature classes or subtypes. The icons indicate the geometry type of the feature classes or subtypes that the rule applies to.

To use the Topology Rules poster for choosing the right topology rule(s) for a given spatial relationship:
- Find the correct geometry category (point, line, or polygon rules).
- Determine whether you want to preserve a spatial relationship within one or between two feature classes or subtypes.
- Identify the spatial relationship that you want to preserve.

Tables 14.1, 14.2, and 14.3 list examples of some of common spatial relationships for polygons, lines, and points, and the topology rules that are used to preserve them.

Geodatabase topology

Table 14.1 Polygons

Spatial relationship	Rule
Polygons are adjacent to each other, e.g., voting districts, soils, parcels	**Must not overlap** — Polygons must not overlap within a feature class or subtype. Polygons can be disconnected or touch at a point or touch along an edge. Polygon errors are created from areas where polygons overlap. *Use this rule to make sure that no polygon overlaps another polygon in the same feature class or subtype.* A voting district map cannot have any overlaps in its coverage. **Must not have gaps** — Polygons must not have a void between them within a feature class or subtype. Line errors are created from the outlines of void areas in a single polygon or between polygons. Polygon boundaries that are not coincident with other polygon boundaries are errors. *Use this rule when all of your polygons should form a continuous surface with no voids or gaps.* Soil polygons cannot include gaps or form voids—they must form a continuous fabric.
Polygons of one feature class contain polygons of another feature class, e.g., states containing counties, census tracts containing census blocks	**Must be covered by** — Polygons in one feature class or subtype must be covered by a single polygon from another feature class or subtype. Polygon errors are created from polygons from the first feature class or subtype that are not covered by a single polygon from the second feature class or subtype. *Use this rule when you want one set of polygons to be covered by some part of another single polygon in another feature class or subtype.* Counties must be covered by states.
Polygons contain points	**Contains point** — Each polygon of the first feature class or subtype must contain within its boundaries at least one point of the second feature class or subtype. Polygon errors are created from the polygons that do not contain at least one point. A point on the boundary of a polygon is not contained in that polygon. *Use this rule to make sure that all polygons have at least one point within their boundaries. Overlapping polygons can share a point in that overlapping area.* School district boundaries must contain at least one school.

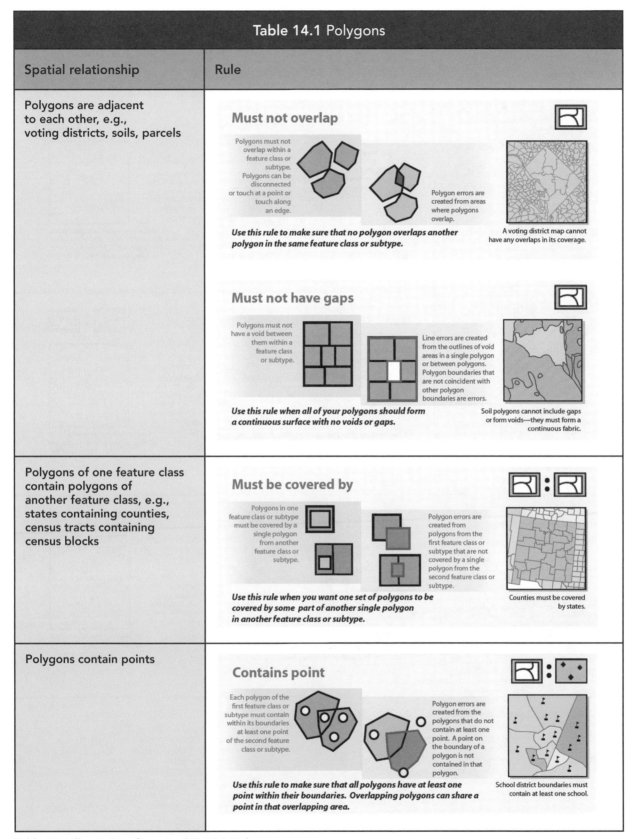

Table 14.1 illustrations from ArcGIS 10.1 Help.

Chapter 14: Editing with topology

Table 14.2 Lines

Spatial relationship	Rule
Line features must be connected with each other, e.g., streets, rivers, telecommunication lines	**Must not have dangles** — The end of a line must touch any part of one other line or any part of itself within a feature class or subtype. Point errors are created at the end of a line that does not touch at least one other line or itself. *Use this rule when you want lines in a feature class or subtype to connect to one another.* A street network has line segments that connect. If segments end for dead-end roads or cul-de-sacs, you could choose to set as exceptions during an edit session.
Line features must be coincident with the boundary of polygons, e.g., lot lines and parcel boundaries	**Must be covered by boundary of** — Lines in one feature class or subtype must be covered by the boundaries of polygons in another feature class or subtype. Line errors are created on lines that are not covered by the boundaries of polygons. *Use this rule when you want to model lines that are coincident with the boundaries of polygons.* Polylines used for displaying block and lot boundaries must be covered by parcel boundaries.
Endpoint of a line must coincide with a point feature	**Endpoint must be covered by** — The ends of lines in one feature class or subtype must be covered by points in another feature class or subtype. Point errors are created at the ends of lines that are not covered by a point. *Use this rule when you want to model the ends of lines in one feature class or subtype that are coincident with point features in another feature class.* Endpoints of secondary electric lines must be capped by either a transformer or meter.

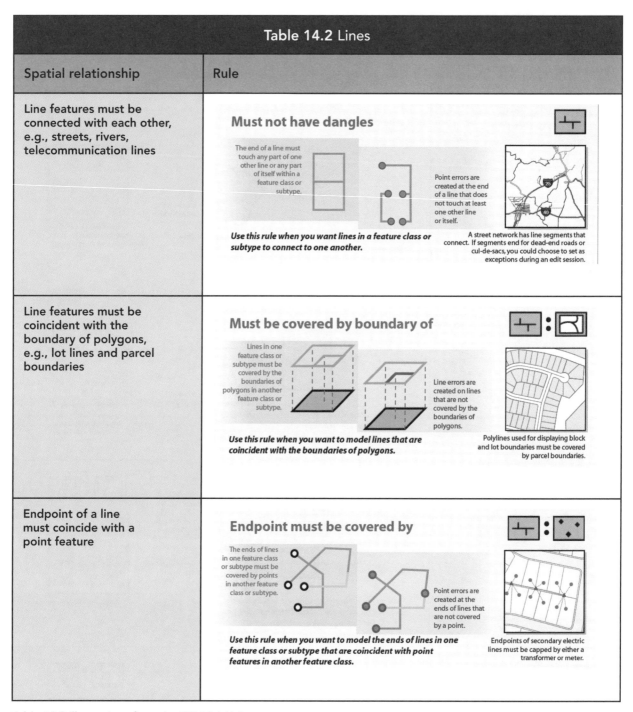

Table 14.2 illustrations from ArcGIS 10.1 Help.

Geodatabase topology

Spatial relationship	Rule
Points must be separate; they must not be stacked, e.g., address points, fitting points in a distribution network	**Must be disjoint** — Points cannot overlap within the same feature class or subtype. Point errors are created where points overlap themselves. Use this rule when points within one feature class or subtype should never occupy the same space. Fittings in a water distribution network should not overlap.
Points must be aligned with points from another feature class or subtype, e.g., service points and electric meters	**Must be coincident with** — Points in one feature class or subtype must be coincident with points in another feature class or subtype. Point errors are created where points from the first feature class or subtype are not covered by points from the second feature class or subtype. Use this rule when points from one feature class or subtype should be aligned with points from another feature class or subtype. Meters must be coincident with service points in an electric utility network.
Points must fall on a line, e.g., water monitoring stations and streams; bus stops and bus routes	**Point must be covered by line** — Points in one feature class or subtype must be covered by lines in another feature class or subtype. Point errors are created on the points that are not covered by lines. Use this rule when you want to model points that are coincident with lines. Monitoring stations must fall along streams.

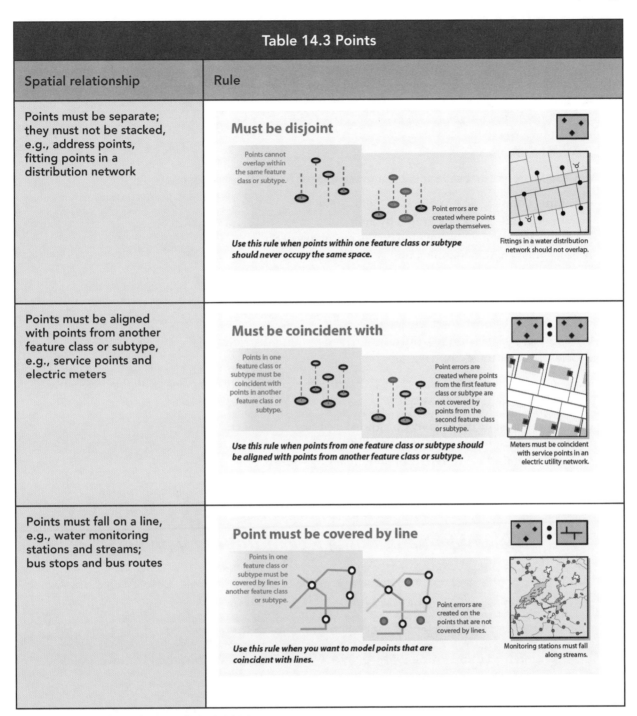

Table 14.3 illustrations from ArcGIS 10.1 Help.

Chapter 14: Editing with topology

Challenge

For each of the following scenarios, choose the appropriate topology rule to model the spatial relationships.

1. You want to ensure that electricity lines are properly connected with each other

    a. Must not have gaps
    b. Must not have dangles
    c. Must be covered by
    d. Must be disjoint

2. You want to ensure that soil Sample Points are located within Soils polygons

    a. Sample Points Must be coincident with Soils
    b. Sample Points Must be covered by Soils
    c. Sample Points must be covered by the boundary of Soils
    d. Soils must contain Sample Points

3. You want to ensure that an area is seamlessly covered with sales districts

    a. Sales districts must be disjoint
    b. Sales districts must not have dangles
    c. Boundary of Sales districts must be covered by Sales districts
    d. Sales districts must not have gaps

4. You want to ensure that line features in a Coastline feature class are coincident with the boundaries of polygons in a Counties feature class

    a. Counties must be covered by Coastline
    b. Counties must be coincident with Coastline
    c. Coastline must be covered by the boundary of Counties
    d. Coastline must be covered by Counties

Designing and building a geodatabase topology

To build a geodatabase topology, you use tools in the Catalog window (or ArcCatalog) or automate the process using geoprocessing tools. Before building a geodatabase topology, you design it by doing the following:

- Identifying feature classes that will participate in the topology

- Organizing them into a feature dataset (which enforces that they all have the same spatial reference)

- Specifying topology rules

- Ranking feature classes by their accuracy: most to least accurate (optional)

Once you have designed the topology, you can build a test geodatabase to test your design, and then refine and adjust it as needed.

For more information about creating a topology refer to the ArcGIS Help topics "Designing a geodatabase topology" and "Creating a topology."

Geodatabase topology errors

To apply the topology rules to the participating feature classes, you validate the geodatabase topology. Initially (before the topology has been validated for the first time), the entire extent of the topology is considered a "dirty area," an area where features have not been checked against the topology rules yet. During the initial validation, vertices that are closer together than the cluster tolerance are snapped together. The dirty area is cleared and any rule violations found are flagged as potential errors.

You have to evaluate these rule violations to see if there are spatial errors that need to be resolved, or if there is a valid spatial relationship that should be marked as an exception. For example, if a topology rule requires that street features are connected with other street features at both ends, a dead-end street or cul-de-sac is a rule violation, but it is a valid exception to the rule.

If you find that a rule violation is indeed a spatial error, you have to determine the best way to resolve it. You can use any editing tool or some predefined fixes to resolve topology errors. In exercise 14b, you will use editing tools; in exercise 14c, you will use predefined fixes.

Exercise 14b: Edit with geodatabase topology (ArcGIS 10.1)

You continue to work on the land trust boundary. You will now use a geodatabase topology to enforce coincidence between the land trust boundary and the boundaries of the adjacent annexation and precincts polygons.

Note: These instructions apply to ArcGIS 10.1 only. If you are working in ArcGIS 10.0, follow the ArcGIS 10.0 instructions for exercise 14b that follow.

1. If necessary, start ArcMap and open the ...\DesktopAssociate\Chapter14\TopoEditing. Display the Editor and Topology toolbars, if necessary.

The map shows land trust, annexation, and precinct boundaries north of the City of Noblesville, Indiana.

2. In the Catalog window, open the Properties of the ...\DesktopAssociate\Chapter14\Noblesville\Landbase\Landbase_Topology. In the Rules tab, examine the rules that have been defined for this topology.

The topology rules enforce that the land trust boundary is covered by the boundaries of the Annexation and Precincts layers.

3. Close the Topology Properties. From the Catalog window, click and drag the Landbase_Topology into your map. Click No when asked if you want to add all the feature classes. (They are already in the map.)

4. Open the Layer Properties of the Landbase_Topology layer. In the Symbology tab, check the box next to dirty areas. Then click OK to close the Layer Properties window.

Currently, the entire extent of the topology is a dirty area. This is because the topology has never been validated since it was built. You will now validate the Landbase_Topology layer and apply topology rules to the participating feature classes.

5. In the Catalog window, right-click the Landbase_Topology layer and choose Validate. Click OK in the message window, and then click the Refresh button in the lower left corner of the map display to clear the dirty areas from the map.

6. Open the properties of the Landbase_Topology layer. In the Errors tab, click Generate Summary.

Five rule violations have been found during validation. You will now investigate these rule violations and evaluate whether they are spatial errors.

7. Click OK to close the Layer Properties window.

8. Start an Edit session. On the Topology toolbar, click the Select Topology button. Select Geodatabase Topology and make sure the Landbase_Topology is selected. Then click OK.

9. Zoom to the Land Trust bookmark. On the Topology toolbar, click the Error Inspector button, and click Search Now.

The five errors that have been found during the validation are listed in the Error Inspector.

10. **In the Error Inspector, select each of the errors (the rows) one after the other, and examine the spatial relationships in the map. You may want to turn the Landbase_Topology, Land Trust, and Annexation layers off and on to see which features are coincident.**

Hint: In the map, selected errors are marked in black.

Not all of the rule violations are real errors. For example, the rule violation at the northern boundary should be marked as an exception, because the land trust extends beyond the precinct and annexation polygons. At the southwestern boundary, there is another possible exception to the rules: the Land Trust boundary is only coincident with the annexation boundary, because the precinct polygon extends further south.

11. **Make sure the Landbase_Topology, Land Trust and Annexation layers are turned on. In the Error Inspector, select the two errors in the lower right land trust boundary, as shown in the following graphic. Right-click any of the selected errors and choose Zoom To.**

Hint: To select multiple errors in the Error Inspector, hold down the Ctrl key.

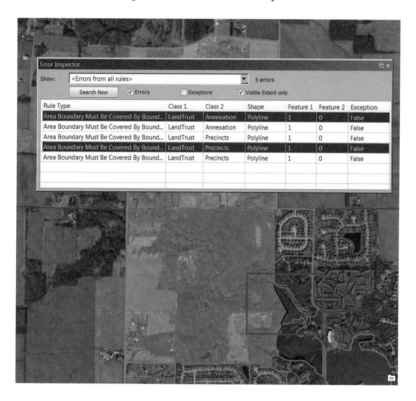

Note: The errors may be in a different order in your error inspector than what is shown in the graphic above.

In this extent, the land trust boundary does not align the with annexation and precinct boundaries. You will fix these two errors by aligning the land trust edge (the edge causing the error) with the edge along the river, which is shared by annexation and precinct polygons.

12. From the Topology toolbar, click the Align Edge tool . With the Align Edge tool active, first click the selected errors, and then click the Annexation boundary along the river, as shown in the graphic below.

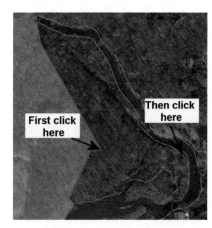

13. Click any other tool (e.g., the Topology Edit tool) to disable the Align Edge tool. Zoom to the Land Trust bookmark. On the Topology toolbar, click the Validate Current Extent button .

14. In the Error Inspector, click Search Now.

The two misalignment errors have been resolved. There are three errors left.

15. Select the three remaining errors in the Error Inspector, right-click them, and choose Mark As Exception.

16. Save your edits and stop editing. If you are continuing with the next exercise, leave ArcMap open. Otherwise, close ArcMap.

Exercise 14b: Edit with geodatabase topology (ArcGIS 10.0)

You continue to work on the land trust boundary. You will now use a geodatabase topology to enforce coincidence between the land trust boundary and the boundaries of the adjacent annexation and precincts polygons.

Note: These instructions apply to ArcGIS 10.0 only. If you are working in ArcGIS 10.1, follow the ArcGIS 10.1 instructions for exercise 14b above.

1. If necessary, start ArcMap and open the …\DesktopAssociate\Chapter14\TopoEditing_100. Display the Editor and Topology toolbars, if necessary.

The map shows land trust, annexation, and precinct boundaries north of the City of Noblesville, Indiana.

2. In the Catalog window, open the Properties of the …\DesktopAssociate\Chapter14\Noblesville\ Landbase\Landbase_Topology. In the Rules tab, examine the rules that have been defined for this topology.

The topology rules enforce that the land trust boundary is covered by the boundaries of the Annexation and Precincts layers.

3. Close the Topology Properties. From the Catalog window, click and drag the Landbase_Topology into your map. Click No when asked if you want to add all the feature classes. (They are already in the map.)

4. Open the properties of the Landbase_Topology layer. In the Symbology tab, check the box next to dirty areas. Then click OK to close the Layer Properties window.

Currently, the entire extent of the topology is a dirty area. This is because the topology has never been validated since it was built. You will now validate the Landbase_Topology and apply topology rules to the participating feature classes.

5. In the Catalog window, right-click the Landbase_Topology layer and choose Validate. Click OK in the message window, and then click the Refresh button in the lower left corner of the map display to clear the dirty area from the map.

Note: If you get an error message about an existing schema lock, make sure that you are outside of an edit session.

6. Open the properties of the Landbase_Topology layer. In the Errors tab, click Generate Summary.

Five rule violations have been found during validation. You will now investigate these rule violations and evaluate whether they are spatial errors.

7. Click OK to close the Layer Properties window.

8. Start an Edit session. On the Topology toolbar, select Landbase_Topology from the drop-down list, if necessary.

9. Zoom to the Land Trust bookmark. On the Topology toolbar, click the Error Inspector button and click Search Now.

The five errors that have been found during the validation are listed in the Error Inspector.

10. In the Error Inspector, select each of the errors (the rows) one after the other, and examine the spatial relationships in the map. You may want to turn the Landbase_Topology, Land Trust, and Annexation layers off and on to see which features are coincident.

Hint: In the map, selected errors are marked in black.

Not all of the rule violations are real errors. For example, the rule violation at the northern boundary should be marked as an exception, because the land trust extends beyond the precinct and annexation polygons. At the southwestern boundary, there is another possible exception to the rules: the Land Trust boundary is only coincident with the annexation boundary, because the precinct polygon extends further south.

11. Make sure the Landbase_Topology, Land Trust, and Annexation layers are turned on. In the Error Inspector, select the two errors in the lower right land trust boundary, as shown in the graphic below. Right-click any of the selected errors and choose Zoom To.

Hint: To select multiple errors in the Error Inspector, hold down the Ctrl key.

Note: The errors may be in a different order in your error inspector than what is shown in the graphic above.

In this extent, the land trust boundary does not align the with annexation and precinct boundaries. You will fix these two errors by reshaping the land trust edge (the edge causing the error) by tracing the edge along the river, which is shared by annexation and precinct polygons.

12. On the Snapping toolbar, click Snapping and turn on Snap To Topology Nodes. From the Snapping menu, open the Snapping Options and turn on SnapTips. If necessary, open the Editor Options. In the Topology tab, check the box to display Unselected Nodes.

13. Using the Topology Edit tool, 🖫, select the edge that contains the two errors. From the Topology toolbar, click the Reshape Edge tool 🗗 .

14. **Refer to the graphic above:**
    - With the Reshape Edge tool active, click the topology node marked A in the graphic.
    - On the Feature Construction toolbar, click the Trace tool 🔁 . Press O on your keyboard to open the Trace Options. Make sure that the Offset is set to 0, and then click OK to close the Trace Options.
    - Click the same node again to start tracing. Trace the Annexation-Precinct edge along the river until you reach the node marked B. Snap to the node marked B and double-click it to finish the sketch.

Note: Make sure that you properly snap to node B before you finish the sketch. Otherwise, you will get an error message that the task could not be completed.

15. Using the Topology Edit tool, right-click the reshaped edge and choose Clear Selected Topology Elements. Zoom to the Land Trust bookmark. On the Topology toolbar, click the Validate Current Extent button 🗹 .

16. In the Error Inspector, click Search Now.

The two misalignment errors have been resolved. There are three errors left.

17. **Select the three remaining errors in the Error Inspector, right-click them, and choose Mark As Exception.**

18. **Save your edits and stop editing. If you are continuing with the next exercise, leave ArcMap open. Otherwise, close ArcMap.**

Using predefined fixes

With geodatabase topology you can either manually fix a topology error, as you did in the last exercise, or you can apply a predefined fix. Most topology rule violations have one or more predefined fixes that automatically resolve the error. When you select an error in the Error Inspector, you have access to the list of predefined fixes. If there are predefined fixes for an error, you choose which one to apply, based on the nature of the error. For example, if two street features are not connected with each other because one of the features is too short (undershoot) or too long (overshoot), you can use the Extend or Trim predefined fix to resolve the error. If there is no predefined fix for an error, you can use any editing tool to manually fix the error. For example, in the previous exercise, you used the Align Edge tool to snap the boundaries of two adjacent polygons together, since there is no predefined fix for this type of error.

Exercise 14c: Edit with geodatabase topology and predefined fixes

You continue your work maintaining the GIS data for a land trust. You will now build a geodatabase topology to validate the spatial relationship between a streams and a water samples feature class.

1. **ArcGIS 10.1:** If necessary, start ArcMap and open the ...\DesktopAssociate\Chapter14\LandTrust_Hydro.mxd.

    ArcGIS 10.0: If necessary, start ArcMap and open the ...\DesktopAssociate\Chapter14\LandTrust_Hydro_100.mxd.

2. **Display the Editor, Snapping, and Topology toolbars, if necessary.**

The map document shows the area of the land trust, streams, and water sample locations and already contains a geodatabase topology layer.

3. **Open the properties of the Hydrology_Topology layer. Examine the rules that have been defined for this topology.**

The topology rules enforce that streams are connected with each other and that water sample points are covered by streams.

4. **In the Symbology tab, turn on the Dirty Areas. Close the Layer Properties window.**

5. **Start an Edit session.**

6. **ArcGIS 10.1:** On the Topology toolbar, click the Select Topology button, check Geodatabase Topology, and make sure the Hydrology_Topology is selected.

    ArcGIS 10.0: On the Topology toolbar, make sure the Hydrology_Topology is selected from the drop-down list.

7. **Make sure that you are zoomed to the full extent. On the Topology toolbar, click the Validate Current Extent button.**

Decision point: How many errors are found in total?

Hint: To answer this question, you can generate a summary report in the Hydrology_Topology Layer Properties or search in the Error Inspector.

8. **Zoom to different areas in the map and examine the errors.**

Decision point: How many of the errors found during validation are real errors (not exceptions)?

All the Must Not Have Dangle errors at the ends of the streams are exceptions to the rule. Of course, a stream feature must begin and end somewhere. There are only two real Must Not Have Dangle errors and one Point Must Be Covered By Line error. You will use the ArcGIS predefined fixes to resolve these errors.

9. Zoom to the Dangle Error 1 bookmark. Using the Measure tool, measure the length of the overshoot.

10. From the Topology toolbar, select the Fix Topology Error tool, right-click the error, and choose Trim. For a Maximum Distance, type in 2 and press Enter. Validate the Topology in the current extent.

The first dangle error caused by the overshoot has been fixed. Next, you will use a different predefined fix to resolve a dangle error caused by an undershoot.

11. Zoom to the Dangle Error 2 bookmark. Using the Measure tool, measure the length of the undershoot.

12. Using the Fix Topology Error tool, right-click the error and choose Extend. For Maximum Distance, type in 30 and press Enter. Validate the Topology in the current extent.

Now the streams are properly connected. Next you will fix the Point Must Be Covered By Line error.

13. Zoom to the Point-Line Error bookmark. Using the Fix Topology Error tool, right-click the error.

There is no predefined fix to move the water sample point on top of the stream feature. Therefore, you will use the regular Edit tool to move the water sample point onto the line.

14. On the Snapping toolbar, make sure that Vertex snapping is enabled. Right-click the error and choose Select Features.

15. Using the Edit tool from the Editor toolbar, hover over the selected water sample point, and drag and snap it to the nearest vertex along the stream, as shown in the graphic below.

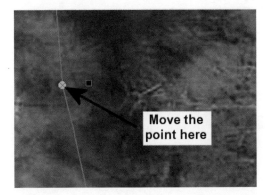

16. Clear the selected feature and zoom to the full extent. Then validate the Topology in the current extent. In the Error Inspector, click Search Now.

The remaining 15 errors are all Dangle errors at the beginnings and ends of the stream features.

17. In the Error Inspector, select all 15 errors and mark them as exceptions.

Hint: Holding down the Shift key, select the errors in the Error Inspector, right-click them, and choose Mark As Exception.

18. Save your edits and stop editing. Close ArcMap.

Answers to challenge questions

1. You want to ensure that electricity lines are properly connected with each other

    a. Must not have gaps
    b. Must not have dangles
    c. Must be covered by
    d. Must be disjoint

2. You want to ensure that soil Sample Points are located within Soils polygons

    a. Sample Points Must be coincident with Soils
    b. Sample Points Must be covered by Soils
    c. Sample Points must be covered by the boundary of Soils
    d. Soils must contain Sample Points

3. You want to ensure that an area is seamlessly covered with sales districts

    a. Sales districts must be disjoint
    b. Sales districts must not have dangles
    c. Boundary of Sales districts must be covered by Sales districts
    d. Sales districts must not have gaps

4. You want to ensure that line features in a Coastline feature class are coincident with the boundaries of polygons in a Counties feature class

    a. Counties must be covered by Coastline
    b. Counties must be coincident with Coastline
    c. Coastline must be covered by the boundary of Counties
    d. Coastline must be covered by Counties

Key terms

Edge: In topology, an edge defines lines or polygon boundaries.

Node: The point representing the beginning or ending point of an edge, topologically linked to all the edges that meet there.

Resources

- ArcGIS Help 10.1 > Desktop > Editing> Editing topology > Editing shared geometry with topology
    - Editing shared geometry
    - Creating a map topology
    - Selecting topology edges
- ArcGIS Help 10.1 > Desktop > Editing> Editing topology > Geodatabase topology
    - Geodatabase topology rules and topology error fixes
    - About fixing topology errors
    - Fixing topology errors
- ArcGIS Help 10.1 > Geodata > Data Types > Topologies
    - Topology in ArcGIS
    - Designing a topology
    - Creating a topology

- ArcGIS Help 10.1 > Desktop > Editing > Fundamentals of Editing > Editing tutorial
  - Exercise 4: Editing Shared features and topologies
- ArcGIS Help 10.1 > Geodata > Geodatabases > Building a geodatabase tutorial
  - Exercise 8: Creating a topology
- Esri Web Course: Getting Started with Geodatabase Topology (for ArcGIS 10)
  - **http://training.esri.com**
- Help 10.1 > Geodata > Data Types > Topologies
  - Topology in ArcGIS
  - Designing a topology
  - Creating a topology
- ArcGIS Help 10.1 > Desktop > Editing > Fundamentals of Editing > Editing tutorial
  - Exercise 4: Editing Shared features and topologies
- ArcGIS Help 10.1 > Geodata > Geodatabases > Building a geodatabase tutorial
  - Exercise 8: Creating a topology
- Esri Web Course: Getting Started with Geodatabase Topology (for ArcGIS 10)
  - **http://training.esri.com**Exercise 4: Editing Shared features and topologies
- ArcGIS Help 10.1 > Geodata > Geodatabases > Building a geodatabase tutorial
  - Exercise 8: Creating a topology
- Esri Web Course: Getting Started with Geodatabase Topology (for ArcGIS 10)
  - **http://training.esri.com**

chapter 15

Geoprocessing for analysis

Geoprocessing tools for analysis   224
    Common geoprocessing tools used for analysis
    Choosing and finding tools
    Challenge
    Executing geoprocessing tools
    Exercise 15a: Execute a tool from its dialog box
    Tool parameters

Geoprocessing models for analysis   230
    Process states
    Storing models
    Environment settings
    Exercise 15b: Set environment settings and complete the geoprocessing model
    Validating models
    Running models
    Setting model parameters
    Exercise 15c: Validate and run the model from ModelBuilder
    Tips for managing intermediate data
    Exercise 15d: Set model parameters and run the Model tool

Using Python scripts for analysis   235
    Python in ArcGIS
    Exercise 15e: Create a script tool from a Python script
    Exercise 15f: Add a script tool to the model

Answers to chapter 15 questions   238

Answers to challenge questions   238

Key terms   238

Resources   239

Chapter 15: Geoprocessing for analysis

In GIS analysis you apply different analytical techniques to your spatial data to answer a geographic question. For example, you may ask, "Where is the best location for a new nature reserve, a new store, a new power plant?" "Why do people prefer to live in certain areas?" "How can you manage urban growth over time?" "What is the impact of a natural disaster, and how can you mitigate it?"

Geoprocessing plays an essential role in GIS analysis: a variety of different geoprocessing tools allows you to extract information from your data, combine the information from different datasets, and create new output data that contains the information you need to answer your questions. Based on the new information, you will be able to answer questions and make more informed decisions (figure 15.1).

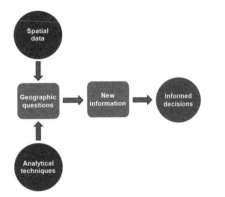

Figure 15.1 GIS analysis helps you answer a geographic questions and make more informed decisions. Esri.

Skills measured

- Given a task, determine the appropriate geoprocessing tool, parameters, and environment settings to accomplish the task.
- Determine the appropriate workflow to complete a given geoprocessing task.
- Compare and contrast options for automating work in ArcGIS Desktop (e.g. scripting, modeling).
- Explain the circumstances under which it is appropriate to embed a Python script within a toolbox.

Prerequisites

- Knowledge of the spatial analysis workflow
- Hands-on experience with ArcGIS geoprocessing, including
  - Geoprocessing tools
  - Basics of geoprocessing models, including
    - Creating geoprocessing models
    - Model processes

Geoprocessing tools for analysis

Using the right geoprocessing tools and workflows is critical for the success of your analysis and ultimately for making more informed decisions based on your analysis results.

Common geoprocessing tools used for analysis

Tables 15.1, 15.2, and 15.3 give you an overview of the most commonly used geoprocessing tools in the analysis workflow.

Geoprocessing tools for analysis **225**

Table 15.1 Tools for extracting features

Task	Tool	How the tool works
1. Extract features or portions of features that are located in your study area 2. Eliminate features that are located outside of a desired area	Clip	Cuts out the area of a feature class that intersects with a clip feature class
Eliminate features or portions of features that do not fall into a desired area	Erase	Creates a feature class from the features or portion of features outside the erase feature class (inverse of the Clip tool).
Divide features into multiple feature classes, e.g. to use them as input for analysis tools	Split	Splits the input feature class into multiple output feature classes,

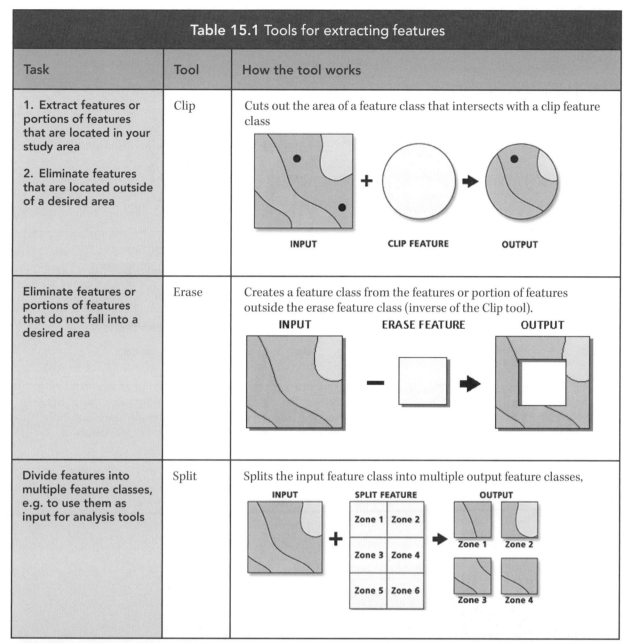

Table 15.1 illustrations from ArcGIS 10.1 Help.

Table 15.2 Tools for combining features

Task	Tool	How the tool works
1. Combine feature classes or tables 2. Create input datasets for analysis tools	Merge	Combines multiple input datasets of the same geometry type into a new output dataset.

Continued on next page

Table 15.2 Tools for combining features (continued)

1. Add additional features or records to an existing feature class or table 2. Create input datasets for analysis tools	Append	Adds input datasets of the same geometry type to an existing dataset.
1. Combine features based on one or more attribute values 2. Generalize datasets before using them in analysis tools	Dissolve	Aggregates features based on specified attributes. INPUT → OUTPUT
1. Locate features or portions of features within a certain distance of other features 2. Determine how much of another layer exists within a certain distance from a feature 3. Create geodesic buffers, i.e., large buffer polygons that accurately represent distances on Earth's surface	Buffer	Creates a new output feature class with polygons around features of the input feature class, based on a buffer distance. The buffer distance will determine the area of the buffer polygons. INPUT → OUTPUT Or with Dissolve type: ALL: INPUT → OUTPUT

Table 15.2 illustrations from ArcGIS 10.1 Help.

Geoprocessing tools for analysis **227**

Table 15.3 Tools for combining geometries and attritbutes

Task	Tool	How the tool works
Identify the area of overlap between features from different feature classes	Intersect	• Calculates the common areas of multiple feature classes in a new output feature class • Input feature classes can have point, line, or polygon geometry. The output feature class has the lowest dimension geometry (point < line < polygon) of the inputs • Combines all the attributes from all input feature classes in the output feature class
Identify the total area covered by features from different feature classes Create an output layer that can be queried for any combination of input attributes	Union	• Combines all features from all input feature classes in the output feature class • Combines all the attributes from all input feature classes in the output feature class
Join features based on their spatial location rather than based on their attribute values	Spatial Join	• Joins attributes from a feature class to a target feature class based on a spatial relationship in a new output feature class • Spatial relationships include e.g., intersection, distance, containment, coincidence, or adjacency

Table 15.3 illustrations from ArcGIS 10.1 Help.

Choosing and finding tools

Choosing the right geoprocessing tool is a crucial part of planning your analysis workflow. As you plan your analysis, you break down your workflow into tasks and match these tasks to the appropriate geoprocessing tool.

Once you have decided on using a particular tool, there are two basics ways to find it:
- Using the Search window
- Browsing using ArcToolbox or the Catalog window

If you know where a geoprocessing tool is located, you can browse to it and open it directly from ArcToolbox or the Catalog window. The six most commonly used geoprocessing tools are also accessible from the Geoprocessing menu.

If you don't know where a tool is located, you can search for it in the Search window. Searching for geoprocessing tools in the Search window is based on the tool name or keywords. If you don't know the name of a tool, you can enter some keywords that describe what you want the tool to do. For example, if you type the term "cookie cutter" in the Search window, it finds the Clip and the Erase tools that both extract features based on a "cookie cutter" feature class.

To search for tools in any of the system toolboxes that come with ArcGIS software, you can immediately type in the search terms. To search for a tool in a custom toolbox that you or another user created, you must first create an index for the custom toolbox.

Challenge

For each of the following scenarios, choose the most appropriate geoprocessing tool.

1. You plan to work with parcels in a metro area that is composed of multiple counties. Each county maintains their own parcel data. Your goal is to combine all county parcel datasets into a new output dataset to be used in your future analysis. What tool would you use to combine the parcel datasets?

    a. Spatial Join
    b. Merge
    c. Erase
    d. Dissolve

2. You have a polygon feature class of wetland protection areas and a polygon feature class of potential residential development areas. You want to identify areas for potential development outside of the wetland protection areas. Which geoprocessing tool would you use?

    a. Union
    b. Merge
    c. Erase
    d. Intersect

Executing geoprocessing tools

The three primary methods to execute geoprocessing tools are:
- Execute individual tools through the tool dialog
- Run a sequence of tools in a model in ModelBuilder
- Execute one or more geoprocessing tools through a Python script

Table 15.4 Methods to execute geoprocessing tools

Method	Uses
Running individual tools sequentially (from the tool dialog box)	The most common methods of executing tools: • For simple types of analysis • For one-time analyses that will not be repeated
Model Builder	Method for stringing together multiple geoprocessing tools into an analysis workflow: • For more complex types of analysis • Steps need planning • Can be tested, repeated, shared
Python	Method for creating custom analysis tools: • For analysis workflows that entail conditional logic • To make use of Python functions • For analysis workflows that need to be shared as a script

Exercise 15a: Execute a tool from its dialog box

Suppose you are working with the City of Fishers on identifying a new park location. The result of your analysis will be a feature class of vacant parcels that are at least one mile from the Interstate 69 highway, zoned as residential, and 0.2 acres or larger in size. To prepare data for this analysis you will now extract the line feature of Interstate 69 from a street centerlines feature class.

1. **ArcGIS 10.1:** Start ArcMap and open the …\DesktopAssociate\Chapter15\FishersTool.mxd.

    ArcGIS 10.0: Start ArcMap and open the …\DesktopAssociate\Chapter15\FishersTool_100.mxd.

    The map document shows a street centerlines layer.

    The Centerlines layer contains many more features than needed for the project. For your project, you need just the features of the Interstate 69 highway, which runs through the southeastern part of the layer. You will extract all the Interstate 69 features and aggregate them into one.

2. **From the Selection menu, click Select By Attributes. In the Select By Attributes dialog, build and execute the following expression: "ROAD_NAME" = 'I 69'.**

Twenty-four features of Interstate 69 were selected. (You could double-check this in the table of contents in List By Selection.)

Decision point: To simplify the data used in the analysis, you would like to aggregate all selected features into one single feature. What geoprocessing tool should you use?

_____

You will use the Dissolve tool to create the Interstate 69 feature class. Like all geoprocessing tools, you can run the Dissolve tool on all features in a layer, or only on selected features. If you have a set of features selected in the input layer (such as the one that you have), the tool automatically runs on the selected set.

3. **Open the Search window and locate the Dissolve tool.**

Hint: You can open the Search window from the Standard toolbar. To search for tools only, select Tools in the upper part of the Search window.

Question 1: What is another way you could locate the Dissolve geoprocessing tool?

Once you have located a tool, you can open it directly from the Search window.

4. **Click Dissolve (Data Management) to open the tool.**

5. **Run the Dissolve tool with the following parameters:**
    - Input Features: Centerlines
    - Output Feature Class: …DesktopAssociate\Chapter15\Fishers.gdb\Interstate69

When the tool is finished, the Interstate69 feature class is automatically added to the table of contents.

6. **Open the attribute table of the Interstate69 layer and confirm that it contains only one feature (i.e., has only one record in the attribute table).**

7. **If you are continuing with the next exercise, leave ArcMap open. Otherwise, close ArcMap.**

> ## *Tool parameters*
> Tool parameters are settings that define the execution of the geoprocessing tool. All geoprocessing tools have a parameter for input datasets and most have a parameter for an output dataset. Some parameters are required to run the tool, whereas others are optional.

Geoprocessing models for analysis

Rather than running the tools individually from the tool dialog box, it may be more efficient to set up your analysis workflow as a geoprocessing model. Models are a visual way of automating your analysis.

Process states

Models are made up of processes (a tool plus its input and output elements) (figure 15.2).

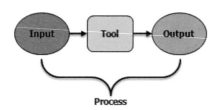

Figure 15.2 A model process consists of a tool and the variables connected to it. Connector lines indicate the sequence of processing. Esri.

Each process in a model is in one of four process states: Not ready-to-run, Ready-to-run, Running, or Has-been-run. The color of the tool element changes to reflect its state, as summarized in table 15.5.

Table 15.5 Process state in a model		
Process state	Tool color	Description
Not-ready-to-run	White	The required tool parameters have not been specified or are invalid.
Ready-to-run	Yellow	All required tool parameters have been specified and are valid.
Running	Red	Input data is processed, output data is created, and messages are added to the Results window.
Has-been-run	Yellow (Drop shadows added to tools and outputs elements when run from within ModelBuilder)	The process has run, and the derived data has been generated.

Storing models

Models are saved in a toolbox that you need write-access to. Since the system toolboxes that come with the ArcGIS software are read-only, you first create a new (custom) toolbox before you can save the model. You can create a custom toolbox as a .tbx file in any folder or you can create it at the root level of a geodatabase. When you save a model in a custom toolbox, it becomes a model tool (figure 15.3).

Figure 15.3 Models are stored in a user-defined (custom) toolbox, which can be created either at the root level of any geodatabase or as a .tbx file inside of any folder. Created by the author.

Environment settings

Besides tool parameters, environment settings are additional parameters that affect the output of geoprocessing tools. Environment settings include parameters such as the output workspace, the output spatial reference, or the processing extent and help you standardize your analysis results. If you establish environment settings before performing your analysis, some tool parameters will already be populated. For example, if you set the output workspace to a specific geodatabase, all of your geoprocessing results are automatically stored in that workspace.

You can apply environment settings at four levels: at the application, tool, model, and model process level (figure 15.4).

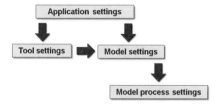

Figure 15.4 The four environment levels form a hierarchy where the application level is highest. In this hierarchy, environment settings are passed down to the next level, as illustrated above. At each level, you can override the passed-down environment settings with another setting. Esri.

- The top level environment setting is the application level. Application level settings are set in the Geoprocessing Options and are system-wide defaults that will be applied to all tools you run.
- Tool-level settings are applied to a single use of a tool and override the application-level settings.
- Model-level settings are specified and saved with a model; they override tool- and application-level settings.
- Model process-level settings are specified at the model process level and are saved with the model; they override model-level settings.

Exercise 15b: Set environment settings and complete the geoprocessing model

You are still working with the City of Fishers on identifying a new park location. You will add a process to an existing model and set different environment settings at the model and the model process level.

1. **ArcGIS 10.1:** If necessary, open the ...\DesktopAssociate\Chapter15\FishersModel.mxd. In the Catalog window, navigate to the ...DesktopAssociate\Chapter15\Fishers.gdb\CityPark toolbox.

    ArcGIS 10.0: If necessary, open the ...\DesktopAssociate\Chapter15\FishersModel_100.mxd. In the Catalog window, navigate to the ...DesktopAssociate\Chapter15\CityPark toolbox.

The map document contains the City of Fishers parcel layer and the Interstate 69 feature that you extracted in the previous exercise.

2. **Right-click the ParkSelection model in the CityPark toolbox and choose Edit to open the Model window.**

The model contains two processes that are not connected at this point. The Buffer model process will buffer the Interstate 69 feature by one mile. The Select Layer By Attributes process will select vacant parcels that are larger than 0.02 acres.

First you will set the workspace environment settings at the model level.

3. **From the Model menu, click Model Properties. Click the Environments tab.**

4. **Check Workspace, then click the Values button to open the Environment Settings dialog box.**

5. **Click Workspace to expose the Current Workspace and the Scratch Workspace environment settings.**

The Current Workspace setting specifies the default location for the inputs and outputs of the geoprocessing tools. The Scratch workspace is used in ModelBuilder to store intermediate data that is created during the execution of a model.

6. **Set both the Current Workspace and the Scratch Workspace to ...DesktopAssociate\Chapter15\Fishers.gdb. Then click OK to close all open dialog boxes.**

Next, you will add a third model process to erase the areas within the Interstate buffer from the selected parcels.

7. Locate the Erase (Analysis) tool and add it to the Model window.

Hint: You can drag and drop the Erase tool directly from the Search window to the Model window.

Once you have selected parcels for potential park locations, you want to add them to a different geodatabase than the rest of your outputs. To override the environment setting that you set earlier, you will set the Current and the Scratch Workspace environment settings at the model process level.

8. Right-click the Erase tool and click Properties. In the Environments tab, check and expand the workspace environments. Click the Values button.

9. Expand Workspace and set both the Current and the Scratch Workspace to …DesktopAssociate\Chapter15\CityParkSiteSelection.gdb. Then, click OK to close all open dialog boxes.

10. In the Model window, select the Connect button ↗ on the ModelBuilder toolbar.

11. Click and drag a connection line from the green Parcels (2) variable (the output of the Select Layer By Attributes tool) to the Erase tool. From the pop-up menu, select Input Features.

12. Drag a connection line from the green Interstate Buffer variable (the output of the Buffer tool) to the Erase tool and choose Erase Features from the pop-up menu.

13. Open the Erase tool dialog box and confirm that the Output Feature Class will be saved to …\DesktopAssociate\Chapter15\CityParkSiteSelection.gdb.

14. Name the Output Feature Class **ParcelsForParks**. Then click OK to close the Erase dialog box.

15. Right-click the ParcelsforParks output variable and click Add To Display.

16. Click the Auto Layout button ▪ on the ModelBuilder toolbar to arrange the tools.

Your model should look like the one in the graphic below.

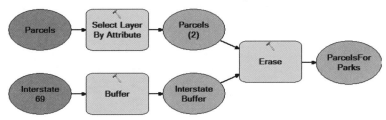

17. Save the model.

18. If you are continuing with the next exercise, leave ArcMap open. Otherwise, close ArcMap.

Validating models

When you validate a model, all model processes, including their data variables and tool parameters, are checked to see if they are valid for the tools in which they are used. If all variables and tool parameters in a model process are validated, the process is set to the Ready-to-run state. If some data variables or tool parameters are not validated (or missing), the process is returned to a Not-ready-to-run state. To fix this, you must open the variable or the tool and supply the correct value.

Running models

You can run a model from within ModelBuilder or from the model tool dialog box, Python window, or in a script. Intermediate data is automatically deleted when you run a model from its dialog box or from the Python window, but not when you run it in the ModelBuilder window.

Running a model from the ModelBuilder application allows you to run single processes or the entire model. Typically, you run single processes initially to make sure they run and produce the desired output. Running a model from the model tool dialog box, Python window, or in a script will execute the entire model.

Setting model parameters

Model parameters are the values that the users will see and will be able to change when they open the model tool's dialog box.

You can create a model parameter from any variable in the model. For example, the Input and Output data variables are typically set as model parameters, so that the user can specify different input datasets from the model tool dialog box and determine where to save the output. Another common workflow is to expose a tool parameter as a variable and set the variable as a model parameter. For example, if you use the Buffer tool in a model, you may want to expose the buffer distance as a variable and designate the variable as a model parameter. This way the user can specify a buffer distance. You will do this in the next exercise.

Exercise 15c: Validate and run the model from ModelBuilder

You will now validate and run the model that you created in the previous two exercises.

Dependency: You need to complete exercise 15b before running this exercise. If you have not run exercise 15b, please do so before starting this exercise.

1. **ArcGIS 10.1:** If necessary, open the ...\DesktopAssociate\Chapter15\FishersModel.mxd. Navigate to the ...DesktopAssociate\Chapter15\Fishers.gdb\CityPark toolbox.

   ArcGIS 10.0: If necessary, open the ...\DesktopAssociate\Chapter15\FishersModel.mxd. Navigate to the ...DesktopAssociate\Chapter15\CityPark toolbox.

2. **Right-click the ParkSelection model and choose Edit to open the Model window.**

First, you will run the model from the ModelBuilder application.

3. **On the ModelBuilder toolbar, click the Run button ▸. Click Close once the model has run.**

The ParcelsForParks layer is added to the map. Drop shadows around the tools and output variables indicate that the model is now in Has-been-run state.

Decision point: Looking at your model, which outputs would be considered intermediate data?

The output feature class of the Buffer tool would be considered intermediate data. The intermediate buffer polygon is not needed once the final output is created.

4. **In the Catalog window, confirm that the Interstate_Buffer feature class is created in the Fishers.gdb that you set as a Scratch workspace at the Model level. Also confirm that the ParcelsforParks feature class is created in the CityParkSiteSelection.gdb that you set as a Current workspace at the Erase process level.**

5. **From the Model menu, click Validate Entire Model.**

Since all tool parameters were validated, the model was reset to Ready-to-run state. Validating the model does not delete the intermediate data.

6. **From the Model main menu, select Delete Intermediate Data.**

7. **From the Model menu, click Run Entire Model to run the model again.**

8. **From the Model menu, select Delete Intermediate Data.**

Decision point: Why was the Select Layer By Attributes process not reset to Ready-to-run state?

Since the Select Layer By Attributes tool does not create intermediate data on disk, but returns a set of selected features, it is not affected when you delete the intermediate data. That's okay. You will use the selected set of features next time you run the model.

9. Save the model and close the Model window.

10. If you are continuing with the next exercise, leave ArcMap open. Otherwise, close ArcMap.

> ## *Tips for managing intermediate data*
> - Do not write intermediate data to a multiuser geodatabase. Every time you want to delete data you will need access to that geodatabase, and you may not have permissions to do that.
> - Do not put intermediate data in a remote dataset or network drive; intermediate data should be kept local. The final output can be stored anywhere.
> - Do not save intermediate data to the same location as final output data.

Exercise 15d: Set model parameters and run the model tool

Before you run a model from the model tool dialog box, you will set some model parameters.

Dependency: You need to complete exercise 15b before running this exercise. If you have not run exercise 15b, please do so before starting this exercise.

1. **ArcGIS 10.1:** If necessary, open the ...\DesktopAssociate\Chapter15\FishersModel.mxd.

   ArcGIS 10.0: If necessary, open the ...\DesktopAssociate\Chapter15\FishersModel_100.mxd.

2. **ArcGIS 10.1:** Right-click the ParkSelection model in the ...\DesktopAssociate\Chapter15\Fishers.gdb\CityPark toolbox and click Open.

   ArcGIS 10.0: Right-click the ParkSelection model in the ...\DesktopAssociate\Chapter15\CityPark toolbox and click Open.

Currently, the model tool has no parameters. You will set some model parameters before you run the model tool.

Note: Do NOT click OK. Clicking OK will run the model tool.

3. **Click Cancel to close the dialog box.**

4. **Right-click the ParkSelection model and click Edit to open the Model window.**

5. **In the Model window, right-click the blue Interstate 69 variable and click Model Parameter.**

A small letter P appears next to the variable indicating that the Interstate 69 variable has been designated as a model parameter.

6. **Right-click the Interstate 69 variable and click Rename. Rename the variable to Buffer Features.**

Next, you will expose the buffer distance parameter as a variable and set it as a model parameter.

7. **Right-click the Buffer tool, point to Make Variable, and click From Parameter. Click Distance [value or field]. Click the Auto-layout button.**

8. **Make the Distance variable a model parameter.**

You will also make the output variable a model parameter.

9. **Right-click the green ParcelsForParks variable and click Model Parameter.**

10. **Rename the ParcelsForParks variable to Output Feature Class.**

Your model should look like the following graphic.

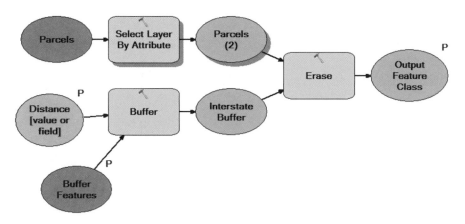

11. Save the model and close the Model window.
12. Right-click the ParkSelection model and click Open.

Now, the model parameters appear in the tool dialog box. If you have run the previous exercise, you see an error icon. This is because the output feature class already exists.

13. Change the buffer distance to 2 miles and rename the output feature class to **ParcelsForParks2**.
14. Click OK to run the model tool. Click Close when the tool is finished.
15. If you are continuing with the next exercise, leave ArcMap open. Otherwise, close ArcMap.

Using Python scripts for analysis

Models are extremely useful if you are unfamiliar with scripting. ModelBuilder even allows you to execute tool-based logic (whether or not a condition is true) and use looping to iterate through sets of data. However, very large models may become complicated and hard to maintain. At this stage in your analysis, it is definitely worth exploring Python scripting as an alternative to maintaining large, complex models (i.e., Python gives you a lot of flexibility in automating your analysis workflows). In Python, you can program pretty much any workflow and most importantly, you can develop your own custom analysis tools that provide beyond out-of-the-box functionality.

> ### *Python in ArcGIS*
>
> ArcMap comes with a built-in Python window that allows you to execute single geoprocessing tools or enter and execute Python code to perform an entire workflow. Once you save your Python code as a file on disk (with a .py extension), it is considered a Python script. You can load and execute Python scripts in the Python window and you can also create a script tool from it. A script tool behaves like any other geoprocessing tool: You can execute it using its dialog box, within models, in the Python window, or within other Python scripts.
>
> Within the Python window, you can also access additional Python functionality using the ArcGIS ArcPy site package. ArcPy provides Python access for all geoprocessing tools including extensions, as well as a wide variety of useful functions and classes that support more complex Python workflows. Using Python and ArcPy, you can develop an infinite number of useful programs that operate on geographic data.

Exercise 15e: Create a script tool from a Python script

Dependency: You need to complete exercise 15c before running this exercise. If you have not completed exercise 15c, please do so before starting this exercise.

1. **ArcGIS 10.1:** Open ArcMap, if necessary, and open the ...\DesktopAssociate\Chapter15\FishersScript.mxd.

    ArcGIS 10.0: Open ArcMap, if necessary, and open the ...\DesktopAssociate\Chapter15\FishersScript_100.mxd.

2. From the Standard tools toolbar, click the Python button to open the Python window.

3. Right-click in the Python window and select Load. Navigate to the ...\DesktopAssociate\Chapter15\CountFeatures folder and open ListFeatureclasses.py.

The script counts the number of features in one or more feature classes in a workspace and provides a text file with the feature class names and feature counts. This is functionality that does not exist as a system geoprocessing tool.

Note: For the ArcGIS Desktop Associate exam, you do not need to know the syntax of a Python script. However, you do need to know about situations when you would use script tools versus system tools and when to use scripts versus models.

4. Scroll down and read through the script. Then, close the Python window.

5. **ArcGIS 10.1:** In the Catalog window, navigate to the ...\DesktopAssociate\Chapter15\CountFeatures\ScriptTool.tbx toolbox.

    ArcGIS 10.0: In the Catalog window, navigate to the ...\DesktopAssociate\Chapter15\CountFeatures\ScriptTool_100.tbx toolbox.

6. **ArcGIS 10.1:** Right-click the ScriptTool.tbx toolbox, point to Add, and then click Script.

    ArcGIS 10.0: Right-click the ScriptTool_100.tbx toolbox, point to Add, and then click Script.

7. Fill out the dialog as follows:
    - Name: ListFeatures
    - Label: List Features
    - Check to Store relative path names, and then click Next.

Now you will add the ListFeatureclasses.py script.

8. Open the ListFeatureclasses.py script file from the ...\DesktopAssociate\Chapter15\CountFeatures folder. Click Next.

Now you will define tool parameters for the input workspace and the result file.

9. Click the first cell under Display Name, and type in **Input Workspace**. Click the first cell under Data Type and choose Workspace or Feature Dataset from the drop-down list.

10. Create another parameter with the Display Name **Result File** and the Data Type Text File. In the Parameter Properties at the bottom of the dialog, click the cell next to Direction and choose Output, then click Finish.

The ListFeatureclasses.py Python script is now available as a geoprocessing tool in the ScriptTools.tbx toolbox.

11. If you are continuing with the next exercise, leave ArcMap open. Otherwise, close ArcMap.

Exercise 15f: Add a script tool to the model

You will now add the script tool that you created in the previous exercise to the ParkSelection model and have it count the number of output features (parcels that are suitable for parks).

Dependency: You need to complete exercise 15e before running this exercise. If you have not completed exercise 15e, please do so before starting this exercise.

1. **ArcGIS 10.1:** Right-click the ParkSelection model in the ...DesktopAssociate\Chapter15\Fishers.gdb\ CityPark toolbox and choose Edit to open the Model window.

    ArcGIS 10.0: Right-click the ParkSelection model in the ...DesktopAssociate\Chapter15\ CityPark toolbox and choose Edit to open the Model window.

2. **ArcGIS 10.1:** In the Catalog window, expand the ...\DesktopAssociate\Chapter15\CountFeatures\ ScriptTool.tbx toolbox.

    ArcGIS 10.0: In the Catalog window, expand the ...\DesktopAssociate\Chapter15\CountFeatures\ ScriptTool_100.tbx toolbox.

3. Drag the ListFeatures script tool into the ModelBuilder window.

4. Double-click the ListFeatures tool to open its dialog box and fill in the parameters as follows:
    - Input Workspace: ...DesktopAssociate\Chapter15\CityParkSiteSelection.gdb
    - Result file: ...DesktopAssociate\Chapter15\FishersParkParcels.txt

Before the ListFeatures tool can run, the ParcelsForParks output feature class must be created. Otherwise, the number of features cannot be counted. To ensure the remaining part of the model executes first, you will set the output feature class of the Erase tool as a precondition for the ListFeatures tool.

5. From the ModelBuilder toolbar, select the Connect button. Click and drag a connection line from the green Output Feature Class variable (the output of the Erase tool) to the ListFeatures tool. From the pop-up menu, select Precondition.

6. From the Model menu select Validate Entire Model.

7. Run the model.

8. Save the model and close the Model window. Close ArcMap.

9. In Windows Explorer, navigate to the ...DesktopAssociate\Chapter15 folder and open the FishersParkParcels.txt file.

The FishersParkParcels text file contains the feature count of the ParcelsForParks feature class that resulted from your analysis. According to the criteria used in the analysis, 1,747 parcels are potential candidates for establishing a new park.

This is a fairly simple example of using the ListFeatures tool. Since there is only one feature class in the CityParkSiteSelection.gdb, the text file is very short. You could have just opened the ParcelsForParks feature class and read out the feature count. However, the tool would be very useful if you had dozens or even hundreds of feature classes in a workspace.

10. Close the text editor.

Answers to chapter 15 questions

Question 1: What is another way you could locate the Dissolve geoprocessing tool?
Answer: You can locate geoprocessing tools by browsing to them in ArcToolbox and the Catalog window under Data Management Tools > Generalization > Dissolve. The Dissolve tool is also available through the Geoprocessing menu as a default tool. This Geoprocessing menu (or any menu or toolbar) can be modified in Customize Mode.

Answers to challenge questions

1. You plan to work with parcels in a metro area that is composed of multiple counties. Each county maintains their own parcel data. Your goal is combine all county parcel datasets into a new output dataset to be used in your future analysis. What tool would you use to combine the parcel datasets?

    a. Spatial Join
    b. Merge
    c. Erase
    d. Dissolve

2. You have a polygon feature class of wetland protection areas and a polygon feature class of potential residential development areas. You want to identify areas for potential development outside of the wetland protection areas. Which geoprocessing tool would you use?

    a. Union
    b. Merge
    c. Erase
    d. Intersect

Key terms

Spatial analysis: The process of examining locations, attributes, and relationships of features in spatial data through overlay and other analytical techniques in order to address a question or gain useful knowledge. Spatial analysis extracts or creates new information from spatial data.

Geoprocessing: A GIS operation used to manipulate GIS data. A typical geoprocessing operation takes an input dataset, performs an operation on that dataset, and returns the result of the operation as an output dataset. Common geoprocessing operations include geographic feature overlay, feature selection and analysis, topology processing, raster processing, and data conversion. Geoprocessing allows for definition, management, and analysis of information used to form decisions.

Model: An abstraction of reality used to represent objects, processes, or events. In geoprocessing in ArcGIS, one process or a sequence of processes connected together, that is created in ModelBuilder.

Parameter: A variable that determines the outcome of a function or operation. In geoprocessing in ArcGIS, a characteristic of a tool. Values set for parameters define a tool's behavior during run time.

Variable: A symbol or placeholder that represents a changeable value or a value that has not yet been assigned. In ModelBuilder, a variable can hold data (e.g., input and output data of a geoprocessing tool) or values (e.g., buffer distances, SQL expressions, Spatial references etc.).

Resources

- ArcGIS Help 10.1 > Geoprocessing > Tool Reference> Analysis Toolbox
  - Extract toolset
    - Clip
    - Split
  - Overlay toolset
    - Erase
    - Intersect
    - Spatial Join
    - Union
  - Proximity Toolset
    - Buffer
- ArcGIS Help 10.1 > Geoprocessing > Finding Tools
  - A quick tour of finding tools
- ArcGIS Help 10.1 > Geoprocessing > Executing tools
  - A quick tour of executing tools
  - Executing tools using the tool dialog > A quick tour of using a tool dialog
  - Executing tools using ModelBuilder > Tutorial: Executing tools in ModelBuilder
- ArcGIS Help 10.1 > Geoprocessing > Creating Tools > Creating tools with ModelBuilder
  - A quick tour of creating tools with ModelBuilder
  - Tutorial: Creating tools with ModelBuilder
- ArcGIS Help 10.1 > Geoprocessing > Creating Tools > Creating tools with Python
  - A quick tour of creating tools with Python
  - What is a script tool?
- ArcGIS Help 10.1 > Geoprocessing > Environment settings
  - What is a geoprocessing environment setting?
  - Environment levels and hierarchy
- ArcGIS Help 10.1 > Geoprocessing > ModelBuilder > Using Model Builder to execute tools
  - Essential vocabulary: Executing tools in ModelBuilder
  - Understanding process state
  - Validating a model

chapter 16

Analyzing and querying tables

Extracting information from attribute tables 242
   Sorting records in a table
   Calculating statistics
   Exercise 16a: Sort records in a table and calculate statistics
   Summarizing a field
   Exercise 16b: Summarize a field and join the output table to another layer

Attribute queries 246
   Query syntax
   Challenge 1

Spatial queries 248
   Spatial selection methods
   Challenge 2

Queries in analysis 250
   Selection methods
   Exercise 16c: Combine attribute and spatial queries

Working with selections 252
   Exercise 16d: Refine the park selection

Answers to chapter 16 questions 253

Answers to challenge questions 253

Key terms 254

Resources 255

Chapter 16: Analyzing and querying tables

You can analyze and query tables to answer questions about your data. Some questions you can answer by simply opening a table and sorting or selecting the records. Other questions can be answered by calculating summary statistics or by running an attribute or spatial query.

These analysis techniques can also support more complex analysis workflows: For example, in a site selection analysis workflow, you may use attribute or spatial queries to specify which locations meet your criteria, and use the selected set as input for geoprocessing tools.

Skills measured

- Manipulate and format the view of attribute data to support a given task or use.
- Given a scenario, determine the appropriate specifications (e.g., target/input data layer(s), selection layer, selection method, selection type) for performing an attribute query.
- Given a scenario, determine the appropriate specifications (e.g., target/input data layer(s), selection layer, selection method, selection type) for performing a spatial query.

Prerequisites

- Hands-on experience extracting information from attribute tables
- Knowledge of table structure, including field types
- Hands-on experience querying features and attribute
- Extracting information from attribute tables

Extracting information from attribute tables

ArcGIS provides some simple methods for exploring attribute values in a table and extracting information from them. You can also use these methods to explore and familiarize yourself with the attributes that you use for analysis.

Methods for extracting information from an attribute table include sorting the records based on one or more fields, calculating statistics, and summarizing the values in a field.

Sorting records in a table

A quick way to extract the minimum, maximum or a range of values from the attribute table is to sort a field in either ascending (A–Z or 1–9) order or descending (Z–A or 9–1) order. For example, to determine the world city with the largest population, you can simply sort the records in the population field in descending order.

For more complex questions, you can sort records based on more than one field. For example, to find the three most populated cities in each world country, you could sort first by country and then by population (figure 16.1).

ObjectID	Shape	CITY_NAME	CNTRY_NAME	POP
1881	Point	Bhairawa	Nepal	63367
1887	Point	Dhangarhi	Nepal	92294
1905	Point	Biratnagar	Nepal	182324
1890	Point	Pokhara	Nepal	200000
1891	Point	Kathmandu	Nepal	1442271
1348	Point	Europoort	Netherlands	-999
1350	Point	Middelburg	Netherlands	46485
1364	Point	Assen	Netherlands	62237
1363	Point	Leeuwarden	Netherlands	91424

Figure 16.1 Advanced Sorting allows you to create a nested structure in a table: In the example above, the records are first sorted by the CNTRY_NAME field, then by the POP field (both in Ascending order). You can sort records in a table based on up to four fields. Created by the author, from Esri Data & Maps, 2010, data courtesy of ArcWorld, ArcWorld Supplement, and DeLorme.

Calculating statistics

To quickly analyze the values in numeric fields in a table, you can calculate statistics. You can determine the minimum and maximum values, the sum (total) value, the mean, and the standard deviation of the values. You can calculate statistics based on all records in a field or selected records. The output statistics are displayed in a dialog. This dialog can be used to copy output statistics into documents, e-mails, layouts, etc. (figure 16.2).

Extracting information from attribute tables **243**

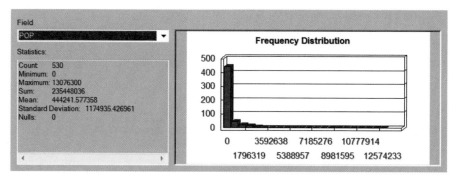

Figure 16.2 The Statistics dialog box gives you some basic summary statistics for a numeric field. A histogram shows how the values are distributed. Created by the author.

Another way to calculate statistics on a field is using the Summary Statistics geoprocessing tool. It allows you to calculate statistics on one or more fields at the same time. In addition to calculating summary statistics for all records or for selected records, the Summary Statistics tool allows you to specify one or more case field(s). The Summary Statistics tool creates a new output table containing the output statistics.

Exercise 16a: Sort records in a table and calculate statistics

In the next two exercises you will work with datasets created by the US Census Bureau to store demographic information: census blocks and the census tracts. A census tract contains several blocks.

1. **ArcGIS 10.1:** Start ArcMap and open the ....\DesktopAssociate\Chapter16\Attributes.mxd.

    ArcGIS 10.0: Start ArcMap and open the ....\DesktopAssociate\Chapter16\Attributes_100.mxd.

The map document shows census blocks for Dane County, Wisconsin.

2. **Open the Census Blocks attribute table.**

3. **Locate the POP2000 field.**

The POP2000 field contains population values for each census block.

Decision point: How can you quickly find out which census block has the highest population?

---

4. **Right-click the POP2000 field and choose Sort Descending.**

The census block with ID number 1759 has the highest population.

5. **Locate the TRACT2000 field.**

The TRACT2000 field contains the census tract number for each census block.

Decision point: Suppose somebody asked you to determine the minimum and maximum population of census blocks within a particular census tract. How can you organize the values in the attribute table to retrieve this information?

---

6. **Right-click the TRACT 2000 field and choose Advanced Sorting.**

7. **Sort the records in the table first by TRACT2000, and then by POP2000 (both in Ascending order).**

The records in the table (representing census blocks) are now sorted by census tract number, and within each census tract, by population. You can now determine the minimum and maximum population values by census block within a given census tract.

Decision point: How can you quickly find the total population of Dane County?

---

8. **Generate statistics of the POP2000 field.**

Hint: Right-click the POP2000 field and choose Statistics.

9. **Calculating statistics gives you more information than sorting. The total population of Dane County is indicated by the Sum.**

10. **Close the statistics dialog box.**

11. **If you are continuing with the next exercise, leave ArcMap open. Otherwise, close ArcMap.**

Summarizing a field

Summarizing the values in an attribute field creates a new output table containing a list of unique values, a count of how many times each value occurs, and optionally some summary statistics such as the minimum, maximum, average, sum, or standard deviation of other fields in the original table. For example, in a Streets attribute table, you can summarize the Street_Name field and include the sum of the Shape_Length field in the output table. The output table would have a record for each unique street name in the field, a Count field indicating how many features with a particular street name occur, and a field containing the total length of all street features with a particular name. A common workflow is to join the new summary statistics table to another layer attribute table, and use the additional fields for further analysis.

Figure 16.3 You can summarize values based on all records or selected records in a field. In the example above, the selected records have been summarized by the StreetName field and the sum of the Shape_Length field is included. The summary table contains a record for each unique street name, a count of how many times each name occurs in the selected set, and the total length of each street. Created by the author.

Note: To summarize a single field, use the field name context menu. To summarize values in more than one field, use the Frequency geoprocessing tool.

Exercise 16b: Summarize a field and join the output table to another layer

You will now see how summarizing a field creates a new table that you can join to a layer's attribute table.

1. **ArcGIS 10.1: If necessary, start ArcMap and open the ....\DesktopAssociate\Chapter16\Attributes.mxd**

    ArcGIS 10.0: If necessary, start ArcMap and open the ....\DesktopAssociate\Chapter16\Attributes_100.mxd

2. **Turn on the Census Tracts layer and open its attribute table.**

The Census Tracts table does not contain any population attributes. To support future analysis, you will summarize population in the Block Groups table by census tract, and then join the output table with the Census Tract layer.

3. Open the Census Blocks attribute table.

4. Summarize the TRACT2000 field using the following parameters:

Hint: Right-click the TRACT2000 field and choose Summarize.
   - Summary statistics to be included in the output table: POP2000 - SUM.
   - Output table: ...DesktopAssociate\Chapter 16\Census.gdb\ Tract_Population

5. Click Yes to add the table to the map.

6. Open the Tract_Population table.

The Tract_Population table contains a count of how many census block are in each census tract and the total population of each census tract. You will now join the Tract_Population table to the Census Tracts layer.

Decision point: What are the key fields in the TractPopulation and the Census Tracts tables that you can use to join the tables?

---

7. Close the table window.

8. Right-click the Census Tracts layer, point to Joins and Relates and choose Join.

9. Run the Join Data tool using the specifications below:

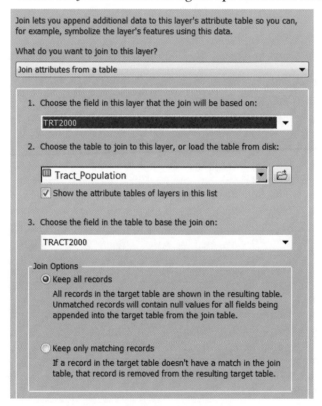

10. Open the Census Tracts attribute table.

The table now contains population data. For more information about table joins, refer to the Table Associations chapter of this book.

11. Close the table window. If you are continuing with the next exercise, leave ArcMap open. Otherwise, close ArcMap.

Attribute queries

Attribute queries allow you to select features and records based on their values in one or more attribute fields. To build an attribute query, you build a query expression: an SQL expression that selects records based on a condition. For example, to select world cities based on the condition that their POPULATION attribute is greater than 50,000, you would build a where clause: SELECT from Cities WHERE: "POPULATION" > 50,000. ArcGIS automatically adds the SELECT from Cities WHERE phrase to the query expression. You can build attribute queries in the Select by Attributes dialog, or use the Select or Select Layer by Attributes geoprocessing tools.

Query syntax

A typical query expression consists of an expression that includes the name of the field being queried, followed by an operator, and a numeric value or string. These are the components of a simple query. You can create complex or compound query expressions by combining conditional statements with AND and OR operators: for example, given a layer of world cities, the query expression "POPULATION" > 50,000 AND "COUNTRY_NAME" = 'Nepal' would select all the cities in Nepal with a population greater than 50,000. The query expression "COUNTRY_NAME" = 'Nepal' OR "POPULATION" > 50,000 would select all world cities with a population greater than 50,000 and all cities in Nepal (no matter what their population).

You can also nest a query within another query expression: The results of the nested or sub-query will be used as input to the main query: for example, the query "COUNTRY_NAME" NOT IN (SELECT "COUNTRY_NAME" FROM indep_countries) would select only countries that are not included in the indep_countries table.

Table 16.1 below summarizes the operators commonly used in query expressions:

Table 16.1 Operators in query expressions

Operator	Use	Example
=	With strings, numbers, and dates	STATE_NAME = 'California'
<, >	With strings, numbers, and dates	"POPULATION" < 50000 "City_NAME" > 'M'
<=, >=	With strings, numbers, and dates	"POPULATION" <= 50000 "City_NAME" >= 'M'
<>	With strings, numbers, and dates	"STATE_NAME" <> 'IOWA'
*,/,+,-	To multiply, divide, add, and subtract values in specified fields	("Shape_Area"/"Shape_Length") <=5
LIKE (with wildcard) NOT LIKE (with wildcard)	To conduct a partial string search—wildcard symbol (i.e., %): anything is acceptable in its place	"STATE_NAME" LIKE 'Miss%'
IS NULL IS NOT NULL	To find NULL values (or values that are not NULL) in a specified field	"POPULATION" IS NULL

Continued on next page

Operator	Use	Example
IN NOT IN	To find records with values in or not in one of several strings or numeric values in a field	"STATE_NAME" IN ('Alabama', 'Alaska', 'California', 'Florida')
BETWEEN x AND y NOT BETWEEN x AND y	To find records with values inside or outside the specified numeric range	"MED_AGE" BETWEEN 18 AND 22
NOT	To find records with values that do not match the expression	NOT "STATE_NAME" = WISCONSIN
AND	To combine two conditions together and select records for which both conditions are true	"SUB_REGION" = 'New England' AND NOT "STATE_NAME" = 'Maine' (Returns all states in the New England subregion, except Maine)
OR	To combine two conditions together and select records when at least one conditions is true	"STATE_NAME" = 'California' OR "STATE_NAME" = 'Nevada' (returns California, Nevada)

Table 16.1 Operators in query expressions (continued)

Challenge 1

1. Which of the following query expressions will select cities with a population larger than 10,000.

   a. "POPULATION" = 10000
   b. "POPULATION" > 10000
   c. "POPULATION" >= 10000
   d. "POPULATION" IS 10000

2. Which of the following query expressions will select all buildings with three or more stories that are zoned Multi-Use (MU).

   a. "BUILDING_FLOORS" >3 AND "ZONE" = 'MU'
   b. "BUILDING_FLOORS" >= 3 OR "ZONE" = 'MU'
   c. "BUILDING_FLOORS" >3 OR "ZONE" = 'MU'
   d. "BUILDING_FLOORS" >= 3 AND "ZONE" = 'MU'

3. Which cities does the following expression select?
   "STATE" = 'CALIFORNIA' AND NOT "CITY_NAME" = 'Los Angeles'

   a. All the cities within California and Los Angeles
   b. All the cities within California except Los Angeles
   c. None of the cities within California except Los Angeles
   d. All the cities outside California and Los Angeles

4. Which parcels does the following query expression select?
   "PrclNo" LIKE '%20%'

   a. All parcels with a parcel number that starts with 20
   b. All parcels with a parcel number that ends with 20
   c. All parcels with a parcel number that contains the numbers 2 and 0
   d. All parcels with a parcel number that contains the number 20

Spatial queries

Spatial queries allow you to select features in different layers based on their spatial relationships with other features. For example, a spatial query can select points that are contained within a polygon or set of polygons, features within a specified distance of other features, or find features that are adjacent. To build spatial queries you define the Target layer, that is, the layer from which features will be selected, and the Source layer, that is, the layer that is used to select features. For example, to select soil sample locations within a distance of 1000 feet of a river, Soil_Samples would be the Target layer and Rivers would be the Source layer. Spatial queries can be performed on all features or selected features in the Source layer. You can build spatial queries in the Select by Location dialog, or use the Select Layer by Location geoprocessing tool.

Spatial selection methods

You can perform spatial queries using a variety of spatial selection methods.

Table 16.2 lists the most commonly used spatial selection methods and their uses.

Table 16.2			
Spatial Selection Method	Use	Spatial Selection Method	Use
Intersect	Selects target features that fully or partially overlap one or more source features	Completely Contain	Selects target features that contain an entire source feature
Are within a distance of	Selects target features within a specified buffer distance of one or more source features	Have their centroid in	Selects target features whose centroid falls inside the geometry of a source feature

Continued on next page

Table 16.2 (continued)

Spatial Selection Method	Use	Spatial Selection Method	Use
Are within	Selects target features that fall inside the geometry of one or more source features	Share a line segment with	Selects target features that share at least two vertex locations with a source feature (must be the same one)
Are completely within	Selects target features that fall inside the geometry of a source feature without touching its boundary	Touch the boundary of	Selects target features that touch the boundary of one or more source features
		Are identical to	Selects target features that share all vertex locations with a source feature
Contain	Selects target features that contain entire source feature(s) or portions of source feature(s).	Are crossed by the outline of	Selects target features that share at least one vertex location with a source feature without sharing a line segment.

Table 16.2 illustrations created by the author based on graphics from the ArcGIS 10.1 Help.

The differences between some of the spatial operators are very subtle. The ArcGIS Help topic "Using Select By Location" explores the differences between the spatial operators in more detail.

Challenge 2

1. Suppose you have a layer of counties and a line layer containing coastlines. You are building a spatial query to select counties that border the coastline. Which spatial selection method would you choose?

    a. Are identical to
    b. Touch the boundary of
    c. Are crossed by the outline of
    d. Intersect

2. Which spatial selection method will select points inside a polygon? (Choose two).

   a. Have their centroid in
   b. Are identical to
   c. Are completely within
   d. Are within
   e. Are within a distance of

3. Which spatial selection methods will select polygons that are adjacent to polygons in another layer? (Choose two.)

   a. Touch the boundary of
   b. Are identical to
   c. Have their centroid in
   d. Share a line segment with
   e. Intersect

4. Suppose you have a layer of rivers and a layer of parks. Which spatial selection methods will select rivers that cross a park? (Choose two.)

   a. Intersect
   b. Completely contain
   c. Are completely within
   d. Are crossed by the outline of
   e. Touch the boundary of

Queries in analysis

Queries are an essential part of many analysis workflows. You may use a single query, a sequence of queries, compound attribute queries, or a combination of spatial and attribute queries to answer geographic questions.

In analysis workflows, you will typically use queries to create subsets of features, for example to use as input for a geoprocessing tool or to narrow down a set of features to the ones that fulfill specific criteria.

> ### *Selection methods*
> Whether you select features using an attribute query, a spatial query or you select features interactively, you can choose from four selection methods. The default selection method is to create a new selection. All other selection methods operate on a current selection:
>
> - Add to current selection
> - Remove from current selection
> - Select from current selection

Exercise 16c: Combine attribute and spatial queries

Suppose you are working with the City of Fishers to identify a parcel for a new park. (If you completed the exercises in the previous chapter, you are familiar with this scenario.) You will begin with a layer of potentially suitable parcels that are larger than five acres and fall into a particular city district. Then, you will narrow down the potential parcels to determine the best park locations.

1. **ArcGIS 10.1:** Open ArcMap, if necessary, and open the ....DesktopAssociate\Chapter16\Selections.mxd

   ArcGIS 10.0: Open ArcMap, if necessary, and open the ....DesktopAssociate\Chapter16\Selections_100.mxd

The map document displays parcel data for the City of Fishers.

2. Open the attribute table of the Parcels for Parks layer and locate the DeedAcres and the PolTwp fields.

The DeedAcres field contains the area of each parcel in acres. The PolTwp field contains a district name for each parcel.

Question 1: Write down the syntax for a query statement that selects the parcels in the Fall Creek district that have acreage larger or equal to 5 acres. Try to come up with the syntax on your own first. If you are not sure if it is correct, you can look it up at the end of the chapter.

---

3. From the Parcels for Parks layer, select the parcels in the Fall Creek district with an acreage of 5 acres or larger.

Twenty-nine parcels should be selected. You can check this in the table of contents in the List By Selection view.

The city wants the new park to be located in the southeast neighborhood of the city.

4. Switch back to List By Drawing order, if necessary. Turn on and extend the Neighborhoods layer.

5. In the Neighborhoods layer, select the polygon for Southeast Fishers.

At this point, you should have one feature from Neighborhoods and 29 features from Parcels for Parks selected.

Note: If you accidentally cleared the selection of the 29 features from Parcels for Parks, run the attribute query from step 3 again.

You will now continue to narrow down your selection of suitable parcels for the park.

Question 2: Which spatial query operator will select the parcels in the Southeast Fishers neighborhood (without selecting the ones that are touching the boundary or are partially outside)? Try to come up with the operator on your own first. If you are not sure if it is correct, you can look it up at the end of the chapter.

Decision point: Which layer is the Target and which one is the Source layer?

---

6. Using Select by Location, narrow down the selection in Parcels for Parks to the parcels that are 5 acres or larger and are inside the selected Neighborhoods polygon (without selecting the ones that are touching the boundary or are partially outside).

Hint: Use Parcels for Parks as the Target and Neighborhoods as the Source layer. Make sure to change the selection method to select from the currently selected features and to check the Use selected features option for the Source layer.

7. Right-click the Neighborhoods layer, point to Selection, and choose Clear selected features.

You have successfully narrowed down the potentially suitable parcels to four.

You will now create a selection layer from the four candidate parcels.

8. Right-click the ParcelsforPark layer, point to Data, and choose Create Layer from Selected Features. Rename the selection layer to **Candidates for Park**.

9. Clear the selected features.

You have narrowed down the search for a new park to only parcels five acres or larger that are located within the Southeast portion of the city.

10. If you are continuing with the next exercise, leave ArcMap open.

Working with selections

A selected set of features can be used in other analysis operations. For example, you can summarize or calculate statistics for a selected set of records, use a selected set as input for a geoprocessing tool, or invert the selected set (by switching the selection). You can save a selected set inside a map document by creating a selection layer, or you can export it to a new feature class or shapefile.

Exercise 16d: Refine the park selection

You will now refine your analysis of potentially suitable park parcels.

1. **ArcGIS 10.1:** Open ArcMap, if necessary, and open the ....DesktopAssociate\Chapter16\ParkSelections.mxd

   ArcGIS 10.0: Open ArcMap, if necessary, and open the ....DesktopAssociate\Chapter16\ParkSelections_100.mxd

Among other layers, the map document contains the Candidates for Parks selection layer that you created in the previous exercise. You will now further narrow down the selection of suitable parcels. For the park parcel, you would prefer a parcel that is more than 1.5 miles away from the Interstate highway.

Decision point: Using a spatial query, how would you select parcels in the Candidates for Parks layer that are 1.5 miles (or more) away from the Interstate highway?

Since you cannot select these features directly, you will first select the candidate parcels that are within a distance of 1.5 miles of the Interstate highway. Then you will switch the selection.

2. Using Select by Location, select the features from Candidates for Park that are within a distance of 1.5 miles of the Interstate layer.

3. In the Candidates for Parks attribute table, click the Switch Selection button.

4. Zoom to the selected features.

You narrowed down the selection to two parcels as the final candidates for the park. Last but not least, you will perform a final reality check and locate the selected parcels on a basemap.

5. Using the Add Data button, add the Imagery Basemap.

Hint: Click the drop-down arrow next to the Add Data button, and choose Add Basemap. Add the Imagery Basemap.

6. Turn off all layers except for the Basemap layer. Then turn the Candidates for Parks layer on and off.

Oops, one of the two selected parcels is actually a lake! You will remove the lake from the selection.

7. In the Selection menu, change the Interactive Selection Method to Remove from Current Selection. Then use the Select Features tool to remove the lower left parcel from the selection.

Next, you will export the final parcels to a new feature class.

8. Export the selected feature from Candidates for Parks using the same coordinate system as the layer's source data. Save the output in your ...DesktopAssociate\Chapter16\Fishers.gdb and name it FinalParcel. Click Yes to add the exported data as a layer.

9. Close ArcMap.

Answers to chapter 16 questions

Question 1: Write down the syntax for a query statement that selects the parcels in the Fall Creek district layer that have acreage larger or equal to 5 acres.
Answer: "PolTwp" = 'Fall Creek' AND "DeedAcres">= 5

Question 2: Which spatial query operator will select the parcels in the Southeast Fishers neighborhood (without selecting the ones that are touching the boundary or are partially outside)?
Answer: Are completely within

Answers to challenge questions

Correct answers shown in bold.

Challenge 1

1. Which of the following query statement will select cities with a population larger than 10,000.

    a. "POPULATION" = 10000
    b. "POPULATION" > 10000
    c. "POPULATION" >= 10000
    d. "POPULATION" IS 10000

2. Which of the following query statements will select all buildings with three or more stories and zoned as Multi-Use (MU).

    a. "BUILDING_FLOORS" > 3 AND "ZONE" = 'MU'
    b. "BUILDING_FLOORS" >= 3 OR "ZONE" = 'MU'
    c. "BUILDING_FLOORS" >3 OR "ZONE" = 'MU'
    d. "BUILDING_FLOORS" >= 3 AND "ZONE" = 'MU'

3. Which cities does the following expression select?
    "STATE" = 'CALIFORNIA' AND NOT "CITY_NAME" = 'Los Angeles'

    a. All the cities within California and Los Angeles
    b. All the cities within California except Los Angeles
    c. None the cities within California except Los Angeles
    d. All the cities outside California and Los Angeles

4. What does the following query statement select?
"PrclNo" LIKE '%20%'

   a. All parcels with a parcel number that starts with 20
   b. All parcels with a parcel number that ends on 20
   c. All parcels with a parcel number that contains the numbers 2 and 0
   d. All parcels with a parcel number that contains the number 20

Challenge 2

1. Suppose you have a layer of counties and a layer containing coastlines. You are building a spatial query to select counties that border the coastline. Which spatial selection method would you choose?

   a. Are identical to
   b. Touch the boundary of
   c. Are crossed by the outline of
   d. Intersect

2. Which spatial selection method will select points inside a polygon? (Choose two.)

   a. Have their centroid in
   b. Are identical to
   c. Are completely within
   d. Are within
   e. Are within a distance of

3. Which spatial selection method will select polygons that are adjacent to polygons in another layer? (Choose two.)

   a. Touch the boundary of
   b. Are identical to
   c. Have their centroid in
   d. Share a line segment with
   e. Intersect

4. Suppose you have a layer of rivers and a layer of parks. Which spatial selection method will select rivers that cross a park area? (Choose two.)

   a. Intersect
   b. Completely contain
   c. Are completely within
   d. Are crossed by the outline of
   e. Touch the boundary of

Key terms

NULL value: The absence of a recorded value for a field. A null value differs from a value of zero in that zero may represent the measure of an attribute, while a null value indicates that no measurement has been taken.

Operator: An operator used to compare logical expressions that returns a result of true or false. Examples of logical operators include less than (<), greater than (>), equal to (=), and not equal to (<>).

Selection: In ArcGIS, a set of selected features (or records)

Query: A request for information. In ArcGIS, a query returns a selected set of features (or records).

Expression: A statement that defines which features will be selected by a query. Expressions are generally part of a SQL statement.

Syntax: The structural rules for using statements in a command or programming language.

Resources

- ArcGIS Help 10.1 > Geodata > Data Types > Tables > Displaying tables
  - Sorting records in a table > About sorting records in a table
  - Summarizing data in a table
  - Viewing statistics for a table
- ArcGIS Help 10.1 > Desktop > Geoprocessing > Tool Reference > Analysis Toolbox > Statistics toolset
  - Frequency
  - Summary Statistics
- ArcGIS Help 10.1 > Desktop > Mapping > Working with layers > Interacting with layer contents
  - Using Select By Location
  - Using Select By Attributes
  - Working with selected features
  - Query expressions in ArcGIS
  - Building a query expression
  - SQL reference for query expressions using in ArcGIS

chapter 17

Performing spatial analysis

Proximity analysis 258
    Exercise 17a: Create service areas and calculate distances

Overlay analysis 262
    Challenge 1
    Exercise 17b: Perform a spatial join

Statistical analysis 266
    Exercise 17c: Perform statistical analysis

Temporal analysis 268
    Visual analysis of temporal data
    Challenge 2
    ArcGIS Tracking Analyst

Answers to challenge questions 270

Key terms 271

Resources 271

Chapter 17: Performing spatial analysis

There are four common types of spatial analysis: overlay analysis, proximity analysis, statistical analysis, and temporal analysis (figure 17.1). Each type of analysis answers different geographic questions, such as the following:

- "Which features are nearest to other features, and what is the distance between them?" Proximity analysis can answer this type of question.
- "What is on top of what?" Overlay analysis can answer this type of question.
- "How much has a city grown in area in the last 20 years?" or "How fast does a hurricane travel across the Caribbean?" Temporal analysis can answer these types of questions.
- "Are there any patterns or relationships in the data? If so, how can they be described, quantified?" Statistical analysis can answer this type of question.

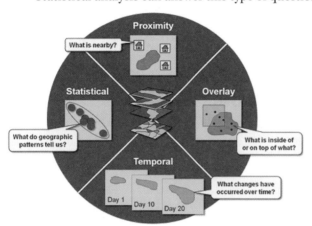

Figure 17.1 The four common types of spatial analysis are proximity, overlay, temporal and statistical analysis. From ArcGIS 10.1 Help.

Any of these four types of analyses can be performed with vector or raster data, using different tools for each data type. Choosing the right geoprocessing tool for a given task is a crucial part of the workflow for any analysis. Here you will focus on the tools for performing these types of analyses with vector data.

Skills measured

- Given a scenario, determine the necessary steps and tools to perform a proximity analysis.
- Given a scenario, determine the necessary steps and tools to perform an overlay analysis.
- Given a scenario, determine appropriate specifications for creating a new spatial join.
- Given a scenario, determine the necessary steps and tools to perform a temporal analysis.
- Given a scenario, determine the necessary steps and tools to perform a statistical analysis.
- Given a scenario, determine the appropriate analysis to answer a problem (i.e., proximity, overlay, temporal, statistical).

Prerequisites

- Knowledge of the spatial analysis workflow
- Hands-on experience with ArcGIS geoprocessing tools, including
  - Common proximity analysis tools, such as Select By Location and Buffer
  - Common overlay tools such as Union and Intersect

Proximity analysis

Proximity analysis evaluates the distances between features to determine which features are nearest to other features, and how close are they. Evaluating the distances between features is important to answer questions such as:

- How far away is the well from the landfill?
- Do any roads pass within 1000 meters of a stream?
- What is the closest cell phone tower from a given location?
- Which area is best covered by a particular cell phone tower?

For example, suppose you are analyzing accident locations. You may ask, "What is the closest hospital to each accident?" Or "What are the distances to all hospitals within a 10 mile radius of each accident?" With this information, you can determine where to send accident victims if the nearest hospital is full. Taking the proximity analysis a step further, you can calculate the service areas covered by each hospital so you can immediately determine where to send victims when an accident happens. Based on these service areas, you may even be able to determine if the current hospital coverage is sufficient, or identify areas where new hospitals are needed.

Table 17.1 lists common proximity analysis tasks, the vector geoprocessing tools that are used to accomplish them, and an explanation of how the tools work.

You may use these tools to perform proximity analysis by itself or use them in combination with other tools to perform other types of analysis (e.g., proximity analysis is often a precursor to overlay analysis).

Table 17.1 Common proximity analysis

Task	Tool	How the tool works
1. Create buffer areas around a feature 2. Create geodesic buffers, i.e., large area buffer polygons that accurately represent distances on the Earth's surface	Buffer	Creates polygons around input features (points, lines, or polygons) based on a buffer distance. The buffer distance determines the size of the polygons.
Create buffer areas at different distances around a feature, e.g., fishing right areas at different distances from an island.	Multiple Ring Buffer	Creates buffer polygons around input features based on multiple buffer distances.
Create areas of influence around point features, e.g., assign hunting territories around different bird of prey nests.	Create Thiessen Polygons	Creates polygons around input point features, where every location in the polygon is closer to its associated point than to any other point in the input feature class.

Continued on next page

Chapter 17: Performing spatial analysis

	Table 17.1 Common proximity analysis (continued)	
Task	Tool	How the tool works
Determine the distance from input features to the nearest feature in another feature class (Near feature class), e.g., calculate distances from a set of schools to the nearest hospital, within a 1 mile radius.	Near	Calculates the distance from each feature in the input feature class to the nearest feature in the near feature class, within a search distance. A new field storing the distance is added to the input attribute table.
Determine the distance between features in two point feature classes, e.g., determine the distance between address points and cell phone towers within a 2 mile radius.	Point Distance	Generates a new table containing the distances between two point feature classes. Distances are calculated from each input point to ALL NEARBY points in the near feature class, within a search distance.
Determine the distance between features (point, line, polygon, multipoint) in two feature classes, e.g., determine the distance between a well and the closest point along each road, within a 2 mile radius.	Generate Near Table	Generates a new table containing the distances between point, multipoint, line, or polygon features in two or more feature classes. Distances are calculated from each input feature to the NEAREST feature, or ALL NEARBY features, within a search distance. Optionally you can write the x, y coordinates and the angle of the nearest feature to the output table.

Table 17.1 illustrations from Esri, ArcGIS 10.1 Help, or created by the author.

Exercise 17a: Create service areas and calculate distances

Suppose you are tasked with recommending which post office should be closed. Criteria for closing a post office are: proximity to the Del Valle distribution center, which also provides post office services, and proximity to another post office. First, you will evaluate the straight line distance from each post office to its closest distribution center. Then you will estimate the catchment area of each post office, that is, the area from which each post office attracts customers.

1. ArcGIS 10.1: Open ArcMap, if necessary, and open the ...DesktopAssociate\Chapter17\Proximity.mxd
   ArcGIS 10.0: Open ArcMap, if necessary, and open the ...DesktopAssociate\Chapter17\Proximity_100.mxd

The map document shows post offices and distribution centers in the City of Austin, Texas.

Decision point: Which tool is appropriate for calculating the straight line distance from each post office to its closest distribution center?

2. Locate the Near geoprocessing tool through the Search window and open it.
3. Run the Near tool with the following parameters:
   - Input Features: Post Offices
   - Near Features: Distribution Centers
4. Open the Post Offices attribute table.
5. Scroll to the right and locate the NEAR_FID and NEAR_DIST fields.

These fields indicate the distance of each post office to its closest distribution center.

Decision point: Which post office is closest to the Del Valle distribution center (NEAR_FID 1)?
Hint: Sort the records in the table first by Near_FID, and then by NEAR_DIST.

Next, you will estimate the catchment area of each post office.

Decision point: Which tool is appropriate for calculating the catchment area of each post office?

6. Locate and open the Create Thiessen Polygon geoprocessing tool.
7. Run the tool with the following parameters:
   - Input Features: Post Offices
   - Output Feature Class: ...DesktopAssociate\Chapter17\Austin.gdb**PO_Thiessen**
   - Output Fields: ALL
8. Open the attribute table of the PO_Thiessen layer.

Each polygon in the PO_Thiessen layer contains all the attributes from its associated post office.

Decision point: Which post offices have a catchment area less than 10 square miles ("Shape_Area" < 278,800,000 square feet) and less than seven miles ("NEAR_DIST" < 36,960 feet) away from the DelValle distribution center ("NEAR_FID" = 1)?

Hint: Use an attribute query to answer this question.

Close ArcMap without saving your map document.

Overlay analysis

Overlay analysis is a type of analysis that combines the geometry and attributes from multiple input feature classes into one single output feature class. Combining information from different feature classes is important to answer questions such as:

- How is vegetation type related to land use?
- Which portion of my parcel is within the 100-year floodplain?
- Which areas zoned for recreational land use are outside of parks?
- What is the total mileage of roads inside each county?

For example, to find the best location to build a new child care facility, you could perform a few overlay operations using feature classes such as parcels, land use, and demographics to find vacant parcels that are not on commercial or industrial land, but are located in or near areas where many young families live.

ArcGIS provides overlay geoprocessing tools that combine geometries and attributes of feature classes in different ways. Table 17.2 lists common overlay tasks, tools used to perform them, and a description of how they work.

Table 17.2 Common overlay analysis tasks

Task	Tool	How the tool works
Combine two or more polygon feature classes, and then query attribute combinations in the output, e.g., find a parcel on public land that is covered by forest, has sandy soil, and has a population of less than one person per square mile.	Union	- Combines features and attributes from all input feature classes in the output feature class. - Input must be polygons.
Determine common areas of all input features, and query attribute combinations in the output, e.g., determine the total mileage of roads that pass through a flood zone.	Intersect	- Combines common areas of multiple feature classes in a new output feature class. - Output feature class contains attributes from all overlapping features.

Continued on next page

Table 17.2 Common overlay analysis tasks (continued)

Task	Tool	How the tool works
Combine areas that are exclusively covered by features from either the input or the update feature class, e.g., find areas that are either recreational land use or lynx habitat, without any areas of overlap.	Symmetrical Difference	• Combines features from an input and an update feature class in a new output feature class and removes common areas. • Output feature class contains attributes from input and update features.
Remove areas from a feature class that overlap another feature class, e.g., remove areas of lynx habitat from areas of recreational land use.	Erase	• Creates a feature class from the features or portion of features outside the erase feature class. • Output feature class contains only the attributes from input features.
Combine input features with features of another feature class and retain features in the extent of the input features, e.g., combine voting districts from the previous elections with a new voting district that partially overlaps the older ones.	Identity	• Combines input features with portions of identity features that overlap. • Output feature class contains attributes from input and identity features.

Continued on next page

Table 17.2 Common overlay analysis tasks (continued)

Task	Tool	How the tool works
Update the features of one feature class with the features of another feature class, e.g., update a parcel feature class with a new parks layer.	Update	Updates the geometry and the attributes of an input feature class with the geometry and attributes of an update feature class. • Output feature class contains the same attribute fields as the input. • Input feature class and update feature class field names must match. • Input and update features must be polygons.
Add the attributes of one feature class to the input feature class, based on spatial relationships between features in the two feature classes, e.g., join counts of an endangered bird species to habitat polygons	Spatial Join	Joins the attributes of a feature class (Join feature class) to another feature class (Target feature class) based on a spatial relationships such as coincidence, proximity, or containment. Output feature class has the same geometry as the Target feature class, but contains attributes from both Join and Target feature classes.

Table 17.2 illustrations from ArcGIS 10.1 Help.

Challenge 1

1. Given a point feature class of vacant buildings (potential shelters) and a polygon feature class representing a floodplain, which overlay tool would you use to remove all vacant buildings located inside the floodplain?

    a. Erase
    b. Identity
    c. Symmetrical Difference
    d. Update

2. Given two feature classes containing nesting area of two bird species, which overlay tool would you use to create a new feature class containing only nesting area common to both species?

    a. Update
    b. Union
    c. Intersect
    d. Spatial Join

3. Given a feature class of census blocks with and demographic attributes for the year 2000, and another feature class containing the same information for the year 2010, which overlay tool would you use to add the 2010 population attributes to the 2000 census blocks?

    a. Update
    b. Spatial Join
    c. Union
    d. Intersect

Exercise 17b: Perform a spatial join

Suppose you are tasked with adding the name and address of the nearest distribution center to each post office to facilitate queries on the Post Offices attribute table.

1. **ArcGIS 10.1:** Open ArcMap, if necessary, and open the ....DesktopAssociate\Chapter17\Overlay.mxd
   ArcGIS 10.0: Open ArcMap, if necessary, and open the ....DesktopAssociate\Chapter17\Overlay_100.mxd

2. Open the attribute table of Distribution Centers.

There are two Distribution Centers in the area, with attributes for their facility ID, name, and address. You will join the attributes of the closest distribution center to the Post Offices attribute table.

3. Close the attribute table. Then, locate the Spatial Join tool and open it.

4. Fill in the following parameters:
   - Target Features: Post Offices
   - Join Features: Distribution Centers
   - Output Feature class: ...\DesktopAssociate\Chapter17\Austin.gdb**PO_SpatialJoin**

Two important steps are choosing the join operation and the match option.

5. Click into the Join Operation field. At the bottom of the dialog, click Show Help.

Decision point: Which Join Operation is appropriate for joining the post office features to the distribution centers?

_____

6. Choose the Join ONE TO MANY option.

Next, you have to choose a match option.

7. Click into the Match Option field.

Decision point: Which Match Option is appropriate for joining the post office features to the distribution centers?

_____

Last, you have to determine a search radius and a distance field.

8. Use the Measure tool to determine the greatest distance between a post office and a distribution center (in miles).

Decision point: What is the largest distance between a post office and a distribution center (in miles)?

_____

9. Click in the Distance Field Name input box and read the tool help about the distance field.

10. Set the remaining parameters as shown below. Then, run the Spatial Join tool.
    - Match Option: CLOSEST
    - Search Radius: 30 miles
    - Distance field: type **DC_Distance**

11. **Open the attribute table of the PO_SpatialJoin feature class and scroll to the end of the attribute table.**

For each post office, the NAME and ADDRESS attributes of the closest distribution center have been joined to the table. The DC_Distance field indicates the distance to the closest distribution center.

12. **Close ArcMap without saving your map document.**

Statistical analysis

Statistical analysis uses spatial statistics to analyze spatial patterns in the data, such as clusters or directional trends. For example, you can use spatial statistics to determine if there are significant clusters of malaria cases in a geographic area. Once identified, you can investigate the causes of clustering. For example, there may be sources of standing water allowing mosquitoes to breed and spread the disease.

Using spatial statistics to analyze patterns in your data reduces subjectivity and helps you explain why features are distributed the way they are. Statistical analysis is often used by itself or in combination with temporal analysis; once you have identified patterns in your data, you can investigate whether or not they are changing over time.

Statistical analysis can help answer questions such as:

- How are occurrences of a disease, wildlife sightings, crime locations, etc., distributed? Where is the geographic center? Can you visualize how occurrences are clustered around the center? Can you visualize the shape and orientation of the distribution to see if there is a directional trend? These types of questions can be answered using Descriptive Statistics tools.

- Are there any clusters in the distribution of occurrences of a disease, wildlife sightings, crime locations, and so on, or are they randomly distributed? If there is clustering or dispersion, is it significant? These types of questions can be answered using Cluster Analysis.

- Where are significant clusters of buildings with a high incidence of reported crimes (hot spots)? Where are clusters of buildings with a low incidence of reported crimes (cold spots)? These types of questions can be answered using Hot Spot analysis.

- What are the driving factors for the clusters of reported crimes? Is the incidence of crime influenced by the poverty rate? These types of questions can be answered through the use of regression analysis.

Table 17.3 Common statistical analysis tools

Task	Tool	How the tool works
Locate the geographic center, e.g., locate elk observations within a park, over several years, to see where elk congregate.	Mean Center / Median Center	Creates a point feature indicating the geographic center of a set of features.
Determine the most centrally located feature, e.g., minimize distances from a set of warehouses to a distribution center by converting the most centrally located warehouse in a warehouse feature class to a distribution center.	Central Feature	Identifies the most centrally located feature in a point, line, or polygon feature class.

Continued on next page

Statistical analysis

Table 17.3 Common statistical analysis tools (continued)

Task	Tool	How the tool works
Determine the overall direction, length, and center of line features, e.g., analyze wildlife migration routes for different species.	Linear Directional Mean	Identifies the mean direction, length, and center of line features. The output will be a line indicating a direction and length, placed in the mean center.
Measure the density of data around a mean, e.g., compare how densely burglaries and auto thefts are distributed around the mean center to develop strategies for preventing these crimes.	Standard Distance	Measures the compactness of a distribution around the mean center. The output will be a new circle polygon placed on the geographic center, whose radius indicates the standard distance between features.
Quantify the direction of spread of a phenomenon, e.g., map the occurrence of burglaries to determine if there is a directional trend.	Directional Distribution	Creates standard deviational ellipses to summarize the central tendency, dispersion, and directional trends of some features.

Table 17.3 illustrations from ArcGIS 10.1 Help.

ArcGIS provides many spatial statistics geoprocessing tools for the different types of statistical analysis. Here, you will focus on some descriptive statistics tools that are useful at the beginning of a statistical analysis to characterize a dataset and gain a better understanding of the distribution of the data. Table 17.3 lists some common descriptive statistics tools, the tools used to perform them, and a description of how each one works. You can run these tools either on the entire input feature class, or on a subset: if you specify a case field, the input features are first grouped according to case field values, then the output is computed for each group separately. For example, if you use a time field as a case field (e.g., YEAR.), the values will be grouped by the different years and you can calculate descriptive statistics separately for several years and observe the change over time.

Exercise 17c: Perform statistical analysis

Suppose you are analyzing population change in Dane County in the US state of Wisconsin, between 2000 and 2010. To start your analysis, you will compare the mean population centers.

1. **ArcGIS 10.1:** Open ArcMap, if necessary, and open the ....DesktopAssociate\Chapter17\Statistics.mxd
   ArcGIS 10.0: Open ArcMap, if necessary, and open the ....DesktopAssociate\Chapter17\Statistics_100.mxd

The map document shows the centroids of each census block in Dane County.

2. Open the Census Points attribute table.

The table contains attributes for the total population in each block for 2000 and 2010. You will use these to calculate the mean center of the population of Dane County for the different years.

3. Close the attribute table and locate and open the Mean Center geoprocessing tool.

4. Run the Mean Center tool with the following parameters:
- Input Feature Class: Census Points
- Output Feature Class: …\DesktopAssociate\Chapter17\Austin.gdb\MeanCntPop
- Weight Field: TOTAL_POP
- Case Field: YEAR

With the YEAR field specified as the case field, the census points are first grouped according to case field values (years), and then a mean center is calculated for each year.

5. Turn off Census Points and zoom to the extent of the MeanCntPop layer.

The points indicating the population mean centers are very close together. This indicates that there has not been much change in the mean center of the population of Dane County between 2000 and 2010, which is a valid result. To confirm this result, you will further describe this population distribution pattern by calculating the directional distribution of the population points.

6. Run the Directional Distribution tool with the following parameters:
- Input Feature Class: Census Points
- Output Feature Class: …\DesktopAssociate\Chapter17\Austin.gdb\DirDistPop
- Weight Field: TOTAL_POP
- Case Field: YEAR

7. Symbolize the DirDistPop layer with Unique values based on the YEAR field.

The directional distribution ellipses for the two years are almost identical. This confirms that there has not been much change in the population distribution of Dane County between 2000 and 2010.

8. Save your map document and close ArcMap.

Temporal analysis

Temporal Analysis allows you to analyze data based on date and/or time information (temporal data). This allows you to determine trends and patterns in your data over time. For example, you can use temporal analysis to determine how long processes take, how often events occur, or how data changes. Temporal analysis could include visualizing temporal data using the ArcMap Time Slider window, or performing statistical analysis using a time field as the case field.

Temporal analysis can answer questions like:
- How much has a city grown in area in the last 25 years?
- How has the mean center of elk observations within a park changed over several years?
- How long does it take for a delivery truck to drive from point A to point B?
- Where is a delivery truck right now?

Visual analysis of temporal data

In order to discover and explore temporal patterns and trends, you may want to visualize how data changes over time. To visualize temporal data, the attribute table must contain a time field that stores the date or time that the information was collected. When a time field is present, you can enable time in the layer properties and create a time animation in the ArcMap Time Slider window. For more information about the Time Slider window, refer to the ArcGIS Help topic "Using the Time Slider window."

Challenge 2

1. Given a feature class of book stores and a feature class of eyewear stores, which analysis type would you use to find the book store closest to an eyewear store?

    a. Proximity
    b. Overlay
    c. Statistical
    d. Temporal

2. Given a feature class of Roads and a feature class of State Parks, which type of analysis would you use to determine the total mileage of roads that runs through a State Park?

    a. Proximity
    b. Statistical
    c. Overlay
    d. Temporal

3. Which type of analysis would you use to evaluate if there was a change in typical summer and winter weather patterns over a period of 50 years?

    a. Proximity
    b. Statistical
    c. Overlay
    d. Temporal

4. Which type of analysis would you use to determine if there was a directional trend in a point feature class of crime locations?

    a. Proximity
    b. Statistical
    c. Overlay
    d. Temporal

ArcGIS Tracking Analyst

Both the ArcMap Time Slider window and the Tracking Analyst extension allow you to create basic time animations to visualize temporal data. Tracking Analyst also provides additional functionality for visualizing and analyzing objects that move or change status over time. For example, Tracking Analyst integrates with Global Positioning System (GPS) units so you can track objects in real time. In addition to real-time tracking, Tracking Analyst also provides options and tools for symbolizing and charting temporal data, processing temporal data, and analyzing change in temporal data.

For more information about ArcGIS Tracking Analyst, refer to the ArcGIS Help document *What is Tracking Analyst?*

Chapter 17: Performing spatial analysis

Answers to challenge questions

Challenge 1

1. Given a point feature class of vacant buildings (potential shelters) and a polygon feature class representing a floodplain, which overlay tool would you use to remove all vacant buildings located inside the floodplain?

   a. Erase
   b. Identity
   c. Symmetrical Difference
   d. Update

2. Given two feature classes containing nesting area of two bird species, which overlay tool would you use to create a new feature class containing only nesting area common to both species?

   a. Update
   b. Union
   c. Intersect
   d. Spatial Join

3. Given a feature class of census blocks with and demographic attributes for the year 2000, and another feature class containing the same information for the year 2010, which overlay tool would you use to add the 2010 population attributes to the 2000 census blocks?

   a. Update
   b. Spatial Join
   c. Union
   d. Intersect

Challenge 2

1. Given a feature class of book stores and a feature class of eyewear stores, which analysis type would you use to find the book store closest to an eyewear store?

   a. Proximity
   b. Overlay
   c. Statistical
   d. Temporal

2. Given a feature class of Roads and a feature class of State Parks, which type of analysis would you use to determine the total mileage of roads that runs through a State Park?

   a. Proximity
   b. Statistical
   c. Overlay
   d. Temporal

3. Which type of analysis would you use to evaluate if there was a change in typical summer and winter weather patterns over a period of 50 years?

   a. Proximity
   b. Statistical
   c. Overlay
   d. Temporal

4. Which type of analysis would you use to determine if there was a directional trend in a point feature class of crime loations?

   a. Proximity
   b. Statistical
   c. Overlay
   d. Temporal

Key terms

Proximity analysis: A type of GIS analysis that determines proximal relationships between features. Proximity analysis answers questions such as "What features are near other features?" or "What is the distance between features?"

Overlay analysis: A type of GIS analysis that combines features and attributes from multiple layers to create new information. Overlay analysis helps answering the question "What is on top of what?"

Temporal analysis: A type of GIS analysis that allows you to track, visualize, and analyze spatial data that move or change status over time. Temporal, or time-based analysis helps answering questions such as " When did things happen?" or "How did things changed over time?"

Thiessen polygons: Polygons generated from a set of sample points. Each Thiessen polygon defines an area of influence around its sample point, so that any location inside the polygon is closer to that point than any of the other sample points.

Statistical analysis: A type of GIS analysis that helps indentifying patterns or relationships in the data. Statistical Analysis extracts additional information from the data that might not be obvious by looking at a map.

Resources

- ArcGIS Help 10.1 > Geoprocessing > Commonly used tools
  - Overlay Analysis
  - Proximity Analysis
  - Statistical Analysis
- ArcGIS Help 10.1 > Geoprocessing > Tool reference
  - Analysis toolbox
    - Proximity toolset > An overview of the Proximity toolset
    - Overlay toolset > An overview of the Overlay toolset
    - Statistics toolset > An overview of the Statistics toolset
  - Spatial Statistics toolbox > Measuring Geographic Distributions toolset
    - An overview of the Measuring Geographic Distributions toolset
  - Spatial Analyst toolbox > Overlay toolset
    - Understanding overlay analysis
- ArcGIS Help 10.1 > Mapping >Time > Visualizing time-enabled data
  - About visualizing temporal data
  - Enabling time on data > About enabling time on data
  - Viewing time-enabled data > Using the Time Slider window
- ArcGIS Help 10.1 > Extensions > Tracking Analyst
  - What is Tracking Analyst?

chapter 18

Organizing layers

Working with layers and data frames 274
  Group layers
  Basemap layers
  Exercise 18a: Organize layers in the table of contents
  Data frame properties
  Coordinate system
  Exercise 18b: Set data frame properties
  Reference scale
  Exercise 18c: Set a reference scale and a fixed scale
  Challenge

Answers to chapter 18 questions 280

Answers to challenge questions 281

Key terms 281

Resources 282

274 Chapter 18: Organizing layers

You start creating a map by adding layers to the table of contents. In the table of contents, layers are organized into one or more data frames. Organizing layers and data frames in your map document is an important part of optimizing your map composition, as well as optimizing the usability and performance of your map.

Skills measured

- Given a map's intended use, determine layer specifications to configure the map's table of contents.
- Given a visualization task, determine the appropriate specifications to complete the task (e.g., data frames, coordinate system settings, scalability of solution).

Prerequisites

- Hands-on experience with data frames and data frame properties
- Hands-on experience with layers and layer properties

Working with layers and data frames

You could use multiple data frames in a map document to show data in different geographic areas side by side, to show one geographic area at different scales, or to show different data layers in the same geographic area and compare them side by side.

Suppose you were making a map of zebra migration in Tanzania's Serengeti National Park. You could use one data frame to show all zebra migration routes at the scale of the entire park and several other data frames to show individual migration routes at larger scales in different areas of the park (inset maps). Another data frame could show Tanzania and the surrounding countries and indicate where Serengeti National Park is located (overview map).

Group layers

Within a data frame, you can improve map usability by organizing several thematically related layers into a group layer. Group layers allow you to manage multiple layers as a single group (figure 18.1). For example, you could organize layers representing national highways, railway lines, navigable rivers and canals, and local paved and unpaved roads into one transportation group layer. In a group layer, you can turn all these layers on and off with one click, set a transparency and visible scale range for the entire group layer, or collapse the group layer to hide the individual layers inside. If you need to, you can even create nested group layers (groups within a group layer).

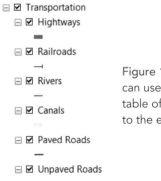

Figure 18.1 A group layer contains other layers. You can use group layers to organize related layers in the table of contents and define drawing options that apply to the entire group layer. Created by the author.

Basemap layers

Zooming and panning in a map should be as seamless and as fast as possible. You can improve the display performance of a map by organizing layers into a basemap layer. Basemap layers are a special type of group layer that use a high-performance drawing engine for fast display (figure 18.2). They "remember" map extents that a user has visited before by creating a local display cache and can therefore draw very fast the next time the user revisits that extent. Since you cannot edit layers inside of a basemap layer, you want to use basemap layers only for layers that don't change frequently and for raster layers that you want to show in the background of your map. In the example of the Serengeti National Park migration map, the basemap layer could contain the park boundary layer, hydrology, and an image showing the terrain. On top of this basemap are displayed the layers representing the migration routes.

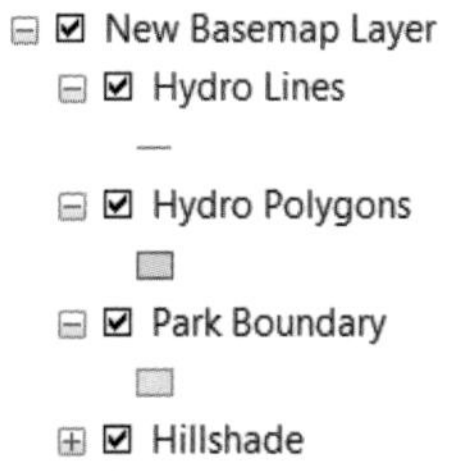

☐ ☑ New Basemap Layer
   ☐ ☑ Hydro Lines
      —
   ☐ ☑ Hydro Polygons
      ☐
   ☐ ☑ Park Boundary
      ☐
   ⊞ ☑ Hillshade

Figure 18.2 A basemap layer is a special type of group layer that is optimized for fast display and seamless panning and zooming. Created by the author.

Exercise 18a: Organize layers in the table of contents

For the exercises in this chapter, suppose you are tasked with creating a population density map of South America that will be included in a report. To get started, you will organize the layers in the table of contents so they are easier to manage.

1. **ArcGIS 10.1:** Start ArcMap and open the ...\DesktopAssociate\Chapter18\OrganizeLayers.mxd.

    ArcGIS 10.0: Start ArcMap and open the ...\DesktopAssociate\Chapter18\OrganizeLayers_100.mxd.

The map document contains a data frame with some basic map layers that you would like to show on the map.

First, you will organize some of the layers into a group layer.

2. **In the table of contents, right-click the South America data frame and choose New Group Layer. Rename the group layer to Transportation.**

3. **Drag the Railroads and Roads layers into the Transportation group layer.**

Hint: Drag the layers below the Transportation group layer.

4. **Move the Transportation group layer below the Cities layer.**

To increase drawing performance of the map when panning and zooming, you will create a basemap layer in the table of contents. You will move layers that won't be modified into the basemap layer.

Note: If a layer added to a basemap layer displays error or warning icons, it means that the layer is not compatible with a basemap layer or has potential drawing performance problems. For more information about these icons, refer to the ArcGIS Help document "Working with basemap layers."

5. **Right-click the South America data frame and choose New Basemap Layer. Rename it Base Features.**

6. **Drag the following layers into the basemap layer:**

    - Rivers
    - Lakes
    - World Countries

Hint: Hold down the CTRL key to select multiple layers at once.

To further organize the table of c ontents, you will insert two more data frames and add layers to them. One of the data frames will be used to show the population density of South America; the other one will be used as an overview map.

7. **From the main menu, click Insert and choose Data Frame.**

8. **Rename the new data frame Population Density.**

9. Select the Countries and World Countries layers, right-click and choose Copy.

10. Right-click one of the selected layers and choose Copy. Right-click the Population Density data frame and choose Paste Layer(s).

11. Zoom to the extent of the Countries layer.

12. Insert another data frame to the table of contents and rename it **Overview Map**.

13. Copy and paste the World Countries layer into the Overview Map data frame.

14. Change the layer symbology as follows:
    - Fill color: Gray 50%
    - Outline Color: No color

Layers in the table of contents are now organized into three different data frames that you will use to show different geographic extents and different aspects of the data.

Note that currently (in data view), you are able to view only the layers in the active data frame.

15. Right-click the South America data frame and choose Activate.

16. Save the map document. If you are continuing with the next exercise, leave ArcMap open. Otherwise, close ArcMap.

Data frame properties

Data frames have properties that apply to all the layers they contain. The following table summarizes some commonly used data frame properties and their uses.

Table 18.1 Data frame properties and their uses

Data frame property	Options	Use
Coordinate system	Can be set to any geographic or projected coordinate system	Defines the coordinate system of the map. All layers in the data frame will be displayed in the coordinate system of the data frame.
Reference scale	Can be set to any map scale	Defines the scale, to which symbols and text will scale (become larger or smaller) when the user zooms in or out.
Label engine	Can be set to Standard or Maplex	Places labels based on the specified label options. Switch from the Standard (default) label engine to Maplex for advanced label options.
Scale/Extent	Can be set to Fixed or Automatic	Specifies whether the user is able to zoom and pan from the current scale/extent of the data frame. Set fixed scale/fixed extent to prevent the user from zooming to any scale other than the one set.
Extent used by the Full Extent command	Can be set to the visible extent, the extent of a feature layer in the data frame, a selected graphic(s), or a custom extent	Redefines full extent when the full extent of all layers in the data frame is too large.

Continued on next page

Table 18.1 Data frame properties and their uses (continued)

Data frame property	Options	Use
Clip to shape	Can be set to the extent of another data frame, the visible extent, the extent of a feature layer in the data frame, a selected graphic(s), or a custom extent	Changes the shape of a data frame, e.g., to a shape other than rectangular.
Extent indicator	Can be set with a border, background and drop shadow, or a leader symbol	Shows the current extent of the data frame (inside another data frame, e.g., an overview map).
Background color	Can be set to any color	Often set to blue to indicate ocean on a world map.

Coordinate system

One of the most important properties of a data frame is its coordinate system: all layers in a data frame are drawn in the coordinate system of the data frame. Choose a projected coordinate system for a data frame based on the spatial properties you want to preserve:

- Conformal projections preserve true shape
- Equal area projections preserve true area
- Equidistant projections preserve true distance
- Azimuthal projections preserve true direction
- Compromise projections attempt to preserve all properties with some distortion to all

The coordinate system also determines the map units and display units of the data frame. While the map units are determined by the coordinate system, the display units can be modified. For example, if the map units are feet and you want to display coordinate values in meters, simply change the display units from feet to meters.

Exercise 18b: Set data frame properties

Now you will refine your South America map by setting the background color, assign an appropriate coordinate system for a population density map, and change the display units in the data frame properties.

1. **ArcGIS 10.1: Open the ...\DesktopAssociate\Chapter18\DataFrameProperties.mxd.**

   ArcGIS 10.0: Open the ...\DesktopAssociate\Chapter18\DataFrameProperties_100.mxd.

The map document shows a South America map similar to the one you created in the previous exercise. To symbolize the ocean around South America you will set the data frame background color to blue.

2. **Open the Properties of the South America data frame.**

3. **In the Frame tab, set the Background color to Blue. Then click OK to close the data frame properties.**

You will apply the same blue background color to the Population Density data frame.

4. **Activate the Population Density data frame and open its properties. Set a blue background color. Leave the data frame properties open.**

Next, you will modify the coordinate system of the data frame. The map will show population density for South America.

278 Chapter 18: Organizing layers

Question 1: What type of projected coordinate system will be appropriate for a population density map?

---

5. In the Coordinate System tab, make a note of the data frame's current coordinate system.

---

6. **ArcGIS 10.1 only:** Set the spatial filter to the Current visible extent.

    If you are working in ArcGIS 10.0, expand the Predefined folder, and then skip to step 8.

Hint: In the upper left corner of the Coordinate System tab, click the Spatial Filter button, and choose Set Spatial Filter.

7. **ArcGIS 10.1 only:** Search for equal area.

8. Expand the Projected Coordinate Systems, Continental and South America folders, and select the South America Albers Equal Area Conic projected coordinate system.

Decision point: What is the geographic coordinate system of the South America Albers Equal Area Conic projected coordinate system? Is it different from the geographic coordinate system of the data frame? Is a geographic transformation required?

The South America Albers Equal Area Conic projected coordinate system is based on a South American 1969 geographic coordinate system. Since the data frame is currently in a WGS 1984 geographic coordinate system, you will have to set a transformation.

9. Click the Transformations button.

10. In the Geographic Coordinate System Transformations dialog box, set the geographic transformation as shown in the graphic below. Then click OK.

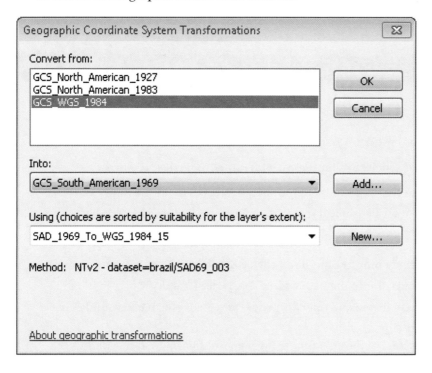

Note: In ArcGIS 10.0, choose SAD_1969_To_WGS_1984_14.

Since the map is intended to be used by an international audience, you will set the display units of the data frame to kilometers (1 kilometer = 1000 meters).

11. **In the General tab, set the Display units to Kilometers.**

12. **Click OK to close the Data Frame Properties.**

The equal area projected coordinate system, the transformation, and the display units are now applied to the Population Density data frame.

13. **Optional: If you would like some more practice, activate the South America data frame and apply the same coordinate system, transformation, and display units to it.**

14. **Save the map document. If you are continuing with the next exercise, leave ArcMap open. Otherwise, close ArcMap.**

Reference scale

Symbols and text in the data frame appear at their specified size at all scales, unless you set a reference scale. For example if you set the size of a marker symbol to 12 pt in the layer properties, it will appear at this size, no matter what scale you zoom to. Setting a reference scale in a data frame "freezes" the size of symbols and text at that scale. At the reference scale, the symbols and text will appear at their "true" scale (i.e., the scale that you have set for them). When you zoom in or out, text and symbols become larger or smaller.

It is a good idea to set a reference scale when creating a map that will be printed. Setting a reference scale ensures that the detail in your data frame will look the same on-screen as when printed. However, when creating an interactive map that will be viewed at different scales, the reference scale should be disabled. As you zoom in and out you do not want the symbols to change in either size or appearance, as this will slow down the map display.

Exercise 18c: Set a reference scale and a fixed scale

You will prepare a copy of the South America map for printing. For the printed version of the map, you will set a reference scale.

1. **ArcGIS 10.1: Open the ...\DesktopAssociate\Chapter18\ReferenceFixedScale.mxd.**

    ArcGIS 10.0: Open the ...\DesktopAssociate\Chapter18\ReferenceFixedScale_100.mxd.

2. **Open the properties of the Cities layer and click the Symbology tab.**

3. **Click the Symbol button and increase the size of the point symbol to 8 pt.**

4. **Click OK to close both the Symbol Selector and the Layer Properties.**

5. **Zoom to any dense urban area in the map (e.g., the Buenos Aires area).**

Without a reference scale set, the marker symbols of the cities layer display as 8 pt font size at all scales.

Next, you will set a reference scale for the South America data frame to scale the cities symbols with the map scale.

6. **Zoom to a scale of 1:40,000,000.**

7. **Right-click the South America data frame, point to Reference Scale and click Set Reference Scale.**

8. **Zoom in to the same area as before.**

When a reference scale is set in the data frame, the marker symbols of the Cities layer are scaled relative to the reference scale.

Next you will set an appropriate scale as a fixed scale for the Population Density data frame.

9. **Activate the Population Density data frame and open its properties.**

10. **In the Data Frame tab, set the Extent to a Fixed Scale of 1:40,000,000.**

The map is now at a fixed scale. The zoom in and zoom out buttons on the Tools toolbar are unavailable. Even if you resized the ArcMap window, the map would stay at a scale of 1:40,000,000.

11. Save the map document. Close ArcMap.

Challenge

1. When should you use group layers?

    a. When you want to improve the drawing performance of the map
    b. When some layers have a different coordinate system than the data frame
    c. When you have many layers that are thematically related
    d. When some layers cannot be put into a basemap layer

2. An ArcGIS user has 20 layers in a map. Four of the feature classes represent different aspects of a sewage system. How can the user organize the sewage system layers in the table of contents without impacting the ability to apply a different color scheme to these layers?

    a. Put the layers into a basemap layer
    b. Put the layers into a group layer
    c. Put the layers into a separate data frame
    d. Put the layers into four different data frames

3. An ArcGIS user received an .mxd file from a co-worker. The user tries to zoom in/out of the current view, but the zoom tools are disabled on the Tools toolbar. What is preventing the user from zooming in/out? (Choose two.)

    a. The map document is set to read-only.
    b. The data frame is not active.
    c. A fixed extent has been set.
    d. A fixed scale has been set.

4. An ArcGIS user has two data frames on a map and needs to add data to only one. How can the user ensure the data is added to the correct data frame? (Choose two.)

    a. Choose the correct data frame in the Add Data geoprocessing tool.
    b. Activate the correct data frame before adding data.
    c. Right-click the correct data frame and choose Add Data.
    d. Choose the correct data frame within the Add Data dialog box.

Answers to chapter 18 questions

Question 1: What type of projected coordinate system will be appropriate for a population density map?
Answer: For a population density map, it is important to show true area. Therefore, you should use an equal area projected coordinate system.

Answers to challenge questions

Correct answers shown in bold.

Challenge

1. When should you use group layers?

   a. When you want to improve the drawing performance of the map
   b. When some layers have a different coordinate system than the data frame
   c. When you have many layers that are thematically related
   d. When some layers cannot be put into a basemap layer

2. An ArcGIS user has 20 layers in a map. Four of the feature classes represent different aspects of a sewage system. How can the user organize the sewage system layers in the table of contents without impacting the ability to apply a different color scheme to these layers?

   a. Put the layer into a basemap layer.
   b. Put the layer into a group layer.
   c. Put the layer into a separate data frame.
   d. Put the layer into four different data frames.

3. An ArcGIS user received an .mxd file from a coworker. The user tries to zoom in/out of the current view, but the zoom tools are disabled on the Tools toolbar. What is preventing the user from zooming in/out? (Choose two.)

   a. The map document is set to read-only.
   b. The data frame is not active.
   c. A fixed Extent has been set.
   d. A fixed scale has been set.

4. An ArcGIS user has two data frames on a map and needs to add data to only one. How can the user ensure the data is added to the correct data frame? (Choose two.)

   a. Choose the correct data frame in the Add Data geoprocessing tool.
   b. Activate the correct data frame before adding data.
   c. Right-click the correct data frame and choose Add Data.
   d. Choose the correct data frame within the Add Data dialog box.

Key terms

Basemap: A map depicting background reference information such as landforms, roads, landmarks, and political boundaries, onto which other thematic information is placed. A basemap is used for locational reference and often includes a geodetic control network as part of its structure.

Inset map: A small map set within a larger map. An inset map might show a detailed part of the map at a larger scale, or the extent of the existing map drawn at a smaller scale within the context of a larger area.

Overview map: A generalized, smaller-scale map that shows the limits of another map's extent along with its surrounding area.

Reference scale: The scale at which symbols appear on a digital page at their true size, specified in page units. As the extent is changed, text and symbols will change scale along with the display. Without a reference scale, symbols will look the same at all map scales.

Resources

- ArcGIS Help 10.1 > Desktop > Mapping > Working with layers > Managing layers
  - Working with group layers
  - Working with basemap layers
- ArcGIS Help 10.1 > Desktop > Mapping > Working with ArcMap
  - Using Data Frames

chapter 19

Displaying layers

Vector layer symbology 284
    Categorical symbology
    Quantitative symbology
    Normalizing attribute values
    Classification methods
    Exercise 19a: Classify and symbolize vector layers

Raster layer symbology 288
    Exercise 19b: Symbolize raster layers

Managing the amount of data viewed in a map 290
    Definition queries
    Scale-dependent display
    Exercise 19c: Apply a definition query
    Exercise 19d: Set scale-dependent display
    Challenge

Answers to chapter 19 questions 293

Answers to challenge questions 293

Key terms 294

Resources 295

You can control a layer's appearance by changing some of its display properties, such as the symbol used to draw it. Different symbolization methods and techniques are used depending on the data source and type of information represented by a layer. For example, vector data is symbolized differently than raster data. You can also control other properties related to a layer's appearance, such as making a layer transparent to see what's beneath it. You can limit the amount of information displayed on a map by hiding some of the features in a layer or by specifying which layers are visible at certain scales.

Skills measured

- Given a scenario, determine appropriate layer properties and settings for a map document.
- Describe how to load data into ArcMap.

Prerequisites

- Knowledge of categorical and quantitative symbology options
- Knowledge of levels of measurement (nominal, ordinal, interval, ratio)

Vector layer symbology

Layers that contain vector data are symbolized with either a single symbol so all features look the same or with symbols that vary based on the values for one or more attributes. The level of measurement represented by that attribute determines whether to use categorical or quantitative symbology.

Categorical symbology

With categorical symbology, features are grouped into categories because they have similar text or numeric attribute values. Each category is then assigned its own symbol using a particular method (renderer) as shown in table 19.1:

Table 19.1 Renderers used for nominal data		
Level of measurement	Method/ renderer	Example
Nominal: Values are qualities, not quantities	Unique values	In a land use layer, each land use type is displayed with a unique color.
	Unique values, many fields	In a buildings layer, categories are based on both ownership (e.g., city, county, private, etc.) and construction year, and then each category is assigned a symbol.

Quantitative symbology

An attribute table may contain numeric attributes that can be used to symbolize a layer so it represents quantities such as a count, a rank, or a ratio. In quantitative symbology, features are grouped into classes based on numeric attribute values, using a classification scheme. Aggregating features into classes allows you to spot patterns in the data more easily. Each class is assigned a symbol using the appropriate method (renderer) as shown in table 19.2.

Table 19.2 Renderers used for ordinal, interval, and ratio data

Level of measurement	Method/renderer	Example
Ordinal: Values are quantities indicating a position or rank, such as first, second, third, without establishing a magnitude	Graduated color	In a layer of neighborhoods that are ranked as more desirable or less desirable, the more desirable ones are symbolized with a darker fill color than the less desirable ones.
	Graduated symbol	In a layer of oil wells, the point size varies based on the depth of the well. Deeper wells are symbolized with a larger point symbol than shallower ones.
Interval: Quantities measured along a linear, calibrated scale, with an arbitrary zero point	Graduated colors	In a layer of archeological sites, the color of the sites varies based on the year of occupation (e.g., 300 AD, 10,000 BPE). Older sites are represented with darker colors; more recent sites are represented with lighter colors.
	Graduated symbols	In a weather stations layer, points size varies based on recorded temperature values. Weather stations recording a higher temperature are represented with a larger point symbol than weather stations recording a lower temperature.
Ratio: Quantities measured along a linear scale with a fixed zero point	Graduated colors	In a building footprints layer, the color of the buildings varies based on age (e.g., 15 years, 20 years). A darker color represents an older building; a lighter color represents a newer one.
	Graduated symbols	When symbolizing pipelines based on the flow volume (cubic feet of gas per day), a thicker line represents higher flow volume than a thinner line.
	Proportional symbols	When symbolizing a point layer of cities based on population, the radius of each point symbol is proportional to the city's population.
	Dot density	When symbolizing a layer of wildlife management units based on caribou sightings, units with more sightings have more dots. Each dot may represent 1,000 sightings; dots are randomly placed within each unit.

Normalizing attribute values

When symbolizing a layer with graduated symbology (graduated color, graduated symbol, and proportional symbol), you can choose another numeric field to normalize the attribute values. The attribute values are divided by the values in the Normalization field. The layer symbology will be based on the ratio of the two fields. For example, instead of symbolizing a layer based on absolute population, you can normalize the population values by area. Then you can symbolize the layer based on population per square mile or population per square kilometer.

Classification methods

Graduated symbology (graduated color and graduated symbol) requires you to classify your data. Choosing the best classification scheme depends on the distribution of the attribute values used to create the classes, as well as what you want to emphasize in the map (table 19.3).

Table 19.3 Classification schemes and their uses			
Classification scheme	Description	Best use	Example
Natural Breaks (Jenks)—default	Similar values are grouped together into classes. Breaks are set to minimize variance within classes and maximize variance between classes.	To emphasize differences between classes when attribute values are clustered and not evenly distributed.	Map block groups to show differences in population.
Equal Interval	Divides the range of values into equal-sized classes, so the difference between high and low values is the same for each class.	To create a map that is easy to interpret when attribute values are evenly distributed.	Map temperatures or precipitation over an area.
Defined Interval	ArcGIS determines the number of classes to create based on the specified interval size and the range of attribute values.	Create class intervals that are easily understood by the map reader.	Map house values with class ranges every $100,000.
Quantile	An equal number of features are placed in each class, there are no empty classes.	Enhance map appearance by distributing uneven data more evenly.	Map block groups with similar populations to emphasize the population differences.
Geometrical Interval	A compromise method between equal interval, Natural Breaks (Jenks), and Quantile. Creates a balance between middle and extreme values.	Create a visually more appealing map when classifying continuous data that is not distributed normally and contains many duplicate values.	Map rainfall where only 15 out of 100 weather stations have recorded data. The rest have none, so their precipitation values are zero.
Standard Deviation	Classes are defined by their distance from the mean value.	Classify data with many values around the mean, so you can easily see values above or below it. Best used with a diverging two-color ramp.	Show how census tract population below the age of 5 compares to the national average (the mean) for that age by symbolizing it with two standard deviations above and below the mean.
Manual	Create your own classification scheme or modify an existing one.	You know your data well and want to emphasize classes that meet specific criteria.	Map watersheds with forest cover less than 50%, 51%–85%, and greater than 85%.

Exercise 19a: Classify and symbolize vector layers

You will now continue working on the South America map. You will apply categorical symbology to a Countries layer and use quantitative symbology in another data frame to create a population density map.

1. **ArcGIS 10.1:** Open the ...\DesktopAssociate \Chapter19\SymbolizeVectorLayers.mxd.
   ArcGIS 10.0: Open the ...\DesktopAssociate \Chapter19\SymbolizeVectorLayers_100.mxd.
2. Open the properties of the Countries layer in the South America data frame.
3. In the Symbology tab, click Categories and confirm the Unique Values renderer is highlighted.
4. For the Value Field, choose CNTRY_NAME. Add All Values and uncheck the symbol for <all other values>.
5. Choose the Muted Pastels color ramp. Then click OK to close the layer properties.

Hint: Right-click the color ramp and uncheck Graphic view to see the names of the color ramps.

Next, you will create the population density map.

6. Activate the Population Density data frame.

Instead of using the Countries layer, you will create the population density map based on a layer of administrative areas. This will give you a more detailed distribution of population densities in South America.

7. In the Catalog window, expand the ...\DesktopAssociate \Chapter19\ South_America geodatabase and drag the Admin feature class to the map above the Countries layer in the Population Density data frame.
8. Open the properties of the Admin layer. In the General tab, change the Layer Name to **Administrative Areas**.
9. Set the following symbology options:
   - Renderer: Quantities, Graduated colors
   - Value field: POP_ADMIN
   - Normalization field: SQKM
   - Color Ramp: Brown Light to Dark

10. Click Apply and move the Layer Properties window to the side so you can inspect the map.

Note: Some features in the Administrative areas were not displayed, so when you applied the graduated colors renderer, you continued to see the Countries layer. This is OK; you will address this issue later in the exercise.

Question 1: Based on the current classification scheme, which class has the most population density values and what effect does this have on the map?

To create a meaningful population density map, you need to distribute the population density values more evenly amongst the classes.

Decision point: Which classification method will distribute the population density values more evenly amongst the classes?

The Quantile method will put an equal number of features in each class, thereby distributing the values more evenly across the map.

11. Click Classify. Change the classification method to Quantile and click OK.
12. Click Apply and examine the effect of the Quantile classification method on the map.

With an equal number of administrative areas in each class, the map looks more visually balanced.

13. Change the number of classes to 7 and click Apply.

Now the map shows more variation in population density. To see the Countries layer underneath, you will make the Administrative Areas layer transparent and change the Countries layer symbology.

14. In the layer properties, Display tab, set a transparency of 30%, and then click OK to close the properties window.

When you applied the graduated colors renderer some features in the Administrative Areas layer were not displayed, and the Countries layer was displayed instead. This is because those areas have a population attribute (POP_ADMIN) value of <null>, which cannot be used for rendering. You will set a white color for the Countries layer to indicate the null values.

15. Set these symbology options for the Countries layer:
- Single Symbol symbology
- Fill Color: White
- Outline Width: 1.25
- Outline Color: Black

Hint: To quickly open the Symbol Selector, you can click the color patch of the layer in the table of contents.

Your population density map is finished now.

16. Close all dialog boxes and save the map document. If you are continuing with the next exercise, leave ArcMap open. Otherwise, close ArcMap.

Raster layer symbology

Similar to vector layers, raster layers can be displayed in many different ways depending on the type of data they contain and which aspect of the raster you want to emphasize. ArcMap chooses an appropriate display method for a given type of raster, which you can adjust as needed. If the raster has a predefined color scheme (a color map), ArcMap automatically uses it to display the raster layer. Table 19.4 lists the most commonly used display methods for raster layers and their uses.

Table 19.4 Renderers used for raster layers		
Renderer	When to use	Example
Stretched	Displays a raster layer by stretching the cell values across a graduated color ramp. Use the Stretched renderer to display a single band, continuous raster layer.	Display a gray-scale aerial photo (single band) with maximum contrast.
RGB Composite	Displays a raster layer with a combination of three bands: the red, the green, and the blue band. Use the RGB Composite renderer to display different band combinations of a multiband raster layer, such as satellite or aerial imagery.	Display a (multiband) satellite image or aerial photo in true color by displaying its red band in red, its green band in green, and its blue band in blue. Use a different band combination (e.g., near-infrared and visible bands) for vegetation or geology analysis.
Classified	Displays a raster layer by grouping cell values into classes. Use the Classified renderer to draw a single-band, continuous raster layer representing phenomena such as slope, elevation, distance, or suitability.	Display a raster of slope percentages to show slope classes (e.g., 0–15%, 15–30%, 30–45%, etc.) in different colors.

Continued on next page

Raster layer symbology **289**

Table 19.4 Renderers used for raster layers (continued)

Renderer	When to use	Example
Unique Values	Displays a raster layer by assigning a different color to each unique cell value (based on a color scheme). Use the Unique Values renderer to display a discrete raster layer with unique colors for each category. Colormap option allows you to specify the color that will be used for each category.	Display a land use raster to show each land use type with a different color. Use colormap option to display a land use raster with standardized, predefined colors for each land use type

Exercise 19b: Symbolize raster layers

You will now add two raster layers to your South America map and optimize their symbology.

1. **ArcGIS 10.1: Open the ...\DesktopAssociate\Chapter19\SymbolizeRasterLayers.mxd.**

    ArcGIS 10.0: Open the ...\DesktopAssociate\Chapter19\SymbolizeRasterLayers_100.mxd.

2. **Activate the South America data frame.**

3. **Turn off all layers in the South America data frame.**

4. **From the ...\DesktopAssociate\Chapter19\SA_Rasters folder, add both raster datasets to the map.**

Hint: In the Catalog window, click the Toggle Contents Panel button to display the contents panel at the bottom of the window. In the catalog tree, select the SA_rasters folder, and then hold down the Shift key and select both raster datasets in the contents panel. Drag both datasets at once into the South America data frame, just above the Base Features layer.

The SA_dem layer represents an elevation model of South America. SA_hillshade is a hillshade raster dataset showing the same area.

5. **In the table of contents, arrange the layers in the order shown in the following graphic.**

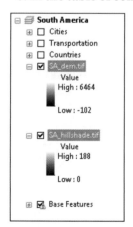

6. **Open the Properties of the SA_dem layer and click the Symbology tab.**

7. **Both the SA_dem and the SA_hillshade layers reference continuous raster datasets, which are displayed by default with a stretched renderer.**

8. **Change the Color Ramp to Elevation #1.**

Hint: Right-click the Color Ramp and uncheck Graphic View to see the names of the color ramps.

9. **In the Display tab, set a transparency of 40% and click OK to close the layer properties.**

Since you have applied a transparency, the hillshade layer is visible underneath the elevation layer.

10. **Turn off the SA_dem layer.**

In some areas, the hillshade raster appears very dark. You will apply a different contrast stretch to improve the display of the SA_hillshade layer.

11. **Open the properties of the SA_hillshade layer.**

12. **In the Symbology tab, for the Stretch Type, select Histogram Equalize. Click Apply and observe the effect that the stretch has on the map.**

The stretch brightens the display of the raster layer and shows more detail in the dark areas.

13. **Close the layer properties.**

14. **Turn the SA_dem layer on and save the map document.**

15. **If you are continuing with the next exercise, leave ArcMap open. Otherwise, close ArcMap.**

Managing the amount of data viewed in a map

During the initial stages of map design, you have to decide how much information is needed to convey the message of your map. You may have to simplify the layers to avoid "cluttering" the map with unnecessary features, especially at smaller scales.

Definition queries

One of the simplest ways to simplify a map is to display only a subset of the features in a vector layer using a definition query (figure 19.1). A definition query is an SQL (Structured Query Language) expression that defines which features in a layer will be displayed in the map. Features that do not satisfy the definition query are not displayed. For example, in a layer of cities with an attribute that distinguishes between capital and non-capital cities, you can create a query to display only the capital cities. A definition query is a layer property that can be edited or removed as needed.

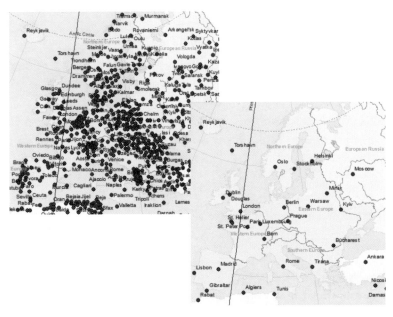

Figure 19.1 Use definition queries to reduce the number of features drawn on a map, without deleting any features from the source data of the layer. From Esri Data & Maps 2010; data courtesy of ArcWorld.

Scale-dependent display

What if you want to display all features in a vector layer, just not at all scales? Sometimes you run into a situation where the display of a layer is only useful in a particular range of map scales. For example, a map of Europe with all countries and all cities in a country displayed at all scales will become unreadable at full

Managing the amount of data viewed in a map **291**

extent because there are far too many cities cluttering the map. Setting a visible scale range for the cities is the solution. In the layer properties, you can specify a minimum and/or a maximum scale for a layer to be displayed. When the map is zoomed outside the specified scale range, the layer is turned off. Displaying a layer within a scale range is also called scale-dependent display.

To manage the amount of data shown on a map at any given scale, you can use definition queries and scale-dependent display in combination. By creating multiple copies of the same layer, you can apply different definition queries to the same data. In the above example of European cities, you may want to display only the capital cities at full extent, and display the medium and small cities only when you zoom in (figure 19.2).

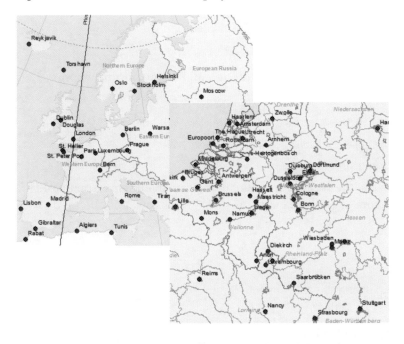

Figure 19.2 The map shown in this example contains two city layers: One layer showing only capital cities, the other showing all major cities in a country. At smaller scales, only the capital cities layer is shown. At larger scales, the major cities layer is shown. Setting different visible scale ranges on multiple copies of the same layer allows you to show different subsets of the features at different map scales. From Esri Data & Maps 2010; data courtesy of ArcWorld.

Exercise 19c: Apply a definition query

To simplify the map, you will reduce the number of visible cities by creating a definition query. You will display only the national and provincial capital cities of each country.

1. **ArcGIS 10.1: Open the …\DesktopAssociate \Chapter19\DefinitionQuery.mxd.**
   ArcGIS 10.0: Open the …\DesktopAssociate \Chapter19\DefinitionQuery_100.mxd.

2. **Open the attribute table of the Cities layer.**

Decision point: Which attribute field tells you whether or not a city is a capital city?

The STATUS field contains this information.

3. **Close the Cities attribute table and open the properties of the Cities layer.**

4. **In the Definition Query tab, click Query Builder.**

5. **Build the following query:**
   "STATUS" = 'National and provincial capital'

6. **Click OK, and then click OK again to close the Layer Properties**

Now that you applied the definition query, only the national and provincial capital cities display on the map.

7. **Save the map document. If you are continuing with the next exercise, leave ArcMap open. Otherwise, close ArcMap.**

Exercise 19d: Set scale-dependent display

Some areas of your South America map contain too many features for the continent-wide scale. To reduce the number of features that are shown at this scale, you will set a scale dependency so that Transportation features show only at regional scales.

1. **ArcGIS 10.1: Open the …\DesktopAssociate \Chapter19\ScaleDependency.mxd.**

    ArcGIS 10.0: Open the …\DesktopAssociate \Chapter19\ScaleDependency_100.mxd.

2. **Zoom in anywhere in the map to a scale of 1:10,000,000.**

3. **Open the properties of the Transportation group layer.**

4. **In the General tab, check the option to not "…show layer when zoomed:" For Out beyond, choose <Use Current Scale>.**

5. **Click OK to close the Group Layer Properties**

Note: You could also control the display of a single layer using this technique.

6. **Zoom to the extent of the Transportation group layer.**

The Transportation group layer does not draw at the continental scale. To indicate that there is some scale dependency set for the layer, the layer's check mark is grayed out in the table of contents.

7. **Save the map document. Close ArcMap.**

Challenge

1. Suppose you just added a layer to a map document showing four South American countries in four data frames. Into which data frame was the layer added?

    a. The data frame specified in the layer properties
    b. The focused data frame
    c. The data frame at the top in the table of contents
    d. The active data frame

2. You have a point feature class of 250 cities with an attribute containing the total population in each city. Population values for the cities vary significantly. Which method is most appropriate to symbolize the city points based on their population?

    a. Graduated color
    b. Graduated symbol
    c. Proportional symbol
    d. Dot density

3. You have a point feature class of soil sample locations with an attribute of soil temperature. What level of measurement does soil temperature represent?

    a. Nominal
    b. Ordinal
    c. Interval
    d. Ratio

4. Which renderer would you use to expand the range of gray values in an aerial photo that appears very dark?

   a. Stretched
   b. RGB Composite
   c. Classified
   d. Unique Values

5. You are working with a very large road dataset, and you are interested in viewing only the highway features within the road dataset. How can you accomplish this?

   a. Set a definition query for the layer
   b. Run the Mask Features geoprocessing tool
   c. Set a visible scale range for the layer
   d. Use Select By Location

Answers to chapter 19 questions

Question 1: Based on the current classification scheme, which class has the most population density values and what effect does this have on the map?
Answer: With the current classification scheme, most population density values fall into the lowest class. Most polygons in the map show the same color.

Answers to challenge questions

Correct answers shown in bold.

1. Suppose you just added a layer to the map document showing four South American countries in four data frames. Into which data frame was the layer added?

   a. The data frame specified in the layer properties
   b. The focused data frame
   c. The data frame at the top in the table of contents
   d. The active data frame

2. You have a point feature class of 250 cities with an attribute containing the total population in each city. Population values for the cities vary significantly. Which method is most appropriate to symbolize the city points based on their population?

   a. Graduated color
   b. Graduated symbol
   c. Proportional symbol
   d. Dot density

3. You have a point feature class of soil sample locations with an attribute of soil temperature. What level of measurement does soil temperature represent?

   a. Nominal
   b. Ordinal
   c. Interval
   d. Ratio

4. Which renderer would you use to expand the range of gray values in an aerial photo that appears very dark?

   a. Stretched
   b. RGB Composite
   c. Classified
   d. Unique Values

5. You are working with a very large road dataset, and you are interested in viewing only the highway features within the road dataset. How can you accomplish this?

   a. Set a definition query for the layer
   b. Run the Mask Features geoprocessing tool
   c. Set a visible scale range for the layer
   d. Use Select By Location

Key terms

Band: A set of adjacent wavelengths or frequencies with a common characteristic. For example, visible light is one band of the electromagnetic spectrum, which also includes radio, gamma, radar, and infrared waves.

Single-band image: Images composed of a single band. Grayscale images are known as single-band images, since all of the information necessary to render an appropriate grayscale image is contained in a single grid of pixels.

Multiband image: Images composed of multiple bands, typically three. One band each is used to describe the color intensity of red, green, and blue at a specific pixel location in a color image. The pixel values are often written as (R, G, B), where R is the red value, G is the green value, and B is the blue value.

Nominal data: Data divided into categories, within which all elements are assumed to be equal to each other, and in which no category comes before another. For example, a group of polygons colored to represent different soil types.

Ordinal data: Data classified by comparative value; for example, a group of polygons colored lighter to darker to represent less to more densely populated areas.

Interval data: Data classified on a linear calibrated scale, but not relative to a true zero point in time or space. Because there is no true zero point, relative comparisons can be made between the measurements, but ratio and proportion determinations are not as useful. Time of day, calendar years, the Fahrenheit temperature scale, and pH values are all examples of interval measurements.

Ratio data: Interval data with a natural zero point. For example, time is ratio since 0 time is meaningful. Degrees Kelvin has a 0 point (absolute 0) and the steps in both these scales have the same degree of magnitude.

Resources

- ArcGIS Help 10.1 > Desktop > Mapping > Working with layers > Displaying layers
  - Drawing a layer using categories
  - Drawing features to show quantities
    - Using graduated colors
    - Using graduated symbols
    - Using proportional symbols
    - Using dot density layers
    - Classifying data
    - Classifying numerical fields for graduated symbology
- ArcGIS Help 10.1 > Geodata > Data Types > Rasters and images > Displaying raster data
  - Renderers used to display raster data
  - Displaying raster layers
    - Displaying a DEM
    - …with hillshading
  - Options for improving the display of raster data
    - Improving the display of raster data
- ArcGIS Help 10.1 > Desktop > Mapping > Working with ArcMap > Navigating Maps
  - Working with data frame reference scales
- ArcGIS Help 10.1 > Desktop > Mapping > Working with layers > Managing layers
  - Displaying layers at certain scales
  - Displaying a subset of features in a layer

Composing map layouts and graphs

Map layout 298
    Exercise 20a: Create an extent indicator in a map layout
    Focus on data frame
    Exercise 20b: Insert a legend
    Exercise 20c: Add a scale bar

Graphs 303
    Graph types
    Creating a graph from a template
    Exercise 20d: Create a graph from a template
    Challenge

Answers to challenge questions 306

Key terms 307

Resources 307

Chapter 20: Composing map layouts and graphs

A map layout is designed for printing. A printed map is often the best method for sharing geographic information. A map layout begins with a page layout, a collection of map elements laid out and organized on a page.

Data frames are the elements that contain geographic data, the layers that you organize and symbolize in the table of contents, and for which you set display options. You can add additional map elements to the layout to help communicate the map's message and to help the audience understand and use the map effectively.

Graphs represent another way to share geographic information. Graphs can show additional information about features on a map, or they can show the same information in a different way. Graphs are created from tabular data, both spatial and nonspatial. A graph can be added as an element to the map layout. Having graphs on the layout can be extremely useful when making posters.

Skills measured

- Given a scenario, determine the most efficient way to change the current view of the map to the desired view.
- Given a task to set up a page layout, determine the legend content and settings, which layout elements to include, number of pages required, and appropriate reference scale.
- Given a scenario, determine appropriate graph/chart specifications (e.g., graph/chart type, required data, graph/chart properties).

Prerequisites

- Hands-on experience working with layouts
- Hands-on experience creating graphs

Map layout

To create a map layout, you switch from data view to layout view. In layout view, you are able to view all data frames of the map document on a virtual page. To compose a map layout, you arrange, resize, and align data frames and add map elements, such as a title, legend, and scale bar. Typically, you arrange and align data frames by using rulers and guides or by snapping them to a grid. Extent indicators can be used to show the extent of one data frame within another data frame. This is useful when you create overview or locator maps.

There are many map elements that you can add to a layout, some of which are dynamically linked to a data frame, and others that are graphics. All map elements have properties that can be configured and modified as needed.

Table 20.1 lists some common map elements.

Table 20.1 Common map elements	
Map element	**Description**
Title	- Communicates the overall map message - Should be the largest text in the map - Can have a subtitle - By default, the name of the map document
Legend	- Explains the symbols used in the map - Associated with one data frame - Can include or exclude layers from the associated data frame

Continued on next page

Map element	Description
Scale bar	- Indicates scale of the associated data frame - Multiple scale bars possible in a layout - Based on the coordinate system of the data frame - Dynamic—automatically adjusts to changes in the scale of the data frame
Inset map	- Can show more detail than the main data frame, or an overview of the geographic location of the main data frame - May include a dynamic extent indicator
North arrow	- Indicates the directional orientation of the map - Should only be used when the coordinate system of the data frame supports a constant direction in the map (e.g., north is up at all points in the map). - Connected to the data frame (when the data frame is rotated, the north arrow rotates with it)
Neatline	- Graphic element that visually groups related elements

Table 20.1 Common map elements (continued)

Exercise 20a: Create an extent indicator in a map layout

Continuing with your South America map, you will add an extent indicator to the inset map to provide a spatial context for the main data frame.

1. **ArcGIS 10.1:** Open the …\DesktopAssociate\Chapter20\ExtentIndicator.mxd.

   ArcGIS 10.0: Open the …\DesktopAssociate\Chapter20\ExtentIndicator_100.mxd.

2. From the View menu, choose Layout View. Dock the Layout Tools toolbar, if necessary.

Hint: Another way to switch to layout view is to click the small Layout View button at the lower left of the map display.

On the virtual map page, you see the South America, Population Density, and Inset Map data frames. They have been arranged on the virtual page to give you a head start with the exercise.

3. Open the properties of the Inset Map data frame and click the Extent Indicators tab.

4. Select Population Density, and then click the right button .

5. Under Options, click Frame, and set these parameters:

   - Border: 2.0 point
   - Color: Tuscan Red

6. Click OK to dismiss the data frame properties window.

Your layout should look like the graphic below:

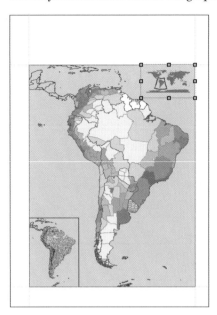

Decision point: Why is the extent indicator in the inset map shown with rounded edges?

The extent indicator is actually a rectangle created by the extent of the Population Density data frame. It appears distorted in the inset map because of the difference between the coordinate systems of the Population Density and Inset Map data frames.

7. Save the map document. If you are continuing with the next exercise, leave ArcMap open. Otherwise, close ArcMap.

> ### Focus on data frame
> While working in layout view, you might need to create, delete, and edit features, graphics, and text just as if you were in data view. Rather than switching to data view, focus on the data frame using the Focus Data Frame button from the Layout toolbar. This allows you to work with the features and elements in that data frame as though you were in data view.

Exercise 20b: Insert a legend

You will now add a legend to your South America map.

1. **ArcGIS 10.1:** Open the …\DesktopAssociate\Chapter20\InsertLegend.mxd.

    ArcGIS 10.0: Open the …\DesktopAssociate\Chapter20\InsertLegend_100.mxd.

The map shows the same layout that you worked with in the previous exercise.

When you create a legend, it is associated with the data frame that is active. You will insert a legend associated with the Population Density data frame.

2. If necessary, select the Population Density data frame.

3. From the Insert menu, choose Legend.

4. In the Legend Items list, select both the Countries and World_Countries layers and click the left arrow button to remove them from the legend, and then click Next.

5. Delete the default Legend Title.

6. Click Next until you reach the end of the legend wizard (accept all other defaults) and click Finish.

7. Move the legend and position it to the lower right side of South America, as shown in the graphic below. Don't worry about the exact position of the legend.

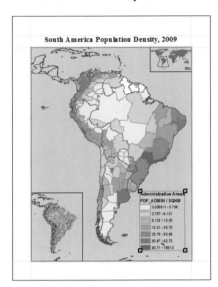

Next, you will refine the text in the legend.

8. Right-click the legend and open its properties. Click the Items tab.

9. **ArcGIS 10.1:** Click the Font drop-down list and choose Apply to the layer name. Set the font to Arial, 12, Normal (uncheck bold).

   ArcGIS 10.0: Click the Change text drop-down list and choose Apply to the layer name. Click Symbol and set the font to Arial, 12, Normal (uncheck bold).

10. **ArcGIS 10.1:** Click the Font drop-down list and choose Apply to the heading. Set the same font properties (Arial, 12, Normal). Then click OK to close the legend properties.

    ArcGIS 10.0: Click the Change text drop-down list and choose Apply to the heading. Click Symbol and set the same font properties (Arial, 12, Normal). Then click OK to close the legend properties.

11. In the table of contents, rename the Administrative Areas layer heading to Population / SQKM. Delete the subheading POP_ADMIN / SQKM.

Hint: To make a heading editable, click on it two separate times (do not double-click).

If the legend and other map elements are not aligned properly, don't worry. You will align them in the next exercise.

12. Save the map document. If you are continuing with the next exercise, leave ArcMap open. Otherwise, close ArcMap.

Exercise 20c: Add a scale bar

Next, you will add a scale bar to your South America map and finalize the map layout.

1. **ArcGIS 10.1:** Open the ...\DesktopAssociate\Chapter20\AddScalebar.mxd

   ArcGIS 10.0: Open the ...\DesktopAssociate\Chapter20\AddScalebar_100.mxd.

2. From the Insert menu, choose Scale Bar. From the Scale Bar Selector, select Scale Line 1, and then click OK. Drag the scale bar underneath the North Arrow.

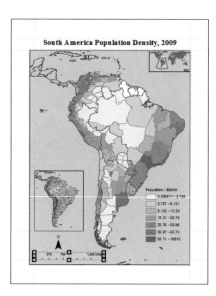

3. Right-click the scale bar and open its properties.

4. Set these parameters:
   - When resizing: adjust number of divisions
   - Division Units: kilometers
   - Label Position: after bar

5. Click Apply.

Decision point: At the current scale bar length, what division value will show a total of 2000 kilometers?

Assuming you are using ArcMap default settings, the current division value is 750 kilometers. The scale bar shows a total of 2,250 kilometers at its current length. With the current resize setting (Adjust number of divisions), ArcMap fits as many divisions as possible into the current scale bar length. Therefore, if your scale bar shows 2,250 kilometers, it fits three divisions of 750 km into the scale bar length.

A division value of 1,000 kilometers will fit into the current scale bar length twice.

6. Set the Division value to 1,000 km and the number of subdivisions to 4. Then click OK.

The scale bar now shows a total of 2,000 kilometers.

7. On the Layout toolbar, click the Toggle Draft mode button.

In draft mode, only the outline of the data frames and map elements in the layout are displayed, not their content. If your map layout has many elements, you might want to speed up your work by using draft mode.

8. Click Toggle Draft Mode button to switch back to the previous mode.

9. Use the graphic below as a guide to align the scale bar.

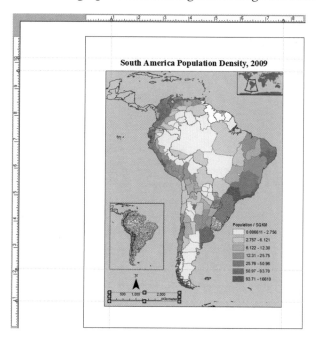

10. Save the map document. If you are continuing with the next exercise, leave ArcMap open. Otherwise, close ArcMap.

Graphs

While maps are an efficient way to visualize the spatial component of geographic data, graphs are a visual way to convey tabular information. Typically, graphs are used for displaying numeric attribute values from a few fields in an easy-to-understand way. While it may be hard to compare numeric values in the attribute table, relationships among them may become obvious when plotting the values on a graph. You can export graphs to various image formats and use them in posters or reports or add them to an ArcMap layout. A graph in the layout can provide additional information about the features on the map and can promote the map's message.

Graph types

The graph type you choose depends on your intent and the nature of the data that you want to display on the graph. For example, a bar graph is most appropriate for comparing population densities in different countries or rainfall amounts at one location across different months of the year. If your intention is to show the amounts of monthly rainfall at a given location as well as the variability over different years, a box plot graph would be a better choice. If your intention is to explore the relationship between two numeric attributes, such as average cancer rates and average household income (without implying that a relationship exists), a scatterplot graph would be the best choice.

Table 20.2 lists the commonly used graph types you can create in ArcGIS and their uses.

Table 20.2 Common graph types and their uses		
Graph type	When to use	Example
Bar	To compare amounts and discover trends in the data	Rainfall amounts collected at monthly intervals
Histogram	To display the range and frequency distribution of values in a field (in a large dataset)	Frequency distribution of population in different census blocks; for example, how often a census block with a low, medium, or high population class occurs in a dataset
Line	To display successive values that have been collected continuously over a specific time period	Stock prices monitored continuously over the last 12 months
Area	To emphasize area in successive values that have been collected continuously over a specific time period	Area burned throughout the time of a wildfire event
Scatterplot	To explore data and look for patterns and relationships between the values in two fields	Crime rates vs. average household income
Box plot	Primarily in scientific applications, to summarize the variability of data values, often in the form of the mean or standard deviation	Measurements of nitrate concentration at different stations along a river; measurements at every station repeated multiple times to capture the variability in nitrate concentrations over a year
Polar	Primarily in mathematical and statistical applications, to simultaneously show magnitude and direction of a phenomenon	Area of runoff flow in each flow direction on different aspects of a slope
Pie	To display relationships between proportions of a whole	Percentage of different ethnicities in a city population

Table 20.2 illustrations from ArcGIS 10.1 Help.

Creating a graph from a template

You can either create a graph "from scratch" by manually specifying the graph type, the fields, and all the parameters that define the appearance of the graph, or you can use a graph template as a starting point. Graph templates contain all the information necessary to create a graph except the actual data. For example, in a graph template, you can specify the graph title, how to configure the horizontal and vertical axes, and other parameters that define the appearance of the graph. For more information about graph templates, refer to the ArcGIS Help topic "Fundamentals of graph templates."

Exercise 20d: Create a graph from a template

To provide additional information for your South America population density map, you will create a graph of population densities in the different South American countries. You will use a graph template as a starting point.

1. **ArcGIS 10.1:** Open the …\DesktopAssociate\Chapter20\CreateGraph.mxd.

   ArcGIS 10.0: Open the …\DesktopAssociate\Chapter20\CreateGraph_100.mxd.

2. In the Population Density data frame, open the attribute table of the Countries layer and locate the POP_SQKM field.

3. Sort the values in the POP_SQKM field in ascending order.

Hint: Right-click the POP_SQKM field heading and choose Sort Ascending.

Two countries have a population density of –99999, which indicate that there is no data value for these countries.

4. Close the table, then locate and open the Make Graph (Data Management) geoprocessing tool.

5. Run the tool with the following parameters:

   - Input graph template or graph: …\DesktopAssociate\Chapter20\PopulationDensity.tee
   - Dataset: Countries
   - X: POP_SQKM
   - Label (optional): CNTRY_NAME
   - Output graph name: PopulationDensity_SA

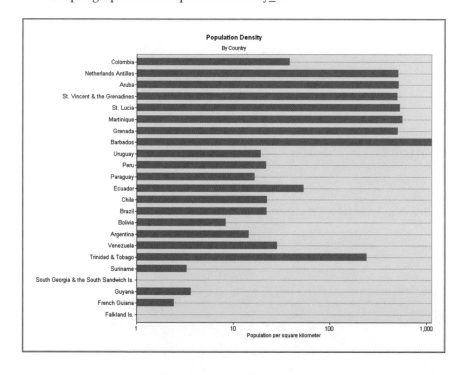

6. The graph created by the Make Graph tool should look like the one in the graphic below.

Note: You might have to expand the graph window to see all countries labeled on the vertical axis.

Notice that population densities for two countries are not shown in the graph. This is because they have negative POP_SQKM values, and the graph template starts the horizontal axis with a value of 1. You will exclude these countries from your graph.

7. Open the Select By Attributes dialog box and select the features from Countries that have a POP_SQKM value greater than 0.

8. Right-click the Population_SA graph and choose Properties. In the Appearance tab, check the option to Show only selected features/records on the graph. Click OK and clear the selected features, if necessary.

9. Right-click the Population_SA graph and choose Add to Layout. Close the graph window.

10. Decrease the size of the graph and arrange the layout elements to make room for the graph.

Hint: You may want to decrease the size of the legend, move the legend up, and move the graph in the lower right corner of the map.

11. Save the map document and close ArcMap.

Challenge

For each of the following situations, choose the most appropriate graph type.

1. Percentage of avian flu cases in different countries (all cases in all countries = 100%)

   a. Box Plot
   b. Bar Graph
   c. Polar
   d. Pie

2. Average wind speed in different directions measured from a weather station

   a. Box Plot
   b. Bar Graph
   c. Polar
   d. Pie

3. GDP for 2000 and 2010 in different countries

   a. Box Plot
   b. Bar Graph
   c. Polar
   d. Pie

Answers to challenge questions

Correct answers shown in bold.

1. Percentage of avian flu cases in different countries (all cases in all countries = 100%)

   a. Box Plot
   b. Bar Graph
   c. Polar
   d. Pie

2. Average wind speed in different directions measured from a weather station

   a. Box Plot
   b. Bar Graph
   c. Polar
   d. Pie

3. GDP for 2000 and 2010 in different countries

   a. Box Plot
   b. Bar Graph
   c. Polar
   d. Pie

Key terms

Layout: The arrangement of elements on a map, possibly including a title, legend, north arrow, scale bar, and geographic data.

Map element: In digital cartography, a distinctly identifiable graphic or object in the map or page layout. For example, a map element can be a title, scale bar, legend, or other map-surround element. The map area itself can be considered a map element; or an object within the map can be referred to as a map element, such as a roads layer or a school symbol.

Resources

- ArcGIS Help 10.1 > Desktop > Mapping > Working with ArcMap > Navigating maps
  - Displaying maps in data view and layout view
- ArcGIS Help 10.1 > Desktop > Mapping > Page layouts
  - A quick tour of page layouts
  - Map elements
  - Working with data frames in page layouts
  - Working with legends
- ArcGIS Help 10.1 > Desktop > Mapping > Graphs
  - A quick tour of graphs
  - Creating graphs
    - The steps to create a graph
    - Types of graphs
    - Creating a graph using geoprocessing tools
- ArcGIS Help 10.1 > Desktop > Mapping > Graphs > Creating graphs > Using graph templates
  - Fundamentals of graph templates
  - Creating a graph template

Creating Data Driven Pages

The index layer   310

Enabling Data Driven Pages   311

   Exercise 21a: Create an index layer and enable Data Driven Pages

   When to use Data Driven Pages

   Challenge

Refining Data Driven Pages   314

   Page definition queries

   Exercise 21b: Create a page definition query

   Dynamic page elements

   Exercise 21c: Create a dynamic legend (ArcGIS 10.1 only) and insert dynamic text

Answers to challenge questions   317

Key terms   317

Resources   318

Chapter 21: Creating Data Driven Pages

Suppose you wanted to make a map of all culturally significant sites in the state of Alaska, and show details such as the outline, trails and structures for each site. You probably think: "It's too much to fit on a single map!" And you are right! What you need in these situations is a series of maps, each showing a culturally significant site or a parcel with a sufficient amount of detail. Data Driven Pages provide you with this functionality.

When you enable Data Driven Pages in the layout, you add the ability to iterate through a series of extents in the main or detail data frame of the map, while the rest of the layout remains the same. The extents in the main data frame are defined by features in an index layer. As you iterate through the different map extents, you will have the impression of stepping through series of map pages. To physically create a series of map pages of the different extents, you can print the map layout with Data Driven Pages enabled, or you can export it to various digital formats. You can export either each page to an individual file, or export the entire map series to a single PDF file.

Skills measured

- Explain the circumstances under which it is appropriate to use Data Driven Pages.
- Describe the components used to create a data-driven page (e.g., necessary components, index layer, dynamic fields, etc.).

Prerequisites

Hands-on experience with map layouts

The index layer

The index layer defines the geographic extent of the main or detail data frame for each page and the total area to be covered by all the pages (figure 21.1). You can either choose an existing layer as an index layer or you can create a new index layer. Any point, line, or polygon layer from the main data frame can serve as an index layer. However, the attribute table of the index layer must contain a field that can be used for naming each page (Name Field) and a field that can be used to determine the order of the pages in the map series (Sort Field).

You use different types of index layers to create different kinds of map series, for example:

- A regular grid of polygons (squares or rectangles) will define pages covering adjacent areas of the same size, for example, 1-by-1 mile areas.
- Irregular polygons will define pages covering different areas, for example, countries, states, or study areas.
- Rectangles that follow a linear feature will define pages in a strip map along a linear feature, for example, a river or a pipeline.

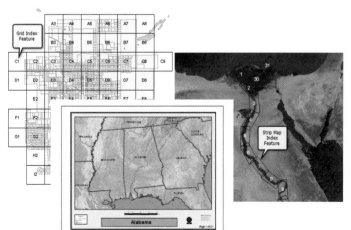

Figure 21.1 Common types of index layers include regular polygon grids, rectangular polygons following a linear feature, or irregular polygons, for example, states. Map by Esri; data from Esri Data & Maps 2010, data courtesy of Esri, derived from TomTom.

Enabling Data Driven Pages

When you enable Data Driven Pages functionality in a map document, you specify which layer to use for the index layer and other parameters that define, for example, the title, the order, and the spatial extent of the maps in the Data Driven Pages series. Based on these parameters, ArcMap will build the Data Driven Pages series, and the navigation tools on the Data Driven Pages toolbar become active. The parameters are customizable and can be modified at any time, as shown in tables 21.1 and 21.2.

Optionally, you can select a method to customize the extent of the detail data frame. By default, ArcMap will apply the Best Fit option (table 21.3).

Table 21.1 Required parameters

Parameter	Description
Data Frame	The data frame that will be used for the detail data frame
Layer	The layer that will be used for the index layer
Name Field	The field in the index layer attribute table that will be used for the page name
Sort Field	The field in the index layer attribute table that defines how Data Driven Pages will be indexed and sorted

Table 21.2 Optional parameters

Parameter	Description
Rotation	The field in the index layer's attribute table that can be used to define the rotation angle for each page—if you decide to apply a map rotation to the detail data frame.
Spatial Reference	The field in the index layer's attribute table that can be used to derive the spatial reference from—if you decide to apply a spatial reference other than the one of the detail data frame to the pages.
Page Number	The field in the index layer's attribute table that can be used to derive the page number.
Starting Page Number	The field in the index layer's attribute table that can be used to start the map series at a page other than the first page.

Table 21.3 Extent parameters (optional)

Parameter	Description
Best Fit (default)	This option will derive the extent of the detail data frame from the index features and applies a margin around it. If the index features are different in size, every page will have a different scale. You are able to modify the margin parameter and specify it either in percent, map, or page units. You may also use margin parameter to create overlap between the pages in the Data Driven Pages series.
Center and maintain current scale	This option will center the detail data frame on the center of the index feature and maintain the scale that is currently set in the detail data frame.
Data Driven Scale	This option will derive the scale of the detail data frame from a field in an index layer's attribute table. For each index feature (i.e., each page) there may be a different scale.

For more information on these parameters, refer to the ArcGIS Desktop Help topic "Creating Data Driven Pages."

Exercise 21a: Create an index layer and enable Data Driven Pages

You will create an index layer and enable Data Driven Pages for a population density map of South America.

1. **ArcGIS 10.1:** Start ArcMap and open the …\DesktopAssociate\Chapter21\CreateDDP.mxd.

   ArcGIS 10.0: Start ArcMap and open the …\DesktopAssociate\Chapter21\CreateDDP_100.mxd.

The map opens in layout view.

2. Open the Grid Index Features geoprocessing tool.

Hint: Use the Search window to find the tool.

3. Run the tool with the following parameters:
   - Output Features Class: …DesktopAssociate\Chapter21\ South_America.gdb\GridIndexFeatures
   - Input Features: SA_Countries
   - Check: Generate PolygonGrid that intersects input features layers or data
   - Check: Use Page Unit and Scale
   - Map scale: 10,000,000
   - Polygon Width: 5 inches
   - Polygon Height: 5 inches

The GridIndexFeatures layer is added to the map.

4. Open the GridIndexFeatures layer attribute table.

The Grid Index Features tool automatically generated the PageName and PageNumber fields that will be used to show the name and number of each data driven page.

5. Close the table, and change the GridIndexFeatures layer symbology to Hollow.

Each grid polygon frames the extent for the detail map that will be displayed in each data driven page.

6. Display the Data Driven Pages toolbar, if necessary. On the Data Driven Pages toolbar, click the Data Driven Pages Setup button and Enable Data Driven Pages.

Hint: To display the Data Driven Pages toolbar, click the Data Driven Pages button on the Layout Tools toolbar.

7. **In the Definition tab, set the following parameters:**
   - Data Frame: Population Density
   - Layer: GridIndexFeatures
   - Name Field: PageName
   - Sort Field: PageNumber

On the right, notice the Optional Fields. You could specify a different starting page number or other optional parameter, but for now, accept the defaults.

8. **Click the Extent tab.**

Since the index features are adjacent to each other, a margin of 125% means that each map in the detail data frame will overlap the adjoining ones by 25% of its extent.

9. **Accept the defaults and click OK.**

When you enabled Data Driven Pages, the map data frame was centered on the first index feature. Since you have set a margin of 125%, the map extent overlaps with the surrounding index features.

10. **In the Data Driven Pages toolbar, click Next Page button .**

Notice that the page name shown on the Data Driven Pages toolbar changed when you navigated to the next page. You could type in a page name here (e.g., C3) and press enter to navigate to that particular page.

To see the difference in the page extents, you will change the index layer to SA_Countries.

11. **Open the Data Driven Pages Setup dialog again and set these parameters in the Definition tab:**
    - Layer: SA_Countries
    - Name Field: CNTRY_NAME
    - Sort Field: CNTRY_NAME
    - Check: Sort Ascending

With the SA_Countries layer set as the index layer, the detail data frame is now centered on Argentina, the first page in the series.

12. **Step through some more pages .**

In the Data Driven Pages toolbar, the page name changes to a different country. This is because you chose the CNTRY_NAME as the Name Field. Also, the map scale changes from one page to the next because each country has a different extent. Remember, the Data Driven Pages series uses the default Best Fit extent option.

13. **Highlight the country name and type Peru, and then press Enter.**

Peru is now centered in the detail frame.

14. **Save the map document. If you are continuing with the next exercise, leave ArcMap open. Otherwise, close ArcMap.**

When to use Data Driven Pages

Data Driven Pages are well-suited for the following situations:

- You want to make a series of maps showing the same features or phenomena in different geographic areas.
- You want to apply the same layout to the map series.

Chapter 21: Creating Data Driven Pages

- You want to divide a large mapping area into sections and show the sections with more detail in a series of maps.
  - Entire map area may be too large to show with sufficient amount of detail (i.e., at an appropriate scale) on a single map.
  - Sections may be non-adjacent geographic areas.

Do not use Data Driven Pages in the following situations:

- You want to make a series of maps showing different features or phenomena in the same geographic area.
- You are able to show the entire mapping area with sufficient amount of detail.

Challenge

For each of the following situations, determine whether or not it is appropriate to use Data Driven Pages.

1. You need to create maps for a report showing the distribution of six species within a national park. Each map must show how a single species is distributed throughout the entire park.

   a. Use Data Driven Pages
   b. Do not use Data Driven Pages

2. Create a map series covering a river system where each map displays a different section of the river.

   a. Use Data Driven Pages
   b. Do not use Data Driven Pages

3. A landscape architect asks you for a map layout showing two different city parks side by side at the same map scale, including the boundary, trails, vegetation, facilities, and topography.

   a. Use Data Driven Pages
   b. Do not use Data Driven Pages

4. You want to make a series of maps of a university campus and in each map you want to display a separate building.

   a. Use Data Driven Pages
   b. Do not use Data Driven Pages

Refining Data Driven Pages

You can use various page effects to refine Data Driven Pages. For example, you can insert additional dynamic elements such as a dynamic page title or a dynamic extent indicator. Also, you can emphasize the map content inside the area defined by the index feature by a page effect called a page definition query.

Page definition queries

A page definition query is a special type of definition query that updates as you navigate through the Data Driven Pages series. It allows you to specify an SQL query to define which features of a layer draw on a page.

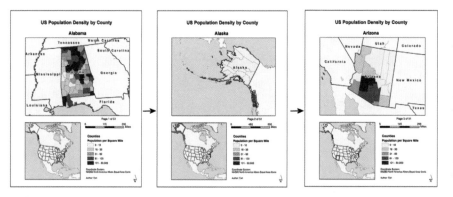

Figure 21.2 In this example, a page definition query has been applied to the Counties layer. States are used as an index layer for the Data Driven Pages series. With the page definition query enabled, only the counties within the current state are displayed on the page. Created by the author using data from Esri Data & Maps, 2010, courtesy of ArcUSA, U.S. Census, Esri (Pop2010 fields).

Exercise 21b: Create a page definition query

You will now create a definition query for your map series of population density in South American countries so that on each page you only see only population density of the administrative areas for the country shown in the detail map.

1. **ArcGIS 10.1:** Open the ...\DesktopAssociate\Chapter21\PageDefinitionQuery.mxd.

   ArcGIS 10.0: Open the ...\DesktopAssociate\Chapter21\PageDefinitionQuery_100.mxd.

2. Open the Administrative Areas properties.

3. In the Definition Query tab, click Page Definition, and then check Enable to enable the page definition query.

4. Accept the other defaults and then click OK.

5. Close the layer properties.

6. Step through a few pages ▶.

With the page definition query set, the map in the detail data frame is greatly simplified and the information for each country is displayed much more effectively.

Note: To remove a page definition query, open the layer properties. In the Definition Query tab, click Page Definition, and then uncheck Enable.

7. Return to the first page.

8. Save the map document. If you are continuing with the next exercise, leave ArcMap open. Otherwise, close ArcMap.

Dynamic page elements

Data Driven Pages integrate well with dynamic page elements that update as you step through the map series. To set up this dynamic behavior in map elements such as the locator map, the legend, scale bar, and north arrow dynamic to the layout, you must associate them with the detail data frame.

- Locator map: Dynamic locator maps provide an overview of where a page is located in the index layer by using a dynamic extent indicator: as you step through the map series, the extent indicator highlights outline the current index feature.

- Dynamic text: In a Data Driven Page series, dynamic text allows for text elements in the page layout to update as you go from one page to another. Dynamic text may, for example, be used for page titles, page numbers, and labels of the neighboring pages.

- Legend: In a Data Driven Page series, the legend of the detail data frame can list only the legend items and categories that are in the current page. To enable this functionality in the legend properties, you set up the map extent option to only show classes that are visible in the current map extent (Items tab).

- Scale bar: In a Data Driven Page series, the scale bar will reflect changes in the map scale of in the detail data frame.

- North arrow: If you apply a rotation to the detail data frame in specific or all pages in the map series, the north arrow is rotated with it. Rotation is often applied to make pages in a strip map more readable.

Exercise 21c: Create a dynamic legend (Arcgis 10.1 only) and insert dynamic text

To refine your Data Driven Pages series, you will set up the legend to update dynamically as you step through the pages and insert some dynamic text elements.

Note: Dynamic legends are a new feature in ArcGIS 10.1.

1. **ArcGIS 10.1:** Open the ...\DesktopAssociate\Chapter21\RefineDDP.mxd.

   ArcGIS 10.0: Open the ...\DesktopAssociate\Chapter21\RefineDDP_100.mxd.

2. **ArcGIS 10.1 only: Open the Legend Properties and click the Items tab.**

If you are working in ArcGIS 10.0, skip to step 5.

3. **Under Map Extent Options, check the "Only show classes that are visible in the current map extent" option. Then click OK to close the Legend Properties window.**

4. **Step through a few more pages.**

The legend only shows the population classes that are currently displayed in the detail data frame. Next you will insert dynamic text to show the page name and number.

5. **On the Data Driven Pages toolbar, click Page Text and choose Data Driven Page Name.**

The country name has been added to the current data driven page, but the text is quite small.

6. **Using the Select Elements tool, right-click the country name and choose Properties. Click Change Symbol and format this text as Times New Roman, 30-point type. Click OK to close all open dialog boxes.**

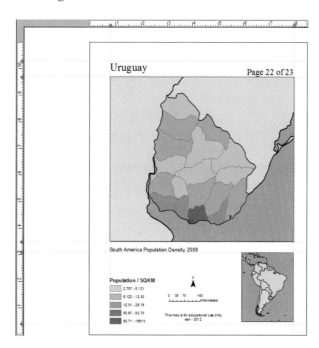

7. **Move to the country name text to the guides in the upper left corner of the map, as shown in the graphic above.**

Next you will insert the dynamic page number.

8. **From the Page Text menu on the Data Driven Pages toolbar, insert the Data Driven Page with Count dynamic text element.**

Note: You can also add dynamic text elements from the Insert menu.

9. **Format this text as Times New Roman, 24 points.**

10. **Move to the text to the guides in the upper right corner of the map, as shown in the graphic above.**

11. Step through a more few pages and observe the elements updating dynamically.

12. Save the map document. Close ArcMap.

Answers to challenge questions

Correct answers shown in bold.

1. You need to create maps for a report showing the distribution of six species within a national park. Each map must show how a single species is distributed throughout the entire park.

   a. Use Data Driven Pages
   b. Do not use Data Driven Pages

Explanation: You don't need to create Data Driven Pages because all maps will have the same extent. You should create separate map documents for each species (6 maps) and then export them as images that will be included in the report.

2. Create a map series covering a river system where each map displays a different section of the river.

   a. Use Data Driven Pages
   b. Do not use Data Driven Pages

Explanation: Use Data Driven Pages to create a strip map where each section of river is displayed on a separate page.

3. A landscape architect asks you for a map layout showing two different city parks side by side at the same map scale, including the boundary, trails, vegetation, facilities, and topography.

   a. Use Data Driven Pages
   b. Do not use Data Driven Pages

Explanation: You don't need to create Data Driven Pages because each park is displayed in a different data frame on the same layout page.

4. You want to make a series of maps of a university campus and in each map you want to display a separate building.

   a. Use Data Driven Pages
   b. Do not use Data Driven Pages

Explanation: Use Data Driven Pages so that each building is displayed on a separate page.

Key terms

Map series: A collection of maps usually addressing a particular theme.

Resources

- ArcGIS Help 10.1 > Desktop > Mapping > Page layouts > Data Driven Pages
  - What are Data Driven Pages?
  - Creating Data Driven Pages
  - Exporting Data Driven Pages
  - Navigating Data Driven Pages
  - Using dynamic text with Data Driven Pages
  - Using page definition queries
- ArcGIS Help 10.1 > Desktop > Mapping > Page layouts
  - Working with legends

chapter 22

Creating map text and symbols

Map text 320
   Challenge 1
   Label classes
   Exercise22a: Create label classes
   Label expressions
   Exercise 22b: Create label expressions
   Comparing label engines
   Challenge 2
   Converting labels to annotation
   Exercise 22c: Use annotation to refine the map

Custom symbols 329
   Symbol layers
   Locking symbol layers
   Storing custom symbols
   Exercise 22d: Create custom marker symbols

Answers to chapter 22 questions 332

Answers to challenge questions 332

Key terms 334

Resources 334

When you create a map you spend a lot of time symbolizing layers to convey the message of your map. But something may still be missing—map text! Map text helps to convey or clarify the message of your map. How can you ensure that your use of map text supports the map's purpose? Think about what message would a population map of Africa convey if the cities and countries were not labeled with their names?

When you symbolize your Africa map, you may or may not find all the symbols you need in the symbol styles provided by ArcGIS. If the perfect symbol is not available, you may want to create your own symbol. You can modify the appearance of any symbol that comes with ArcGIS, or you can create custom symbols from any available picture or font character, or by combining existing symbols.

Skills measured

- Determine appropriate specifications for adding text to a map.
- Given a task to create a layer symbol, determine the geometry type of the feature, where to create the symbol, and what should be specified in each layer of the symbol.

Prerequisites

Hands-on experience with map text, including
- Labels
- Placement rules

Knowledge of geodatabase and map annotation
Knowledge of ArcGIS symbols and styles, including
- Types of available symbols
- Elements contained in a style

Map text

You use map text to support your map's message, and perhaps to draw attention to particular features or areas in the map. For example, on a map of Africa, you may want to label countries, cities, rivers, and lakes with their names, and emphasize the general location of the Sahara desert with map text across that region. In a map layout, you use map text to title the map, provide information about the author, data source, map projection, and date the map was created. ArcGIS provides four types of map text: labels, geodatabase annotation, map annotation, and graphic text (figure 22.1):

- Labels are descriptive map text based on one or more attributes and dynamically placed on or near map features, based on predefined placement rules.
- Annotation is map text that can be individually positioned and modified. Annotation can be created from labels and can be stored as a geodatabase feature class or as map annotation (in the map document). There are two types of annotation: map annotation that is stored in the map document (.mxd) and geodatabase annotation that is stored as a feature class in a geodatabase.
- Graphic text is map text that appears only in layout view. It can by either static text such as the map title, or dynamic text such as the name of a page in a series of Data Driven Pages.

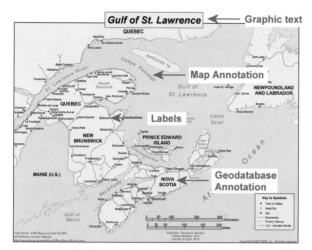

Figure 22.1 It would be possible to create the map above with four kinds of map text: graphic text is used for the map title, labels can be used for the city names, geodatabase annotation for the province names, and map annotation for the regions in the ocean. Map by Esri; data from Esri Data & Maps, 2001, data courtesy of ArcWorld Supplement and DMTI Spatial Inc.

Note: The map in figure 22.1 has been created to demonstrate the three types of map text. Cartographically it is not best practice to mix labels and annotation in the same map. Table 22.1 provides more details about types of map text.

Table 22.1 Types of map text

Type of map text	Generated from	Positioning	Storage options	When to use
Labels	One or more fields in a feature attribute table	Dynamically placed based on placement rules	Not stored, but generated on the fly • Labeling properties are stored in the map document, .lyr file, or layer package • Label text is dynamically generated from field(s) in the attribute table	• For automatic conflict detection and resolution • When attribute values in label fields change frequently • For adding text to a map to support editing workflows
Geodatabase annotation	Labels or user input	Text positions stored; position and appearance of individual text pieces can be edited	Stored as a feature class in a geodatabase • Stand-alone: no connection between features and their annotation • Feature linked: when features are moved or updated, annotation moves or updates as well	• For cartographic control over map text; to edit and move individual pieces of text • When control over text appearance and placement is desired • For re-using map text in different map documents

Continued on next page

322 Chapter 22: Creating map text and symbols

Table 22.1 Types of map text (continued)				
Type of map text	Generated from	Positioning	Storage options	When to use
Map annotation	Labels or user input	Text positions stored; position and appearance of individual text pieces can be edited	Map document or map package	• For adding a few pieces of map text • For adding map text without corresponding vector features, e.g., name of an ocean without an ocean polygon
Graphic text	User input (static text) or other sources (dynamic text, e.g., current date and time from the computer's clock.)	Text positions stored in the map layout	Map document or map package	For adding text to a layout

Challenge 1

For each of the scenarios below, choose the most appropriate type of map text.

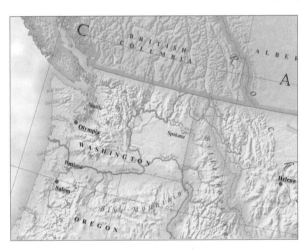

From Esri Data & Maps 2010; data courtesy of Esri, TomTom, ArcWorld Supplement, ArcWorld, DeLorme, NAVTEQ, Intermap, AND, USGS, NRCAN, Kadaster NL, and the GIS User Community.

1. Refer to the graphic above. Suppose you are creating a map of the Pacific Northwest United States with vector layers of states, provinces (Canadian), cities, rivers, and a raster shaded relief layer for the background. Which type of map text is most appropriate for indicating names of mountain ranges shown in the raster layer? (Choose two.)

   a. Labels
   b. Graphic Text
   c. Map annotation
   d. Geodatabase annotation

2. Suppose you are creating a city map with layers of streets and building footprints. You would like to emphasize public buildings, such as museums, historic buildings, large hotels, popular restaurants, and so on. The buildings' attribute table does not contain this information, but your coworker created some map text for the buildings and stored it in a server location that you can access. Which type of map text allows you to most efficiently re-use your coworker's map text for the buildings (without conversion)?

   a. Labels
   b. Graphic text
   c. Map annotation
   d. Geodatabase annotation

3. Suppose you are making an interactive map that you share within your organization that will be used at many different scales. Which type of map text is best to use for this kind of map?

   a. Labels
   b. Graphic text
   c. Map annotation
   d. Geodatabase annotation

Label classes

Label classes allow you to specify labeling properties for groups of features within a layer. By default, every layer has one label class (called Default), which defines the label properties of the layer. If you create additional label classes in a layer, you can use different font sizes and colors, visible scale ranges, or placement rules for each label class. To define which features belong to each label class, you typically use an SQL query. For example, for a cities layer, you can label cities based on their population attribute, using different font sizes. Cities with a large population would be labeled with a larger font than cities with a smaller population. You can also derive label classes from existing symbology classes in a layer: for each symbol class, ArcMap can generate a label class.

If you decide to convert your labels to annotation, label classes will be automatically converted into annotation subtypes. Another reason to use label classes is to control which features in a layer are labeled and potentially reduce label conflicts in your map. For example, you can create several label classes for a buildings layer and define different visible scale ranges for each label class. At full extent, only the most important buildings in the layer will be labeled; as you zoom in, more and more building labels become visible.

Exercise 22a: Create label classes

You will make a map of Peru showing administrative areas and cities symbolized and labeled. To be able to better distinguish between Lima, the national capital, and all other cities, you will apply label classes to the cities.

1. **ArcGIS 10.1: Start ArcMap and open the ...\DesktopAssociate\Chapter22\ LabelClasses.mxd.**
   ArcGIS 10.0: Start ArcMap and open the ...\DesktopAssociate\Chapter22\ LabelClasses_100.mxd.

2. **Zoom to the Peru 1:12 M bookmark and turn on the Labeling toolbar.**

You will work in the Label Manager window to set up the label classes.

3. **On the Labeling toolbar, click the Label Manager button.**

4. **In the list of Label Classes on the left, click Peru Cities.**

You could create the label classes manually, but in this case, you will create them from the existing symbology categories.

Note: For more information on creating label classes manually, refer to the ArcGIS Help topic: "Using label classes to label features from the same layer differently."

   5. **In the Add label classes from symbology categories area, make sure both categories are checked, then click Add.**

You will replace the Default label class with new ones you will create.

   6. **Click Yes in the warning message to overwrite label classes.**

You will enter the parameters for each of the new label classes.

   7. **Click the National capital label class and click Symbol. Then click Edit Symbol.**

   8. **Enter the parameters shown in the table below.**

General tab	Font: Arial, Size: 10
	Style: Bold
	Color: Tuscan Red
	Y Offset: 2
Formatted Text tab	Character spacing: 30
Mask tab	Style: Halo
	Size: 1
	Click Symbol: make sure the Fill color is white and Outline Width is 0

   9. **Click OK three times to return to the Label Manager window.**

   10. **Apply the technique you just used and the information in the table below to specify the parameters for the Other city label class.**

General tab	Font: Arial, Size: 8
	Style: Bold
	Color: Black
	Horizontal Alignment: Center

Because you created the label classes based on existing symbology categories, you do NOT need to create an SQL query to select the features to be labeled in each label class.

   11. **Select each label class of Peru Cities, click SQL Query, and confirm that the SQL Queries have been automatically created. Click OK to return to the Label Manager.**

   12. **Click OK to close the Label Manager.**

Lima is now labeled with the parameters you specified in the National capital label class, and all other cities are labeled with the ones for Other city. The map is much improved now, and you can easily distinguish between the national capital and other cities.

   13. **If you are continuing with the next exercise, leave ArcMap open.**

Label expressions

Label expressions are statements in VBScript, Python, or Java script that determine the label text. Each label class can have its own label expression. You typically use label expressions to create labels based on more than one attribute field, insert your own text in the label, or format portions of your label text differently. For example, you may create a label expression for labeling weather stations based on the average annual amount of rainfall and average wind speed attributes (stored in separate fields). The label expression could include a function to make the average rainfall values appear on one line of text, and wind speed values appear on a second line. You can also use ArcGIS text formatting tags to show the average rainfall values in blue, italicized text, and the average wind speed values in regular, black text (as shown in figure 22.2).

Figure 22.2 Label expressions allow you to concatenate label fields, format the labels, and add your own text. They can be built as VBScript, Python, or Java script statements. Esri.

Question 1: Refer to the following code. Which of the following VB Script label expressions will create the labels shown in the graphic above?

Hint: To look up the syntax of label expressions refer to the ArcGIS Help topics "Building Label Expressions" and "Using text formatting tags."

a. "<CLR blue='255'>" & "<BOL>" & [AVR_RAIN] & " " & "in" & "</BOL>" & "</CLR>" & vbnewline & [AVR_WIND] & " " & "MPH"
b. "<CLR blue='255'>" & "<ITA>" & [AVR_RAIN] & " " & "in" & "</ITA>" & "</CLR>" & vbnewline & [AVR_WIND] & " " & "MPH"
c. "<CLR blue='255'>" & "<ITA>" & [AVR_RAIN] & "</ITA>" & "</CLR>" & vbnewline & [AVR_WIND]
d. "<CLR blue='255'>" & [AVR_RAIN] & " " & "in" & "</CLR>" & vbnewline & [AVR_WIND] & " " & "MPH"

Exercise 22b: Create label expressions

You will refine the map of Peru by adding some label expressions to the label classes. For exercise purposes, you will use the Python and VBScript parsers.

1. **ArcGIS 10.1:** If necessary, start ArcMap. Open the …\DesktopAssociate\Chapter22\ LabelExpressions.mxd.
   ArcGIS 10.0: If necessary, start ArcMap. Open the …\DesktopAssociate\Chapter22\ LabelExpressions_100.mxd.

2. If necessary, turn on the Labeling toolbar. Zoom to the Peru 1:12 M bookmark.

Below the label for Lima, the capital city, you will add another label showing the city's population that is stored in the attribute table's POP field.

3. **ArcGIS 10.1:** In the Label Manager, select the National capital label class and click Expression.

   - For the Parser, choose Python
   - In the Expression area, type [CITY_NAME] + '\n' + [POP]

Similar to the example in the question above, this Python label expression labels the capital city with its name and population (in a new line).

   ArcGIS 10.0: In the Label Manager, select the National capital label class and click Expression.

- Make sure that the Parser is set to VBScript
- In the Expression area, type [CITY_NAME] + vbnewline + [POP]

This VB script label expression labels the capital city with its name and population (in a new line).

4. Click Verify to preview a sample label. Then click OK to dismiss the verification window.

You can make the result look even better by reducing the font size of the population label. You can do this using the Python parser, but instead, you will load a saved expression in a VBScript.

5. Change the parser to VBScript, if necessary, and click Load.

6. Browse to your …\DesktopAssociate\Chapter22 folder, and open CapitalCity.lxp.

7. Click OK to close the Label Expression window and click Apply in the Label Manager.

Lima is now labeled with its name and population.

To relieve congestion in the map, you will replace the labels for the administrative areas with a three-letter abbreviation, using another saved expression.

8. Open the PeruAdminAreas attribute table and examine the values in the ISO_CODE field.

The first two characters in the ISO_CODE values (PE) stand for Peru. The last three characters represent the abbreviations of the administrative areas. The VB Script label expression you will load extracts the last three characters from the ISO_CODE field and uses them to label the administrative areas.

9. Close the attribute table.

10. In the Label Manager, select the PeruAdminAreas layer's Default label class.

11. Load the PeruAdminAreas.lxp expression.

To extract the last three characters from the ISO_CODE field, the label expression uses the VBScript Mid function.

12. Click OK to close the Label Expression window and the Label Manager.

Administrative areas are now labeled with their three-letter abbreviation.

13. Save the map document. If you are continuing with the next exercise, leave ArcMap open. Otherwise, close ArcMap.

Comparing label engines

Note: In ArcGIS 10.0 Maplex is implemented as an extension, not as a label engine. In ArcGIS 10.0, the Maplex extension is included in the ArcInfo license of ArcGIS Desktop.

Labels are dynamically placed by a label engine. That means every time you zoom and pan to a different location in the map, the label engine generates new text locations based on the placement rules set for the labels. The standard label engine provides basic placement rules for points, lines and polygons. You use the standard label engine for simple maps with only a few layers, when you do not need cartographic placement quality in your labels.

Besides the default Standard label engine, ArcGIS provides a special set of tools that allows finer control of how labels are placed: the Maplex label engine. In addition to everything you can do with the Standard label engine, the Maplex label engine provides many additional strategies for label positioning. For example, the Maplex label engine provides these specific strategies:

- Street labels are spread along the length of the street, avoiding street intersections.
- Contour labels are placed in a laddered stack, aligned uphill or to the orientation of the page.
- River labels are placed smoothly along the bends of a river.
- Parcel labels are placed in the center of the parcel, avoiding building footprints.

- Labels of neighboring states or countries are placed together on opposite sides of the border, repeating labels at a specified distance.

In addition Maplex offers strategies for fitting labels in very dense areas and resolving conflicts between labels. For example, you can use abbreviation dictionaries, automatic font reduction, and key numbering (e.g., in dense areas, labels that don't fit are replaced with a number and placed in a list).

For more information on how to label features using the Maplex label engine, refer to the ArcGIS Help tutorial "Introduction to the Maplex Label Engine."

Challenge 2

Identify which label engine is best to use for each of the following scenarios:

1. Suppose you are labeling a map of parcels and building footprints, and the parcels range in size. You don't want the map text for parcels to overlap the buildings. You will need to manage map text in dense areas with many small parcels.

    a. Standard
    b. Maplex

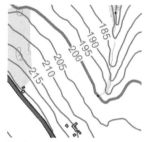
Esri.

2. Suppose you need to label a map of contour lines and you want to achieve the label effect shown in the graphic above.

    a. Standard
    b. Maplex

3. Suppose you have a map with many label classes and label expressions that you have used for a number of years (without changing the labels). When you initially created the map, you did not have any extensions available. Now you want to add a new layer to the map, and you want to label that layer.

    a. Standard
    b. Maplex

Converting labels to annotation

Once you have placed labels in your map and defined all their properties, you have the option to convert the labels to annotation. Annotation allows you to edit the positions of map text and gives you control over all aspects of their appearance. Typically, you will refine label placement and label appearance as much as possible, using label properties in either the Standard or Maplex label engine, before converting labels to annotation.

Exercise 22c: Use annotation to refine the map

You will now further improve the map of Peru by repositioning and modifying some of the map text. To have control over individual labels for city names, you will convert the city name labels to annotation.

1. **ArcGIS 10.1:** If necessary, start ArcMap. Open the ...\DesktopAssociate\Chapter22\ AnnoFeatures.mxd.

    ArcGIS 10.0: If necessary, start ArcMap. Open the ...\DesktopAssociate\Chapter22\ AnnoFeatures_100.mxd.

2. Zoom to the Peru 1:12 M bookmark.

The map should be zoomed to a scale of approximately 1:12,000,000. Before you convert the labels to annotation, you will set a reference scale for the data frame.

3. Open the data frame properties. In the General tab, set the Reference Scale to 1: 12,000,000.

4. Zoom in and out and observe how the labels scale relative to the reference scale. Then zoom back to the Peru 1:12 M bookmark.

5. Right-click the Peru Cities layer and choose Convert Labels to Annotation. Convert the labels using these parameters:

- Store the annotation in a database

Notice that the reference scale for the annotation is derived from the reference scale of the data frame

- Create annotation for All features.
- Create feature linked annotation.
- Accept the default annotation feature class destination and name.
- Check the option to convert unplaced labels to unplaced annotation.

The PeruCitiesAnno layer is added to the table of contents. A new annotation feature class called PeruCitiesAnno is added to the SouthAmerica.gdb (you might have to refresh the geodatabase to see it). The labels of Peru Cities have been turned off.

6. **Zoom in and out and observe how the annotation scales proportionally with the features.**

This is because you have set the reference scale in the data frame properties.

7. Zoom back to the Peru 1:12 M bookmark.

8. Turn on the Editor toolbar and start an edit session.

Hint: From the Editor toolbar, click Start Editing.

9. Using the Edit tool, select any city feature and move it.

Since you created feature linked annotation, the corresponding annotation moves with the feature.

10. Delete the city feature.

Hint: To delete the city feature, right-click it and choose Delete.

When you deleted the feature, the annotation was deleted, as well.

11. Click Undo several times until the city feature that you moved and deleted is back at its original position.

12. Use the Edit Annotation Tool to select some annotation and reposition it to a cartographically better location, or delete it.

Note: For example, in the upper right you can move the annotation for Iquitos so it doesn't overlap the river. Or, to the right of Lima, delete the annotation for Huancavelica because it's too long and flows into adjoining administrative areas.

13. From the Editor toolbar, click Save Edits, then Stop Editing.

14. Save the map document. If you are continuing with the next exercise, leave ArcMap open. Otherwise, close ArcMap.

Custom symbols

In addition to adding map text, you may want to refine your map by creating custom symbols. You can create a custom symbol when you want to use a special symbol that is not available in the styles provided by ArcGIS.

Symbol layers

Custom symbols consist of one or more symbol layers, for which you can specify different properties. To create a new custom symbol, you start with an existing one. For example, to create a new marker symbol, you start by modifying the size and color of an existing marker symbol, then add layers to it. You can create a custom symbol from existing marker symbols, font characters, or from pictures. You compose a custom symbol by modifying the properties of each symbol layer.

To create a custom symbol for polygon features you can specify different fill patterns for each layer. You can fill polygon symbol layers with marker symbols, line patterns, color gradients, or pictures. See figure 22.3 for an example of custom symbols.

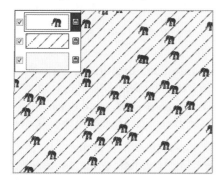

Figure 22.3 The polygon symbol in the illustration consists of three polygon layers: one filled with a solid yellow fill color (bottom layer), another one with a green hashed line pattern (middle layer), and a third layer filled with an elephant marker symbol randomly distributed (top layer). Created by the author.

Line symbols can consist of a combination of line and marker symbols. For example, to indicate a bicycle path, you could create a line symbol with two line layers and a marker symbol layer (figure 22.4).

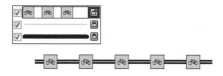

Figure 22.4 The line symbol in the example above consists of two line layers, a thicker black one (bottom layer) and a green one (middle layer), and a bicycle marker symbol (top layer). Created by the author.

Locking symbol layers

There are two places where you can change the color of a symbol in ArcMap: in the Symbol Selector that you open by clicking a symbol in the table of contents, or in the Symbol Property Editor that you open to create a custom symbol. When you create a custom symbol from different layers, you can allow the user to change the color of a symbol layer in the Symbol Selector, or you can lock the symbol layers. If all layers in a symbol are locked, the color control in the Symbol Selector is unavailable. The color can only be modified from the color control in the Symbol Property editor. Whether or not a layer is locked is indicated by an icon next to the layer in the Symbol Property Editor. If you unlock one or more symbol layers, the color control in the Symbol Selector becomes available.

Storing custom symbols

If you create a new custom symbol and symbolize a layer with it, the symbol is automatically stored in the map document. To make custom symbols available outside of the current map document, you store them in style files. You can create a new style file in any folder and store your custom symbols in it. You can share style files by storing them on a central server, e-mailing them, or burning them on a DVD. To have your custom symbols available any time you login to your computer, you can store custom symbols in your personal style file in your user profile. If you choose to share map layers as a layer file (.lyr), layer package (.lpk) or map package (.mpk), custom symbols will also be saved in these files.

Exercise 22d: Create custom marker symbols

To help plan for a trip to Peru, you will make a map that shows UNESCO World Cultural or Natural Heritage sites. To emphasize the ruins of Machu Picchu, you will create your own custom marker symbol.

1. **ArcGIS 10.1: If necessary, start ArcMap. Open the …\DesktopAssociate\Chapter22\ CustomMarkerSymbols.mxd.**
   ArcGIS 10.0: If necessary, start ArcMap. Open the …\DesktopAssociate\Chapter22\ CustomMarkerSymbols_100.mxd.

The map document contains a layer of UNESCO World Cultural and Natural Heritage sites. You are not satisfied with the predefined marker symbol in any of the symbol styles that come with ArcGIS. Instead you will use a set of custom symbols that a coworker made for you.

2. **From the Customize menu, open the Style Manager. Click Styles.**

3. **ArcGIS 10.1: Click Add Style to List. From the …\DesktopAssociate\Chapter22 folder, open Peru.style and click OK to return to the Style Manager. Expand Peru.style and click Marker symbols.**
   ArcGIS 10.0: Click Add Style to List. From the …\DesktopAssociate\Chapter22 folder, open Peru_100.style and click OK to return to the Style Manager. Expand Peru_100.style and click Marker symbols.

The style contains marker symbols for the UNESCO heritage sites and other sites of historical or natural interest. You will use these to symbolize the layers for UNESCO World Cultural and Natural Heritage sites. But first, you will create a new marker symbol for Machu Picchu, which is both a UNESCO cultural and natural site.

4. **Right-click in the white space in the right panel of the Style Manager and choose New > Marker Symbol.**

5. **For the first layer of the new symbol, change the Type to Character Marker Symbol and set the following parameters:**

Font	ESRI Default Marker
Unicode	34
Size	16
Color	Flame Red

6. **Click the layer in the Layers list to lock it.**

The icon next to the layer now indicates that it is locked. Next, you will add another layer to the symbol.

7. **Click the Add Layer button.**

The new layer is locked by default.

8. Specify these parameters for the new layer:

Font	ESRI Conservation
Unicode	99
Size	12
Color	White

Click OK to close the Symbol Property Editor.

9. In the Style Manager, right-click the new symbol and rename it to **Machu Picchu**. Then close the Style Manager.

10. In the table of contents, click the symbol for the UNESCO World Cultural and Natural Heritage sites. Select the Machu Picchu symbol from Peru.style.

11. Notice that the symbol color cannot be changed. This is because you locked the layers of the Machu Picchu symbol. Click OK.

12. Next, symbolize these sites with the appropriate symbols from the Peru style: UNESCO World Cultural Heritage sites, UNESCO World Natural Heritage sites, and Other sites.

Looking at the map now, you decide to add a drop-shadow to the Machu Picchu symbol to make it stand out.

13. In the table of contents, click the Machu Picchu symbol to open the Symbol Editor. Click Edit Symbol.

14. Add another layer to the Machu Picchu symbol with these parameters:

Font	Character Marker Symbol, ESRI Default Marker
Unicode	34
Size	16
Color	Black
Offset	X: 1.00 and Y: −2.00

15. Click the down-arrow twice ⬇ to move the layer to the bottom. Then click OK to return to the Symbol Selector.

There is still more that could be done to complete the map. For example, you could label the sites of interest that you just symbolized, add Peru cities, and resolve any resulting label conflicts.

16. Save the map document. Close ArcMap.

Answers to chapter 22 questions

Question 1: Which of the following VB Script label expressions will create the labels shown in figure 22.2?

    a. "<CLR blue='255'>" & "<BOL>" & [AVR_RAIN] & " " & "in" & "</BOL>" & "</CLR>" & vbnewline & [AVR_WIND] & " " & "MPH"
    b. "<CLR blue='255'>" & "<ITA>" & [AVR_RAIN] & " " & "in" & "</ITA>" & "</CLR>" & vbnewline & [AVR_WIND] & " " & "MPH"
    c. "<CLR blue='255'>" & "<ITA>" & [AVR_RAIN] & "</ITA>" & "</CLR>" & vbnewline & [AVR_WIND]
    d. "<CLR blue='255'>" & [AVR_RAIN] & " " & "in" & "</CLR>" & vbnewline & [AVR_WIND] & " " & "MPH"

Answers to challenge questions

Challenge 1

For each of the scenarios below, choose the most appropriate type of map text.

1. Suppose you are creating a map of the Pacific Northwest United States with vector layers of states, provinces (Canadian), cities, rivers, and a raster shaded relief layer for the background. Which type of map text is most appropriate for indicating names of mountain ranges shown in the raster layer? (Choose two.)

    a. Labels
    b. Graphic Text
    c. Map annotation
    d. Geodatabase annotation

2. Suppose you are creating a city map with layers of streets and building footprints. You would like to

emphasize public buildings, such as museums, historic buildings, large hotels, popular restaurants, and so on. The buildings' attribute table does not contain this information, but your coworker created some map text for the buildings and stored it in a server location that you can access. Which type of map text allows you to most efficiently re-use your coworker's map text for the buildings (without conversion)?

   a. Labels
   b. Graphic Text
   c. Map annotation
   d. Geodatabase annotation

3. Suppose you are making an interactive map that you share within your organization that will be used at many different scales. Which type of map text is best to use for this kind of map?

   a. Labels
   b. Graphic Text
   c. Map annotation
   d. Geodatabase annotation

Challenge 2

1. Suppose you are labeling a map of parcels and building footprints, and the parcels range in size. You don't want the map text for parcels to overlap the buildings. You will need to manage map text in dense areas with many small parcels.

   a. Standard
   b. Maplex

2. Suppose you need to label a map of contour lines and you want to achieve the label effect shown in the following graphic.

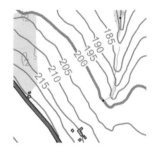

   a. Standard
   b. Maplex

3. Suppose you have a map with many label classes and label expressions that you have used for a number of years (without changing the labels). When you initially created the map, you did not have any extensions available. Now you want to add a new layer to the map, and you want to label that layer.

   c. Standard
   d. Maplex

Explanation: Since the map has been created in version 9.0, without access to any extensions, the labels must have been placed using the Standard label engine. If you would now switch to the Maplex label engine to labels buildings, and then switch back to the Standard label engine, to modify properties of the existing labels, the Maplex label settings would be lost. Therefore, it is better to use the same Standard label engine now—and maybe consider changing all labels to use the Maplex label engine later.

Key terms

Label: In ArcGIS, descriptive text, usually based on one or more feature attributes. Labels are placed dynamically on or near features based on user-defined rules and in response to changes in the map display. Labels cannot be individually selected and modified by the user. Label placement rules and display properties (such as font size and color) are defined for an entire layer.

Annotation: In ArcGIS, text or graphics that can be individually selected, positioned, and modified. Annotation may be manually entered or generated from labels. Annotation can be stored as features in a geodatabase or as map annotation in a data frame.

Resources

- ArcGIS Desktop Help > Desktop > Mapping > Adding text to the map > Displaying labels > Changing how labels are displayed
  - About displaying labels
  - Using label classes to label features from the same layer differently
  - Building label classes from symbology classes
  - Setting scale ranges for label classes
  - Changing the appearance of labels by changing the label symbol
- ArcGIS Desktop Help > Desktop > Mapping > Adding text to the map > Displaying labels > Specifying the text for labels
  - About specifying text for labels
  - Building label expressions
- ArcGIS Desktop Help > Desktop > Mapping > Adding text to the map
  - Using text formatting tags
- ArcGIS Desktop Help > Desktop > Mapping > Adding text to the map > Displaying labels >Specifying the text of labels
  - Building label expressions
- ArcGIS Desktop Help > ArcGIS Tutorials > Introducing the Maplex Label Engine tutorial
- ArcGIS Desktop Help > Desktop > Mapping > Symbols and styles
  - What are symbols and styles?
  - A quick tour of symbols and styles
- ArcGIS Desktop Help > Desktop > Mapping > Symbols and styles > Creating new symbols
  - About Creating new symbols
  - Creating marker symbols
  - Creating line symbols
  - Creating new fill symbols

Preparing maps for publishing

Optimizing maps for the web 336

Sharing a map as a service 337
   Analyzing a map
   Previewing a map service
   Exercise 23a: Prepare a map for publishing and create a service definition file (ArcGIS 10.1)
   Exercise 23a: Prepare a map for publishing and create a map service definition files (ArcGIS 10.0)
   Publishing a map

Preparing temporal data for a web map 342
   Creating time-aware layers
   Exercise 23b: Create a map with time-aware layers

Key terms 344

Resources 344

Chapter 23: Preparing maps for publishing

Interacting with maps over the Internet is an everyday activity for anyone with web-enabled devices, web browsers, or smart phones. People want to see where there friends are, what the weather looks like, and how to get to the nearest gas station. Maps you create with ArcGIS for Desktop can also be shared and then accessed over the web. Before sharing, maps are optimized to perform with the efficiency Internet users have come to expect. You carry out the map optimization process by using map optimization workflows in conjunction with a few ArcGIS for Desktop tools.

Publishing maps with time-aware layers adds another dimension to your map services. In addition to showing where things are located and how they relate to each other, you can visualize how things change over time.

Skills measured
- Determine how to optimize a map for publishing to ArcGIS for Server.
- Describe how to make maps time-aware and display dynamic features using the time slider.

Prerequisites
Knowledge of what map service is

Optimizing maps for the web

Optimizing a map for use on the web requires that you follow certain principles and practices during each phase of the authoring and publishing process. Figure 23.1 shows the workflow.

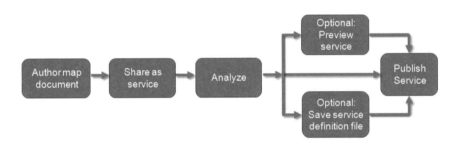

Figure 23.1 Before sharing map documents as a service they are analyzed to identify any issues that might negatively impact their performance. Also before publishing a service, you can preview it or save it to a service definition file. Created by the author.

To author a map document that is optimized for rapid display, there are a number of workflows (summarized in table 23.1), to manage data frames, layers, and source data.

Table 23.1 Creating map documents for the web	
Prepare data frames	• Remove unused layers rather than turning them off • Set custom full extent • Use the Standard label Engine rather than Maplex for faster label display
Prepare layers	• Simplify symbols or use symbols from the ESRI Optimized style • Set scale dependent display for features and labels • Apply definition queries when appropriate • Build pyramids for raster layers
Prepare layer source data	• Project source data into the coordinate system of their data frame to avoid projection-on-the fly • Generalize source data if they contain more detail than displayed on the map • Move source data to the computer that hosts the map service to avoid network delays

Sharing a map as a service

When you share a map as a service, you set up the properties to optimize the drawing performance of the map service. One of these properties is anti-aliasing, which is a graphics technique that blends foreground and background pixels near edges of objects, making borders appear smoother. The anti-aliasing property has options to prevent display artifacts and optimize the drawing quality of the map service. Anti-aliasing impacts drawing speed but has options that allow you to balance drawing quality with drawing speed.

For more information on optimizing drawing quality, refer to the ArcGIS Help topic "Drawing behaviors of the map service."

Analyzing a map

Analyzing a map evaluates it for any issues that could either prevent the map from publishing or negatively impact drawing performance. Functionality and data designed to work in ArcGIS for Desktop may not be supported when publishing the map as a service. Unsupported functionality and data are identified as errors that you have to fix before publishing is possible.

Table 23.2 lists some common issues that may be identified as errors when you analyze a map, along with their suggested solutions.

Table 23.2 Issues identified as errors when analyzing a map

Issue	Solution
Unsupported data source for layers • ArcInfo or PC ARC/INFO coverage • CAD • Excel file • ArcIMS or ArcGIS feature service • In-memory feature class	Convert to geodatabase format
Unsupported Layer type • Basemap layers from ArcGIS Online • Topology layers • LAS dataset layers	Remove before sharing the map as a service
Unsupported symbols, e.g., 3D symbols, texture fill symbols	Replace with supported symbols
Graphics text, map annotation	Convert to geodatabase annotation
Feature selections	Clear selections before sharing the map as a service
Dynamic hillshading in raster layers	Turn off hillshade effect in layer properties; compute a physical hillshade layer instead
Clip to Shape option set in data frame	Turn off Clip to Shape option
Basemap layers cannot be published directly as a map service	Remove basemap layers; publish separately as a basemap service

Chapter 23: Preparing maps for publishing

Previewing a map service

Before you publish a map service, it is a good idea to preview it from within ArcMap to see how it looks and how fast it draws. A map service is streamed to client applications, such as a web application that you open in your browser, as a series of images. When you preview the map service, you can experiment with different image formats that will be used to stream the map service to the client applications. The image format is important. It affects the drawing quality of the map service, whether or not the map service supports transparency (enabling visualization of other layers or services underneath it), and the amount of network traffic caused by the map delivery. Experimenting with different image formats in the Preview ArcGIS Server window will give you a better understanding of how a map service will perform when client applications request it.

Table 23.3 lists the different available image formats, their limitations, and uses.

For more information on the different image formats, refer to the ArcGIS Help topic "Previewing your map."

Table 23.3 Image formats used by map services		
Format	Limitation	Use
BMP	Uncompressed format, produces large-size images	For establishing baseline drawing performance when previewing a map service (generally the slowest)
JPG	Does not support transparency	For services with a large color variation that do not need a transparent background, e.g., basemap services that contain only raster imagery
PNG8	Supports up to 256 colors	For services that need a transparent background and have no more than 256 colors, e.g., roads and boundaries
PNG24	Does not work well with Internet Explorer 6 or earlier.	For services that have a transparent background and more than 256 colors (if fewer than 256 colors, use PNG 8), e.g., roads and boundaries
PNG32	Larger files than PNG24 and PNG8; can cause increased network traffic	For services that have transparent map features with more than 256 colors, e.g., imagery, hillshades, gradient fills, and transparent polygon features.
GIF	Supports up to 256 colors	For simple maps that must be transmitted quickly over the network or Internet

Exercise 23a: Prepare a map for publishing and create a service definition file (ArcGIS 10.1)

The ArcGIS for Server expert in your group has asked you to prepare a map document for publishing as a map service. Since you may not have ArcGIS for Server installed on your computer, you will create a service definition file to hand over to the server expert, who will then publish it to ArcGIS for Server.

Note: These instructions apply to ArcGIS 10.1 only. If you are working in ArcGIS 10.0, follow the ArcGIS 10.0 instructions for exercise 23a.

1. **Start ArcMap and open the ...\DesktopAssociate\Chapter23\PrepareToPublish.mxd.**

The map document shows world countries symbolized and labeled by the year when they joined the United Nations. Ocean names are shown as map annotation.

Note: Countries labeled with 9999 are either not members of the UN (such as Antarctica) or there is no data for these countries.

2. **To prepare the map for publishing, do the following:**
   - Remove the Basemap layer from the table of contents.
   - In the Countries layer properties, remove the halo around the labels.

Hint: To remove the halo, on the Labels tab, click Symbol > Edit Symbol > Mask tab and set the Style to None.
   - Convert the Countries labels to feature-linked geodatabase annotation.

Hint: Right-click the Countries layer and choose Convert Labels to Annotation.

Next, you will create the service definition file.

3. **From the File menu, point to Share As and choose Service. Check the option to Save a service definition file and then click Next.**

Since you are creating a service definition file that will be published later, you do not need to choose an ArcGIS for Server connection.

4. **Check the No available connection option. Then check the option to Include data in service definition when publishing. Ensure that the Service name will be PrepareToPublish, and then click Next.**

5. **Save the PrepareToPublish service definition file in your ...\DesktopAssociate\Chapter23 folder. Then click Continue.**

Before you can create the service definition file, you need to analyze the map.

6. **In the Service Editor window, click Analyze.**

The Prepare window lists errors, warnings, and messages about issues in the map that could either prevent the map from publishing (errors) or negatively impact drawing performance.

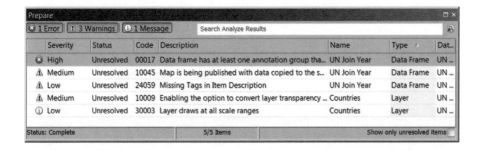

The High Severity error listed in the Prepare window refers to the map annotation used for Ocean names. Map annotation is not supported in a map service.

7. **Right-click the error and choose Convert Graphics to Features. Save the output annotation feature class in your ...\DesktopAssociate\Chapter23\World_UN.gdb and change the name to Oceans_text.**

Note: Important: Check the option to Automatically delete graphics after conversion.

8. **Click Yes to add the exported feature class as a layer.**

In the Prepare window, one of the Medium Severity warnings (Code 10045) alerts you that the map is being published with data copied to the server. Since the data referenced by the current map is relatively small in size and will not be updated any time soon, it is okay to copy the data to the server. You can ignore this warning.

Chapter 23: Preparing maps for publishing

Next, you will address the layer transparency and missing tags warnings.

9. **Right-click the Code 10009 transparency warning and select Use Color Transparency.**

10. **At this point, you could set a visible scale range for the countries layer, which would improve its drawing performance. However, since it is the only feature layer in this map, you will ignore this message.**

11. **Right-click the Code 24059 warning about Missing tags in the Item description and choose to Show the item description page.**

12. **In the Service Editor, for Tags, type World, UN, United Nations.**

13. **Analyze the map.**

The issues that you addressed are now resolved.

Before you save the service definition file, you will preview the map's performance and responsiveness when published to ArcGIS for Server.

14. **In the Service Editor window, click Preview.**

15. **Zoom and pan to different areas in the map and watch the display performance shown (refresh time in seconds) on the toolbar at the top of the Preview ArcGIS for Server window.**

The reported number of seconds represents the time it takes for the service to draw an image of the map when accessed through a client application.

16. **Click the arrow button at the right edge of the toolbar to view more properties of the map service.**

Additional service properties include anti-aliasing options and different image formats used by the map service. Anti-aliasing improves the quality of the map service by blending foreground and background pixels to smooth the borders of features and text.

Decision point: What are the default anti-aliasing options specified in the current service definition file?

_____

Decision point: Which image formats are available for the map service?

_____

17. **If you would like, pan and zoom the map with different image formats set in the Preview window. When you are finished, set the image format back to the default PNG8.**

18. **Click Stage to create the service definition file and package the data.**

The packaging process takes a few moments. When you see the Service Publishing Result message, the service definition file is ready for publishing.

Click OK to close the message. If you are continuing with the next exercise, leave ArcMap open.

Exercise 23a: Prepare a map for publishing and create a map service definition file (ArcGIS 10.0)

The ArcGIS for Server expert in your group has asked you to prepare a map document for publishing as a map service. Since you may not have ArcGIS for Server installed on your computer, you will create a service definition file to hand over to the server expert, who will then publish it to ArcGIS for Server.

Note: These instructions apply to ArcGIS 10.0 only. If you are working in ArcGIS 10.1, follow the ArcGIS 10.1 instructions for exercise 23a earlier in this chapter.

1. **Start ArcMap and open the ...\DesktopAssociate\Chapter23\PrepareToPublish_100.mxd.**

The map document shows world countries symbolized and labeled by the year when they joined the United Nations. Ocean names are shown as map annotation.

Note: Countries labeled with 9999 are either not members of the UN (such as Antarctica) or there is no data for these countries.

2. **To prepare the map for publishing, do the following:**
   - Remove the Basemap layer from the table of contents.
   - In the Countries layer properties, remove the halo around the labels.

Hint: To remove the halo, on the Labels tab, click Symbol > Edit Symbol > Mask tab and set the Style to None.

   - Convert the Countries labels to feature-linked geodatabase annotation.

Hint: Right-click the Countries layer and choose Convert Labels to Annotation.

Next, you will analyze the map, address any issues that come up in the analysis, and then save the map as a map service definition file.

3. **Display the Map Service Publishing toolbar. On the Map Service Publishing toolbar, click the Analyze Map button.**

The Prepare window lists errors, warnings, and messages about issues in the map that could either prevent the map from publishing (errors) or negatively impact drawing performance.

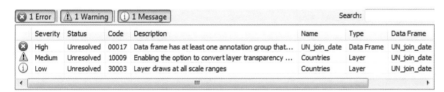

The High Severity error listed in the Prepare window refers to the map annotation used for Ocean names. Map annotation is not supported in a map service.

4. **Right-click the error and choose Convert Graphics to Features. Save the output annotation feature class in your ...\DesktopAssociate\Chapter23\World_UN.gdb and change the name to Oceans_text.**

Note: Important: Check the option to Automatically delete graphics after conversion.

5. **Click Yes to add the exported feature class as a layer.**

Next, you will address the layer transparency and missing tags warnings.

6. **Right-click the Code 10009 transparency warning and select Use Color Transparency.**

At this point, you could set a visible scale range for the countries layer, which would improve its drawing performance. However, since it is the only feature layer in this map, you will ignore this message.

7. **Reanalyze the map.**

The issues that you addressed are now resolved.

Before you save the map service definition file, you will preview the map's performance and responsiveness when published to ArcGIS for Server.

8. **On the Map Service Publishing toolbar, click the Preview Map button.**

9. **Zoom and pan to different areas in the map and watch the display performance shown (refresh time in seconds) on the toolbar at the top of the Preview ArcGIS Server window.**

The reported number of seconds represents the time it takes for the service to draw an image of the map when accessed through a client application.

10. **Click the arrow button at the right edge of the toolbar and click Properties.**

Additional service properties include anti-aliasing options and different image formats used by the map service. Anti-aliasing improves the quality of the map service by blending foreground and background pixels to smooth the borders of features and text.

Decision point: What anti-aliasing options are currently specified?

___

Decision point: Which image formats are available for the map service?

___

11. If you would like, pan and zoom the map with different image formats set in the Preview window. When you are finished, set the image format to the PNG. Then click OK and close the Preview ArcGIS Server window.

Next you will save the map document as a map service definition (MSD) file. In ArcGIS 10.0, the MSD file can be published as an MSD-based map service.

12. Save the map document. On the Map Service Publishing toolbar, click the Save Map Service Definition button. Save the PrepareToPublish_100 map service definition file, in your ...\DesktopAssociate\Chapter23 folder.

The map service definition file is now ready for publishing.

13. If you are continuing with the next exercise, leave ArcMap open.

Publishing a map

After analyzing the map, publishing the map as a service is the final step in the workflow. There are two ways to publish a map as a service: use your organization's own server and ArcGIS for Server installation, or publish the map to ArcGIS Online. If your organization has ArcGIS for Desktop 10.1 and an ArcGIS Online for Organizations subscription, you can publish hosted map services (hosted by Esri) without actually having ArcGIS for Server at your site. This makes it easy to turn GIS data into web services because there is no need to set up and administer ArcGIS for Server.

Preparing temporal data for a web map

Temporal data shows how features change over time. Web maps can be used for visualizing temporal data. Temporal data contains information about the state of the data at a particular time. The area and population of a city in a given year, or the position of the eye of a hurricane at a given date and time are examples of temporal data.

A temporal feature class contains at least one attribute field containing time values, such as dates, months, years, for each feature in that layer. To represent the duration of a process (e.g., duration of a wildfire at a particular area), the attribute table must contain fields representing the start and end times of the process.

You can create time-aware layers by displaying features in a time sequence based on the time field.

Creating time-aware layers

To create time-aware layers, you add one or more temporal feature classes to ArcMap and confirm that the time information is present. You make the layer(s) time-aware by enabling time in their layer properties and specifying which field in the attribute table contains the time information. When enabling time in the layer properties, you also specify a time step interval, which represents how often the time values will be refreshed. Once time is enabled in the layer properties, you can display the layer as an animation in the Time Slider window. The Time Slider allows you to set additional properties that determine how the animation will display. For example you can specify a particular time window to be displayed in the animation. Animations of time-enabled layers that you create in ArcMap are preserved when you share your map as a service.

Preparing temporal data for a web map **343**

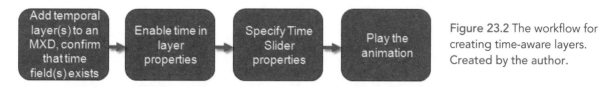

Figure 23.2 The workflow for creating time-aware layers. Created by the author.

Exercise 23b: Create a map with time-aware layers

In preparation for visualizing temporal data in a web map, you will enable time in a layer and set up an animation using the ArcMap Time Slider.

1. **ArcGIS 10.1: Start ArcMap and open the …\DesktopAssociate\Chapter23\TimeAwareMap.mxd**

   ArcGIS 10.0: Start ArcMap and open the …\DesktopAssociate\Chapter23\TimeAwareMap_100.mxd

2. **Open the Countries attribute table.**

In the attribute table, the UN_join_date and the UN_join_year fields contain time information.

3. **Close the attribute table.**

4. **Open the Countries layer properties. In the Time tab, set these parameters:**
   - Check Enable time on this layer
   - Layer Time: Each feature has a single time field
   - Time Field: UN_join_year
   - Field Format: YYYY
   - Time Step Interval: 1 Years
   - Check Display data cumulatively

Click OK.

5. **In the Tools toolbar, click Time Slider.**

The map now shows the group of features that have the earliest time attribute, that is, the countries that joined the United Nations in 1945. You will change the display to a more suitable one by changing the time period.

6. **On the Time Slider, click the Enable time on map button, if necessary.**

7. **On the Time Slider, click the Options button and make the following changes:**
   - In the Time Extent tab, change Restrict full time extent to <Undefined>.
   - For Start time enter **1/1/1944**.
   - For End time enter **1/1/2012**.
   - In the Playback tab, drag the Speed slider to about the middle of the scale, as shown in the graphic. Then click OK.

8. On the Time Slider, set the date to 1/1/1944.

Now the Countries layer is not displayed because there were no member countries in 1944.

9. Click the Play button ▶ to play the animation.

The countries appear as the time slider moves through the years.

10. Close the Time Slider.

You could now share the map as a service as you did in the previous exercise. In the published map service, the default view will display the features of time-enabled layers that fall within the specified time extent.

11. Save the map document. Close ArcMap.

Key terms

Map service: A map document published to ArcGIS for Server.

Temporal data: Data that specifically refers to times or dates. Temporal data may refer to discrete events, such as lightning strikes; moving objects, such as trains; or repeated observations, such as counts from traffic sensors.

Web map: In ArcGIS Online, a web-based, interactive map that allows you to display and query the layers on the map. A web map contains one or more ArcGIS Server map services that are referenced to ArcGIS Online.

Resources

- ArcGIS Desktop Help > Desktop > Mapping > Publishing map services
  - What is a map service?
- ArcGIS Desktop Help > Desktop > Mapping > Publishing map services > Publishing your map as a service
  - Publishing a map service
  - Setting map service properties
  - Previewing your map
- ArcGIS Desktop Help > Desktop > Mapping > Publishing map services > Map authoring for GIS servers
  - Map service planning
  - Map authoring considerations
- ArcGIS Desktop Help > Desktop > Mapping > Publishing map services > Technical guide for map services
  - Supported functionality in map services
  - Drawing behaviors of a map service
  - Serving time-aware layers
- ArcGIS Desktop Help > Desktop > Mapping >Time
  - What is temporal data?
  - A quick tour of temporal data management and visualization
  - Temporal data management > How time is supported in spatial data
- ArcGIS Desktop Help > Desktop > Mapping >Time > Visualizing temporal data > Viewing time-enabled data
  - Using the Time Slider window

chapter 24

Sharing maps and data

Exporting and printing maps   346
  Export formats
  Troubleshooting options when exporting to vector formats
  Exercise 24a: Export from layout view and data view

Printing maps   349
  Exercise 24b: Set up a map for printing
  Troubleshooting common printing problems
  Challenge 1

Sharing maps and data through packaging   353
  Exercise 24c: Work with layer and map packages (ArcGIS 10.1)
  Exercise 24c: Work with layer and map packages (ArcGIS 10.0)

Sharing map documents   357
  Types of pathnames
  Exercise 24d: Repair data sources and set relative pathnames
  Challenge 2

Answers to chapter 24 questions   360

Answers to challenge questions   360

Key terms   361

Resources   361

Chapter 24: Sharing maps and data

As GIS professionals, we share our maps and data to collaborate with team members, clients, contractors, and others. What is the best method for sharing maps and data? The answer to this question depends entirely on what you want to share and whom you want to share it with.

- Suppose you want to share only your map (without the data) with clients, managers, decision makers, or even the general public. The best way to do that may be to export the map to an image format or print it.
- What if you want to share both maps and data with coworkers or other GIS professionals? Then you may want to publish a map or feature service, or create a map or layer package that you can share on ArcGIS Online.
- To quickly share your map with coworkers who share the same internal network and have access to the same data, you could share a map document.

In this chapter, you will take a closer look at methods for sharing maps and data through exporting, printing, packaging, and using relative pathnames.

Skills measured

- Given a scenario, determine the best option for sharing specified content.
- Compare and contrast export formats available in ArcGIS for Desktop.
- Explain the differences between exporting from layout view and data view.
- Given a desired output (e.g., tiling, rasterization, color vs. black and white), determine appropriate map and printer settings.
- Describe how packages are generated and shared from ArcGIS for Desktop.
- Explain considerations for sharing MXDs between workstations (e.g., absolute paths, relative paths with file structure).

Prerequisites

- Hands-on experience exporting and printing maps
- Hands-on experience creating a layout
- Knowledge of relationships between layers and source data

Exporting and printing maps

Technically, exporting and printing are similar processes: to export or print a map, ArcMap creates an intermediate raster from it. Once the raster has been created, ArcMap sends it to the printer engine or export driver, which in turn converts it to the data format the printer understands or to the desired export format. The Output Image Quality (OIQ) or the resolution (dots per inch, or dpi) settings control the quality of the intermediate raster.

You can export or print your map from both data view and layout view, although for printing, layout view is recommended.

- Exporting (or printing) a map from layout view will export/print all data frames and map elements (e.g., legend, scale bar, dynamic text) visible on the layout.
- Exporting (or printing) from data view will export/print only the visible area of the map in the active data frame, without any map elements.

Export formats

You can export your maps to several vector and raster formats. EMF, EPS, AI, PDF, and SVG support a mixture of vector and raster data. BMP, JPEG, PNG, TIFF, and GIF support only raster data. When you export your map (from data view) to any raster formats, you have the option to generate a world file, which will preserve the georeferencing information of the map. In addition, these raster formats (except for GIF) support exporting with a large variety of colors (24-bit true color), which makes them especially suitable for exporting rasters or images.

Table 24.1 lists the formats available for exporting a map and when to use each format.

Table 24.1 Map export formats

Format	Why use this format?	Example
Portable Document Format (.pdf)	Exchange a map with non-GIS users Option to preserve layers and attributes With Data Driven Pages ability to export multiple pages to a single document	Create a digital publication, while preserving the ability to interact with layer visibility and attributes
Adobe Illustrator (.ai)	Refine appearance of a map in preparation for printing Vector layers can be edited effectively in .ai	Separate colors (define in CMYK or RGB) in preparation for offset printing (where CMYK colors are printed separately)
Tagged Image File Format (.tiff)	Create a high-quality, compressed (lossless) image from a map document Supported in a large variety of image processing applications	Use a raster image of a map as a background in another map document
Joint Photographic Experts Group (.jpeg)	Create a compressed image from a map document	Include a raster image of a map in a report
Graphic Interchange Format (.gif)	Create compressed image from a map document Supports transparency	Display a map image as a thumbnail on a website with a limited number of colors
Microsoft Windows bitmap (.bmp)	Create a high-quality (uncompressed) image from a map document	Use a high-quality map image in a printed atlas
Portable Network Graphics (.png)	Create compressed image from a map document Lossless compression: keeps text and line work legible Supports transparency	Include a high-quality map containing vector and raster layers as an image on a website
Scalable Vector Graphics (.svg)	Create an XML-definition from a map document	Create a map image for embedding in a website
Encapsulated PostScript (.eps)	Export a map for high-quality printing	Print high-quality, poster-size maps
Windows Enhanced Metafile (.emf)	Embed a map in a Windows document Option to convert marker symbols to polygons to display them correctly on any machine	Embed a map with picture marker symbols in an MS Word document and resize the marker symbol without loss of quality

For more information on exporting your map in different formats, refer to the ArcGIS Help topic "Exporting your map."

> ### Troubleshooting options when exporting to vector formats
>
> When layers with picture fill symbols or picture marker symbols are exported, these types of symbols sometimes cause the map layers underneath to be rendered striped or pixelated. Exporting your map to any of the vector formats (EPS, PDF, EMF, AI, or SVG) provides options for dealing with picture symbols: you can either vectorize (convert to vector) picture symbols, or rasterize (convert to raster) them. Vectorizing picture symbols converts the raster markers/fills to polygons, resolving the rendering issue.
>
> When layers containing character marker symbols or map text using Esri-specific fonts are exported, the fonts may not be rendered correctly unless they are installed on the machine where the exported file is viewed. The PDF, EPS, and SVG formats allow you to embed these fonts so that the symbology can be displayed correctly.
>
> For more information on exporting your map with different export formats, refer to the ArcGIS Help topic "Exporting your map."

Exercise 24a: Export from layout view and data view

Suppose your manager asked you to provide some examples of your maps to include in a web page and a digital flyer. For the digital flyer, you will export an entire map layout, including all data frames and map elements. For the web page, you will export a single elevation map from one data frame without any additional map elements.

1. **ArcGIS 10.1:** Start ArcMap and open the ...\DesktopAssociate\ Chapter24\ExportMap.mxd.

    ArcGIS 10.0: Start ArcMap and open the ...\DesktopAssociate\Chapter24\ExportMap_100.mxd.

You see a map document with three data frames. One data frame shows an elevation map, another shows a population density map, and a third shows an overview map. The South America population density map is similar to one you prepared in an earlier chapter. You will first export the entire layout with all three data frames from layout view and then export the elevation map from data view.

2. From the File menu, choose Export Map. For Save as type, choose PDF and save the file in your ...DesktopAssociate\Chapter24 folder. Replace the default File name with **SA_Population.pdf**.

3. At the bottom of the Export Map dialog box, expand the Options pane.

The tabs in the Options pane allow you to control the image quality.

4. Accept the default Resolution of 300 dpi and set the Output Image Quality to Best.

The Output Image Quality controls how many pixels in the map document will be used to create the file sent to the export driver. With the Output Image Quality option set to BEST, the ratio is 1:1, which means that no resampling will occur, and every pixel used to display the map layout will be used to generate the intermediate raster that will be exported.

Major cities are represented with a character marker symbol from the ESRI Cartography font. You will ensure that these marker symbols can be displayed on a computer with limited fonts.

5. Click the Format tab and check the option to Embed All Document Fonts.

You will include all the layers in the ArcMap table of contents in the exported PDF document.

6. Click the Advanced tab. For Layers and Attribute, select Export PDF Layers Only. Make sure the Export Map Georeference Information option is checked.

7. Click Save to export the map. Open Windows Explorer and navigate to the SA_Population. pdf.

8. **If you have Adobe Reader installed on your computer, open the SA_Population.pdf document and experiment with turning layers off and on.**

For more information on exporting a map to a PDF document, refer to the ArcGIS Help topic "Exporting to PDF."

Next, you will export the Elevation map in the South America data frame. You will switch to data view to export only the symbolized layers in the data frame, without any additional map elements.

9. **Make sure that the South America data frame is active. In ArcMap, switch to data view.**

Decision point: Which format is appropriate for exporting the map in the South America data frame to a high-quality image that supports transparency?

You will export the raster layers in the South America data frame to PNG format.

10. **From the File menu, choose Export Map. For Save as type, choose PNG and save the file in your …DesktopAssociate\Chapter24 folder. Replace the default File name with SA_Elevation.png.**

11. **Set the Resolution to 300 dpi and check the option to Write a World File.**

You will export the map with 24-bit color (supporting more than 256 colors) and select the color of the blue data frame background to be marked as transparent.

12. **Click the Format tab. For Color mode, choose 24-bit True Color. For both the Background and the Transparent Color, select Sugilite Sky (first row, 10th column). Then click Save.**

13. **Open the SA_Elevation.png in an image viewer (e.g., Paint).**

The exported map contains an image of the layers in the South America dataframe. The blue background color is transparent in the image viewer.

Note: Some image viewers display a transparent color as black or white.

14. **If you are continuing with the next exercise, leave ArcMap open. Otherwise, close ArcMap.**

Printing maps

Prior to printing a map, you must set up the printer and the map page in the map document. Setting up the printer involves selecting the installed printer you want to use and specifying the paper size and orientation. When you set up the page, you can choose the same settings for the size and orientation of the map page, or you can specify your own custom page dimensions.

- To fill the entire paper used by the printer, you can use the Use Printer Paper Settings option. ArcMap then determines the dimensions of the map page based on the printer driver of the selected printer.

- To print your map at a custom size or to share the map with users that may not have access to a particular printer, you can specify your own page dimensions.

For more information on printing your map, refer to the ArcGIS Help topic "About map printing" and the Esri Support Technical Article "HowTo: Set up ArcMap page and printer settings."

Exercise 24b: Set up a map for printing

Suppose you are tasked with setting up a map to be printed in a 4 by 5 inch brochure. You received a map document with a map that was designed to be printed on letter size (8.5 by 11 inch) paper.

1. **ArcGIS 10.1: Start ArcMap, if necessary, and open the …\DesktopAssociate\ Chapter24\ PreparePrinting.mxd.**

    ArcGIS 10.0: Start ArcMap, if necessary, and open the …\DesktopAssociate\ Chapter24\ PreparePrinting_100.mxd.

You see a South America map similar to the one you worked with in an earlier chapter.

2. From the File menu, open Page and the Print Setup dialog box. Click the Name dropdown arrow and inspect the available printers.

Depending on the kind of printers you have installed on your computer, different printer names may appear in the Name dropdown menu. At the minimum, you may have the Microsoft XPS Document Writer installed, which comes with the Windows operating system. Your default printer is marked with a checkmark.

3. Select your default printer. Make sure that Use Printer Paper Settings is unchecked.

After selecting the printer, you specify the paper size and orientation that the printer should use. Since the map was designed to be printed on letter size paper (8.5 by 11 inches), you will set Letter for the paper size. (Later, you will scale the map for the 4 by 5 inch size that you need.)

4. In the Paper section, Size dropdown menu, select Letter. For Orientation, make sure that Portrait is selected.

Note: The available paper sizes in the Size drop-down menu are derived from the printer driver of the selected printer. If your selected printer does not support letter (8.5 by 11 inch) size paper, either select another printer, or select the paper size that comes closest to letter size.

Although the map was designed to be printed on letter size (8.5 by 11 inch) paper, the specifications you received for the brochure require that you print the map at 4 by 5 inches.

5. In the Map Page Size section, specify 4 inches for the Width and 5 inches for the Height. Make sure that Portrait is selected for the Orientation of the page.

6. Check the option to Scale Map Elements proportionally to changes in Page Size. Your Page and Print dialog box should look similar to the one shown in the graphic below.

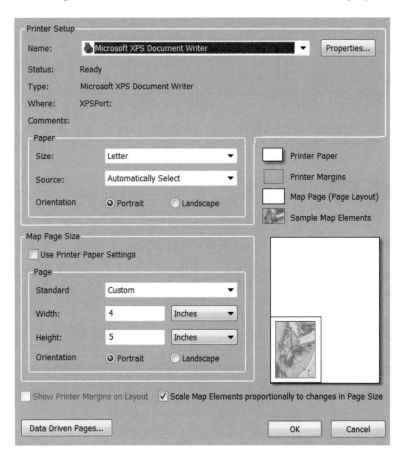

7. Click OK to close the Page and Print Setup dialog box. From the File menu, choose Print Preview.

Now that you changed the size of the map page, the printed map will cover about one quarter of the paper. This is the intended size for the brochure.

8. Click Print to open the Print dialog box.

9. In the Print dialog box, make sure the Windows Printer engine is selected. Set the Output Image Quality option to Best, as shown in the following graphic.

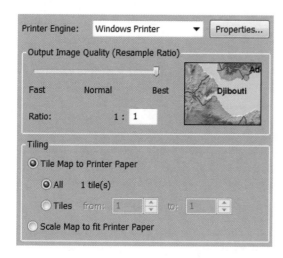

The Output Image Quality controls the number of pixels in the map document that will be used to create the file sent to the printer. With the Output Image Quality option set to BEST, the ratio is 1:1, which means that no resampling will occur, and every pixel used to display the map layout will be used to generate the intermediate raster that is sent to the printer.

By default, the Tile Map to Printer Paper option is selected, which allows you to print the map on multiple pages when it is too large to fit on one page. You will accept this default, although your map will fit on a single piece of (letter size) paper.

Note: Selecting the Scale Map to fit Printer Paper option scales the printed map back to 8.5 by 11 inches.

10. If you are connected to a printer, click OK to print the map; if not, click Cancel to close the Print dialog.

Note: Since you set the Output Image Quality to Best, it may take a few moments until the intermediate raster is generated and sent to the printer.

11. If you are continuing with the next exercise, leave ArcMap open. Otherwise, close ArcMap.

Troubleshooting common printing problems

Since printing a map is a much more complex process than printing a word processing document, there are a number of typical problems that can occur (table 24.2).

Table 24.2 Common problems with map publishing

Problem	Reason	How to resolve	Reference
Printed map displays with low quality (banded or pixilated)	Low output resolution or low Output Image Quality (OIQ) Picture/bitmap marker symbols or fill symbols Transparent symbology	Increase output resolution and OIQ settings Replace picture marker or fill symbols with vector symbols Remove transparency	Knowledge Base—Technical Articles 17332, 18082, 17845 ArcGIS Help topics "About Map Printing" and "Printing a map in ArcMap"

Continued on next page

Table 24.2 Common problems with map publishing (continued)

Problem	Reason	How to resolve	Reference
Printout only shows part of the map	Map layout is too large to fit on one page	Print map on multiple pages using Tile Map to Printer Paper option (preserves map scale) Use Scale Map to fit Printer Paper option (changes map scale)	ArcGIS Help topics "About Map Printing" and "Printing a map in ArcMap"
Map does not scale properly with different paper sizes	By default, maps are printed at their true scale (i.e., the scale set when the map has been created).	Use Scale Map to fit Printer Paper option Set reference scale	ArcGIS Help topic: "About Map Printing" Chapter 18, exercise 18c
When printing a map at different paper sizes, map elements scale improperly	The size of layout elements is relative to the page size that had been set when they were inserted.	Use Scale Map Elements Proportionally to changes in map size option	Knowledge Base— Technical Article 32406
Raster layers print with poor quality or not at all	Raster layers derived from image services have certain print limitations	Increase Output Image Quality (OIQ) setting, or remove image service layer	ArcGIS Help topic "Printing and exporting maps containing service layers"

Challenge 1

For each of the following print or export scenarios, choose the correct troubleshooting option.

1. You want to print a poster size map (60 by 84 inches), but the printer only prints a maximum width of 42 inches. You want to preserve the map scale.

    a. Use Tile Map to Printer Paper option
    b. Use Scale Map Elements proportionally option
    c. Use printer paper settings option
    d. Use Scale Map to fit printer paper option

2. You created a map at a scale of 1: 24,000 and printed it at letter size (8.5 by 11 inches). Now you print the same map at a larger size, and the map symbols and labels appear very small on the printout.

    a. Use Scale Map Elements proportionally option
    b. Use printer paper settings option
    c. Set a reference scale
    d. Increase Output Image Quality (OIQ) setting

3. You created a map at a page size of 34 by 44 inches. You had to decrease the page size to 11 by 17 inches for printing. Now the map title, legend and scale bar do not display at the correct locations in the layout.

    a. Increase output resolution
    b. Increase Output Image Quality (OIQ) setting
    c. Use Scale Map Elements Proportionally option
    d. Use Scale Map to fit printer paper option

4. You exported a map containing a Parks layer on top of a Hillshade layer with 30% transparency to EPS format. The output image displays with very low quality.

   a. Remove transparency
   b. Increase output resolution
   c. Increase Output Image Quality (OIQ) setting
   d. Choose the option to "Vectorize picture symbols"

Sharing maps and data through packaging

Packaging is the best solution for sharing your maps and data with other GIS professionals, such as colleagues in your work group, departments in your organization, or other ArcGIS users (via ArcGIS Online). A package is a single compressed file containing GIS data. You can save a package on a thumb drive, attach it to e-mail, FTP, or upload it to the cloud. Once the recipients of your package have unpacked it, they can immediately work with its contents.

You can package layers in a map document either individually as layer packages, or you can package the entire map document with all its layers as a map package. You can include source data from any storage format, including data stored in an ArcSDE geodatabase, in a layer or map package. Map and Layer packages also allow you to include additional files such as .doc, .txt, .pdf.

Starting with ArcGIS 10.1, you can create additional packages such as geoprocessing, locator, and tile packages. Table 24.3 lists the package types available with ArcGIS 10.1 and higher and example use cases for each type.

Table 24.3 Package types

Package type	Description	Examples
Map package (.mpk)	Contains map document (.mxd) and source datasets referenced by its layers	Package a map document with layers referencing an ArcSDE geodatabase and share it with a group on ArcGIS Online Package a map document and only the schema referenced by its layers as a template Create an archive of a map reflecting the current state of the data
Layer package (.lpk)	Contains layer Properties and source datasets referenced by its layers Option to package only the schema	Share the symbology, labels and field properties that you set up for your map with a co-worker. At ArcGIS 10.1, package a layer and its data sources and share it with a co-worker that uses ArcGIS 9.3.1.
Geoprocessing package (.gpk)	Contains data and tools to create a geoprocessing result Can be created and used only with ArcGIS 10.1 and higher	Share the standard workflow used by your organization to select suitable sites for a project Consolidate data sources used in a geoprocessing model, from different network locations, to a single file geodatabase, so you can run the model on your laptop at a conference

Continued on next page

Table 24.3 Package types (continued)

Package type	Description	Examples
Address locator package (.gcpk)	Contains a single address locator or a composite locator Can be created and used only with ArcGIS 10.1 and higher	Create a composite locator that is optimized for geocoding addresses in a city, and share it with other city departments via ArcGIS Online
Tile package (.tpk)	Contains a set of tiles (images) Can be created and used only with ArcGIS 10.1 and higher	Create a tile package from a map layer and use it as a basemap in ArcGIS Runtime applications, such as lightweight mobile applications

Exercise 24c: Work with layer and map packages (ArcGIS 10.1)

Ever since you and a colleague discovered layer and map packages, you use them frequently to share maps and data for different projects. You will now package your South America map with the data it references and then unpack a layer package that a colleague sent you.

Note: These instructions apply to ArcGIS 10.1 only. If you are working in ArcGIS 10.0, follow the ArcGIS 10.0 instructions for exercise 24c that follow.

1. **Start ArcMap, if necessary, and open the ...\DesktopAssociate\Chapter24\PackageMap.mxd.**

You see the South America map that you worked with in the previous exercise. You will package the map document and the source data into a map package. To package a map document, you need to enter a description in the Map Document Properties.

2. **From the File menu, open the Map Document Properties. For the Description, enter South America: Population per square mile. Then click OK to close the Map Document Properties. Save the map document.**

   - In the Search window, locate the Package Map tool.
   - Run the tool with the following parameters
     - Input Map Document: ...\DesktopAssociate\Chapter24\PackageMap.mxd
     - Output File: ...\DesktopAssociate\Chapter24\SouthAmerica.mpk
     - Extent: Default
     - Package version: CURRENT, 10.0
     - Summary: South America Population Density
     - Tags: Population, South America

Note: It will take a few moments to package the map.

To test if the map package has been properly created, you will open a new blank map document and extract the SouthAmerica map package.

3. **Open a new blank map document. In the Search window, locate and open the Extract Package tool.**

4. **Run the Extract Package tool with the following parameters**
   - Input Package: SouthAmerica.mpk
   - Output folder: ...\DesktopAssociate\Chapter24

5. **When the tool is finished, locate and open the PackageMap.mxd in the Catalog window. You do not need to save the changes in the blank map document.**

Sharing maps and data through packaging

Hint: Navigate to the …\DesktopAssociate\Chapter24\SouthAmerica folder and open the v101 folder.

Next, you will unpack a layer package containing a transportation group layer that a colleague e-mailed to you, and then add the group layer to the current map. Since you frequently unpack map and layer packages, you will specify a default folder location for them.

6. From the Customize menu, open the ArcMap Options, Sharing tab. In the Packaging section, check the option to Use user specified location for unpacking packages. Set the path for unpacking packages to your …\DesktopAssociate\Chapter24 folder and click OK to close the ArcMap Options dialog box.

When you unpack the layer package, the layer is automatically added to the active data frame.

7. Make sure that the Population Density data frame is active. In the Catalog window, navigate to the DesktopAssociate\Chapter24 folder. Right-click the Transportation.lpk and choose Unpack.

The Transportation.lpk layer package is unpacked and automatically added to the Population Density data frame.

Decision point: What would be another way to unpack the Transportation.lpk to the DesktopAssociate\Chapter24 folder?

Suppose a colleague sees the Elevation group layer displayed on your computer screen and likes the symbology. You decide to share the group layer as a layer package.

8. Activate the South America data frame. Then locate and open the Package Layer tool.

9. Run the Package Layer tool with the following parameters:
   - Input Layer: Elevation Group Layer (click the upper one)
   - Output File: …\DesktopAssociate\Chapter24\Elevation.lpk
   - Extent: Default
   - Schema only: leave unchecked
   - Package version: CURRENT, 10.0
   - Summary: South America Elevation and Hillshade
   - Tags: Elevation, Hillshade, South America

The Elevation.lpk layer package is created in your Chapter24 folder.

10. If you like, you can test to see if the layer package has been properly created. Insert a new data frame and use a method of your choice to unpack the layer package and add it to the new data frame.

11. If you are continuing with the next exercise, leave ArcMap open. Otherwise, close ArcMap.

Exercise 24c: Work with layer and map packages (ArcGIS 10.0)

Ever since you and a colleague discovered layer and map packages, you use them frequently to share maps and data for different projects. You will now package your South America map with the data it references, and then unpack a layer package that a colleague sent you.
Note: These instructions apply to ArcGIS 10.0 only. If you are working in ArcGIS 10.1, follow the ArcGIS 10.1 instructions for exercise 24c above.

1. Start ArcMap, if necessary, and open the …\DesktopAssociate\Chapter24\PackageMap_100.mxd.

You see the South America map that you worked with in the previous exercise. You will package the map document and the source data into a map package. To package a map document, you need to enter a description in the Map Document Properties.

2. From the File menu, open the Map Document Properties. For the Description, enter **South America: Population per square mile**. Then click OK to close the Map Document Properties. Save the map document.

- In the Search window, locate the Package Map tool.
- Run the tool with the following parameters
    - Input Map Document: ...\DesktopAssociate\Chapter24\PackageMap_100.mxd
    - Output File: ...\DesktopAssociate\Chapter24\SouthAmerica.mpk
    - Extent: Default

Note: It will take a few moments to package the map.

To test if the map package has been properly created, you will open a new blank map document and extract the SouthAmerica map package.

3. Open a new blank map document. In the Search window, locate and open the Extract Package tool.

4. Run the Extract Package tool with the following parameters
    - Input Package: SouthAmerica.mpk
    - Output folder: ...\DesktopAssociate\Chapter24**SouthAmerica**

Hint: Browse to the ...\DesktopAssociate\Chapter24 folder. For Name, type South America. Then, click Save.

5. When the tool is finished, locate and open the PackageMap_100.mxd in the Catalog window. You do not need to save the changes in the blank map document.

Hint: Navigate to the ...\DesktopAssociate\Chapter24\SouthAmerica folder and open the v10 folder.

Next, you will unpack a layer package from the Catalog window. When you unpack the layer package, the layer is automatically added to the active data frame.

6. Make sure that the Population Density data frame is active. In the Catalog window, navigate to the DesktopAssociate\Chapter24 folder. Right-click the Transportation_100.lpk and choose Unpack.

7. In the table of contents, click List By Source.

The source data of the Transportation.lpk layer package is unpacked to the ... ArcGIS\Packages folder in your user profile.

Decision point: How would you unpack the Transportation.lpk to the DesktopAssociate\Chapter24 folder?

Suppose a colleague sees the Elevation group layer displayed on your computer screen and likes the symbology. You decide to share the group layer as a layer package.

8. Activate the South America data frame. Then locate and open the Package Layer tool.

9. Run the Package Layer tool with the following parameters:
    - Input Layer: Elevation Group Layer (click the upper one)
    - Output File: ...DesktopAssociate\Chapter24\Elevation.lpk
    - Extent: Default
    - Schema only: leave unchecked
    - Package version: CURRENT

The Elevation.lpk layer package is created in your Chapter24 folder.

10. If you like, you can test to see if the layer package has been properly created. Insert a new data frame and use a method of your choice to unpack the layer package and add it to the new data frame.

11. If you are continuing with the next exercise, leave ArcMap open. Otherwise, close ArcMap.

Sharing map documents

Sometimes you want to quickly share a map document, for example, to ask a coworker for a second opinion on some symbology. You can either share a copy of the map document (.mxd) and the source datasets referenced by its layers, or if a coworker has access to the same source data, share only the map document, and set relative pathnames to the source datasets. If a coworker uses an earlier version of ArcGIS, you can even save the map document in an earlier version.

When you copy a map document (.mxd) to a different location, or move the data sources referenced in a map, ArcMap can no longer locate the source data referenced by its layers. For example, if a layer references source data in the C:\DesktopAssociate\Chapter24\SouthAmerica geodatabase, and the map document is moved to an E: drive, the pathname to the source data that is stored in the layer properties isn't valid any more.

Types of pathnames

There are two ways that ArcMap layers can reference their source datasets: they can store the full or absolute pathname to the source data, which is the default, or they can store a relative pathname. An absolute pathname begins with the drive letter followed by the folder structure that contains the data source. For example, the absolute pathname to the source data of a Countries layer could be C:\DesktopAssociate\Chapter24\SouthAmerica.gdb\Countries. Absolute pathnames can break, if the data is copied or moved, for example from a C: to an E: drive.

Storing the location to the source data as a relative pathname can alleviate this problem (figure 24.1).

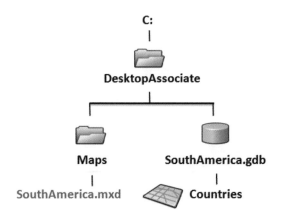

Figure 24.1 Relative pathnames refer to a location relative to the current folder. They use a dot symbol (.) to indicate the current directory and a double-dot (..) symbol to indicate a step up in the folder hierarchy. For example, for a Countries layer in a map document stored in the C:\DesktopAssociate\Maps folder, the relative pathname to the source data would be ..\ SouthAmerica.gdb\Countries. Created by the author.

As long as you maintain the relative location of layers and source data, relative pathnames allow you to move or copy the data to a different folder without breaking the path between the layers and the data source(s). However, relative pathnames cannot span disk drives: the map document and the source data must be stored on the same drive.

Note: You don't enter relative pathnames explicitly in ArcMap. Instead you specify the use of relative pathnames in the map document properties, which will set relative paths for all existing and new layers in the map document.

For more information on absolute and relative pathnames in ArcMap, refer to the ArcGIS Help topic "Paths explained: Absolute, relative, UNC, and URL."

Exercise 24d: Repair data sources and set relative pathnames

Suppose you receive a map document that uses absolute (rather than relative) pathnames to reference the source datasets.

1. **ArcGIS 10.1:** Start ArcMap, if necessary, and open the ...\DesktopAssociate\Chapter24\RelativePaths.mxd.

   ArcGIS 10.0: Start ArcMap, if necessary and open the ...\DesktopAssociate\Chapter24\RelativePaths_100.mxd.

The map contains a layer of world countries symbolized and labeled by the year they joined the United Nations, an annotation layer for the oceans, and a basemap layer. However, both the annotation and the Countries layers cannot be drawn because ArcMap cannot find the source data. Red exclamation marks next to the layers in the table of contents indicate that pathnames to the source data are broken and need to be repaired.

2. Open the Layer Properties of the Countries layer. Click the Source tab.

Decision point: Where does ArcMap expect to find the source data for the Countries layer?

___

The pathname to the Countries feature class, which is the source data of the Countries layer, is stored as C:\DesktopAssociate\Chapter24\World_UN.gdb\Countries.

3. Close the Layer Properties. In the Catalog window, navigate to the …\DesktopAssociate\Chapter24\World folder and expand the World_UN.gdb.

The World_UN.gdb contains both the Countries and the Oceans_Anno feature classes. You will repair the pathname stored in the corresponding layers to point to this location.

4. Right-click the Countries layer, point to Data, and choose Repair Data Source. Add …\DesktopAssociate\Chapter24\World\World_UN.gdb\Countries as a data source. In the table of contents, click List By Source.

Since the Oceans_Anno feature class is stored in the same location as Countries, the path to the Oceans_Anno layer has been automatically repaired, as well. Now that you repaired the path to the source data, ArcMap can draw the Oceans_Anno and the Countries layers.

Question 1: Suppose a data source that affects more than 100 map documents has been moved. How can you automate the repair of the pathnames to the data source for all of the map documents at once?

___

Hint: Refer to the ArcGIS Help topic "Repairing broken data links to answer this question."
To help prevent broken pathnames in the future, you will specify relative pathnames in the map document.

5. From the File menu, open the Map Document Properties. At the bottom of the dialog box, check the option to Store relative pathnames to data sources. Click Apply, and then OK to close the Map Document Properties dialog box. Save the map document.

Note: To store relative paths for all new map documents that you create, check the option to Make relative paths the default for new map documents in the ArcMap Options, General tab.
To test the effect of relative pathnames, you will now move the Chapter24 folder.

6. Open a new blank map. In the Catalog window, copy the Home-DesktopAssociate\Chapter24 folder and paste it to a different location anywhere on your computer (e.g., directly under your C: drive).

7. **ArcGIS 10.1:** Expand the Chapter24 folder at the new location and open the RelativePaths.mxd. In the table of contents, click List By Source, if necessary.

ArcGIS 10.0: Expand the Chapter24 folder at the new location and open the RelativePaths_100.mxd. In the table of contents, click List By Source, if necessary.

The Oceans_Anno and Countries layers now derive their source data from the new location of the Chapter24\World\World_UN.gdb. Setting relative pathnames helps prevent broken paths if the location of the World_UN.gdb is changed.

If you would like to, you can now test the limits of relative paths.

8. Optional: **ArcGIS 10.1:** In the Catalog window, copy only the RelativePaths.mxd file and paste it to any other location on your computer (e.g., to your C:\Users\<your user name>\Documents folder).

   ArcGIS 10.0: In the Catalog window, copy only the RelativePaths_100.mxd file and paste it to any other location on your computer (e.g., to your C:\Users\<your user name>\Documents folder).

9. Optional: **ArcGIS 10.1:** Open the RelativePaths.mxd from the new location.

   ArcGIS 10.0: Open the RelativePaths_100.mxd from the new location.

By moving the map document, you effectively changed the relative path from the map document to the data sources, so the pathnames are broken again. Creating a map package and saving it to a new location will prevent this from happening.

Your last task for today is to clean up and delete the copies of the Chapter24 folder and the RelativePaths.mxd.

10. Open a new blank map document. Since RelativePaths.mxd is a copy of the original, you do not need to save the changes. In the Catalog window, delete the copy of the Chapter24 folder and, if necessary, the copy of the RelativePaths.mxd or Relative paths_100.mxd.

11. Close ArcMap.

Challenge 2

For each of the following scenarios, choose the best sharing method.

1. What is the best method for sharing map and source data with your ArcGIS Online group?

   a. Set relative paths and share the map document
   b. Create a map package
   c. Export map to a PDF document
   d. Export map to a TIFF image

2. What is the best method for sharing the symbology of a layer for use in another map document? (Choose two.)

   a. Create a layer package
   b. Export to an EMF file
   c. Create a map package
   d. Create a layer file
   e. Export to an XML recordset document
   f. Create a tile package

3. Which of the following formats is suitable for sharing data with a user of a GIS software package other than ArcGIS?

   a. Layer package
   b. Shapefile
   c. Map package
   d. Geodatabase feature class

4. Which of the following formats is suitable for sharing a map in ArcReader?

   a. Publish the map to a service
   b. Create a map package
   c. Set relative paths and share the map document
   d. Publish map to a PMF file

Hint: Refer to chapter 1 for answering this question

360 Chapter 24: Sharing maps and data

Answers to chapter 24 questions

Question 1: Suppose a data source that affects more than 100 map documents has been moved. How can you automate the repair of the pathnames to the data source for all of the map documents at once?
Answer: The arcpy.mapping module contains Python methods for updating and fixing data sources in many map documents at once. For more information about these methods, refer to the ArcGIS Help document "Updating and fixing data sources with arcpy.mapping."

Answers to challenge questions

Challenge 1

Correct answers shown in bold.

1. You want to print a poster size map (60 by 84 inches), but the printer only prints a maximum width of 42 inches. You want to preserve the map scale.

    a. Use Tile Map to Printer Paper option
    b. Use Scale Map Elements proportionally option
    c. Use printer paper settings option
    d. Use Scale Map to fit printer paper option

2. You created a map at a scale of 1: 24,000 and printed it at letter size (8.5 by 11 inches). Now you print the same map at a larger size, and the map symbols and labels appear very small on the printed map.

    a. Use Scale Map Elements proportionally option
    b. Use printer paper settings option
    c. Set a reference scale
    d. Increase Output Image Quality (OIQ) setting

3. You created a map at a page size of 34 by 44 inches. You had to decrease the page size to 11 by 17 inches for printing. Now the map title, legend, and scale bar do not display at the correct locations in the layout.

    a. Increase output resolution
    b. Increase Output Image Quality (OIQ) setting
    c. Use Scale Map Elements Proportionally option
    d. Use Scale Map to fit printer paper option

4. You exported a map containing a Parks layer on top of a Hillshade layer with 30% transparency to EPS format. The output image displays with very low quality.

    a. Remove transparency
    b. Increase output resolution
    c. Increase Output Image Quality (OIQ) setting
    d. Choose the option to "Vectorize picture symbols"

Challenge 2

1. What is the best method for sharing map and source data with your ArcGIS Online group?

    a. Set relative paths and share the map document
    b. Create a map package
    c. Export map to a PDF document
    d. Export map to a TIFF image

2. What is the best method for sharing the symbology of a layer to be used in another map document? (Choose two.)

   a. Create a layer package
   b. Export to an EMF file
   c. Create a map package
   d. Create a layer file
   e. Export to an XML recordset document
   f. Create a tile package

3. Which of the following formats is suitable for sharing data with a user of a GIS software package other than ArcGIS?

   a. Layer package
   b. Shapefile
   c. Map package
   d. Geodatabase feature class

4. Which of the following formats is suitable for sharing a map in ArcReader?

   a. Publish the map to a service
   b. Create a map package
   c. Set relative paths and share the map document
   d. Publish map to a PMF file

Key terms

Lossy compression: Lossy compression averages out differences with an approximate rendition of the image resulting.

Lossless compression: Lossless compression keeps all of the detail of the original image, which allows you to uncompress the image without loss of information.

Resources

- ArcGIS Help 10.1 > Desktop > Mapping > Working with ArcMap
  - Referencing data in the map
  - Printing a map in ArcMap
  - Navigating maps
    - Working with data frame reference scales
  - Sharing data through packaging
    - About packaging
    - Creating a map package
    - Creating a layer package
    - Creating a tile package > About tile packages
- ArcGIS Help 10.1 > Desktop > Mapping > Working with layers > Managing layers
  - Repairing broken data links

Chapter 24: Sharing maps and data

- ArcGIS Help 10.1 > Desktop > Geoprocessing
  - Sharing workflows with geoprocessing packages
    - What is a geoprocessing package?
  - ArcPy > Mapping module
    - Updating and fixing data sources with arcpy.mapping
  - Tool reference > Geoprocessing tools supplementary topics
    - Paths explained: Absolute, relative, UNC, and URL
- ArcGIS Help 10.1 > Guide Books > Geocoding > Sharing your address locator
  - About sharing your address locator as a locator package
- ArcGIS Help 10.1 > Desktop > Mapping > Map export and printing
  - About map printing
  - Exporting your map
  - Exporting to PDF
  - Exporting Data Driven Pages
- ArcGIS Help 10.1 > Desktop > Mapping > Using web maps and GIS services
  - Printing and exporting maps containing service layers
- Esri Support (support.esri.com) > Knowledge Base - Technical Articles > ArcInfo Desktop > ArcMap > Printing and Exporting
  - Map print or export blocky, chunky, low quality, or raster banded from ArcMap (Article ID: 17332)
  - HowTo: Modify output resolution (dpi) for ArcMap (Article ID: 18082)
  - HowTo: Use the Output Image Quality option (Article ID: 17845)
  - FAQ: What's the difference between the map page and the printer paper size? (Article ID: 29237)
  - Problem: When changing layout size and/or printers, the map elements scale improperly (Article ID: 32406)
  - Vector data is exported to EPS, PDF, and AI as raster in ArcMap (Article ID: 20051)
  - HowTo: Set up ArcMap page and printer settings (Article ID: 29238)
- Esri Support > White Papers > ArcView
  - ArcGIS and Printing: An Introduction To The Concepts Used When Printing From ArcGISTM

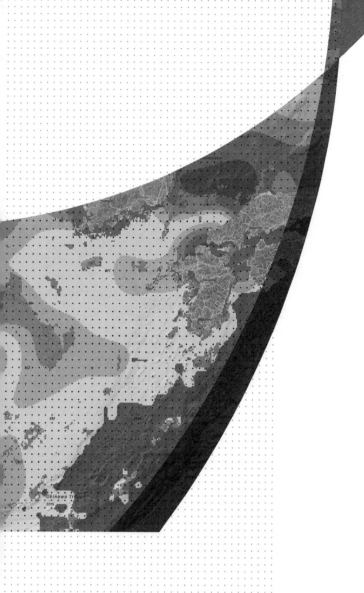

Image and data credits

All screenshots created by the author using exercise data.

Chapter 2

\EsriPress\DesktopAssociate\Chapter02\Boundary.DWG, Yellowstone National Park, 2008.

\EsriPress\DesktopAssociate\Chapter02\yellostlakes, from Esri Data & Maps, 2010; courtesy of U.S. Geological Survey, Esri.

\EsriPress\DesktopAssociate\Chapter02\Yellowstone.tif, data courtesy of U.S. Geological Survey.

\EsriPress\DesktopAssociate\Chapter02\YellowstoneRivers.shp, from Esri Data & Maps, 2010; courtesy of U.S. Geological Survey, Esri.

Chapter 3

\EsriPress\DesktopAssociate\Chapter03\YellowstoneRivers.shp, from Esri Data & Maps, 2010; courtesy of U.S. Geological Survey, Esri.

Chapter 4

\EsriPress\DesktopAssociate\Chapter04\NewData\Rivers\Gya_Cities.shp, Yellowstone National Park, 2008.

\EsriPress\DesktopAssociate\Chapter04\NewData\MajorRoads.shp, from Esri Data & Maps, 2010; courtesy of DeLorme.

\EsriPress\DesktopAssociate\Chapter04\NewData \ParkArea.shp, Yellowstone National Park, 2008.

\EsriPress\DesktopAssociate\Chapter04\NewData \Peaks.shp, Yellowstone National Park, 2008.

\EsriPress\DesktopAssociate\Chapter04\NewData \Trails.shp, Yellowstone National Park, 2008.

\EsriPress\DesktopAssociate\Chapter04\Rivers\Rvs_Fremont_Cnty.shp, from Esri Data & Maps, 2010; courtesy of U.S. Geological Survey, Esri.

\EsriPress\DesktopAssociate\Chapter04\Rivers\Rvs_Gallatin_Cnty.shp, from Esri Data & Maps, 2010; courtesy of U.S. Geological Survey, Esri.

\EsriPress\DesktopAssociate\Chapter04\Rivers\Rvs_Madison_Cnty.shp, from Esri Data & Maps, 2010; courtesy of U.S. Geological Survey, Esri.

\EsriPress\DesktopAssociate\Chapter04\Rivers\Rvs_Park_Cnty.shp, from Esri Data & Maps, 2010; courtesy of U.S. Geological Survey, Esri.

\EsriPress\DesktopAssociate\Chapter04\Rivers\Rvs_Teton_Cnty.shp, from Esri Data & Maps, 2010; courtesy of U.S. Geological Survey, Esri.

\EsriPress\DesktopAssociate\Chapter04\Camps.mdb\Campgrounds, Yellowstone National Park, 2008.

\EsriPress\DesktopAssociate\Chapter04\Yellowstone.gdb\Transportation\BusRoutes, created by the author.

\EsriPress\DesktopAssociate\Chapter04\Yellowstone.gdb\Transportation\BusStops, created by the author.

\EsriPress\DesktopAssociate\Chapter04\Yellowstone.gdb\YellowstoneLakes, from Esri Data & Maps, 2010; courtesy of U.S. Geological Survey, Esri.

\EsriPress\DesktopAssociate\Chapter04\Yellowstone.gdb\YellowstoneRivers, from Esri Data & Maps, 2010; courtesy of U.S. Geological Survey, Esri.

\EsriPress\DesktopAssociate\Chapter04\yellowstlakes, from Esri Data & Maps, 2010; courtesy of U.S. Geological Survey, Esri.

\EsriPress\DesktopAssociate\Chapter04\Highways.shp, from Esri Data & Maps, 2010; courtesy of DeLorme.

\EsriPress\DesktopAssociate\Chapter04\Roads.lyr, from Esri Data & Maps, 2010; courtesy of DeLorme.

\EsriPress\DesktopAssociate\Chapter04\YellowstoneRivers.shp, from Esri Data & Maps, 2010; courtesy of U.S. Geological Survey, Esri.

Chapter 5

\EsriPress\DesktopAssociate\Chapter05\Oceania.gdb\Continents, from Esri Data & Maps, 2010; courtesy of ArcWorld Supplement.

Image and data credits **365**

\EsriPress\DesktopAssociate\Chapter05\Oceania.gdb\Countries, from Esri Data & Maps, 2010; courtesy of ArcWorld Supplement.

\EsriPress\DesktopAssociate\Chapter05\Oceania.gdb\Pop_places, from Esri Data & Maps, 2010; courtesy of DeLorme.

\EsriPress\DesktopAssociate\Chapter05\Hydro_lines.shp, from Esri Data & Maps, 2010; courtesy of DeLorme.

\EsriPress\DesktopAssociate\Chapter05\Hydro_lines_1972.shp, from Esri Data & Maps, 2010; courtesy of DeLorme.

Chapter 6

\EsriPress\DesktopAssociate\Chapter06\Oceania.gdb\Admin_line, from Esri Data & Maps, 2010; courtesy of ArcWorld Supplement.

\EsriPress\DesktopAssociate\Chapter06\Oceania.gdb\Admin_poly, from Esri Data & Maps, 2010; courtesy of ArcWorld Supplement.

\EsriPress\DesktopAssociate\Chapter06\Oceania.gdb\Continents, from Esri Data & Maps, 2010; courtesy of ArcWorld Supplement.

\EsriPress\DesktopAssociate\Chapter06\Oceania.gdb\Countries, from Esri Data & Maps, 2010; courtesy of ArcWorld Supplement.

\EsriPress\DesktopAssociate\Chapter06\Oceania.gdb\gtopo_1km, from Esri Data & Maps, 2010; courtesy of DeLorme.

\EsriPress\DesktopAssociate\Chapter06\Oceania.gdb\Pop_places, from Esri Data & Maps, 2010; courtesy of DeLorme.

\EsriPress\DesktopAssociate\Chapter06\Oceania.gdb\Urban_areas, from Esri Data & Maps, 2010; courtesy of DeLorme.

\EsriPress\DesktopAssociate\Chapter06\OceaniaData.mxd, created by the author.

\EsriPress\DesktopAssociate\Chapter06\OceaniaData_100.mxd, created by the author.

\EsriPress\DesktopAssociate\Chapter06\UrbanAreas.gif, created by the author.

Chapter 7

\EsriPress\DesktopAssociate\Chapter07\Oceania.gdb\Admin_poly, from Esri Data & Maps, 2010; courtesy of ArcWorld Supplement.

\EsriPress\DesktopAssociate\Chapter07\Oceania.gdb\Continents, from Esri Data & Maps, 2010; courtesy of ArcWorld Supplement.

\EsriPress\DesktopAssociate\Chapter07\Oceania.gdb\Countries, from Esri Data & Maps, 2010; courtesy of ArcWorld Supplement.

\EsriPress\DesktopAssociate\Chapter07\Oceania.gdb\Country_info, from Esri Data & Maps, 2010; courtesy of ArcWorld Supplement.

\EsriPress\DesktopAssociate\Chapter07\Oceania.gdb\Gazetteer, from Esri Data & Maps, 2010; courtesy of DCW.

\EsriPress\DesktopAssociate\Chapter07\OceaniaTables.mxd, created by the author.

\EsriPress\DesktopAssociate\Chapter07\OceaniaTables_100.mxd, created by the author.

Chapter 8

\EsriPress\DesktopAssociate\Chapter08\Aerials\Woodside_1991.tif, courtesy of the U.S. Geological Survey.

\EsriPress\DesktopAssociate\Chapter08\Aerials\Woodside1948.tif, courtesy of the U.S. Geological Survey.

\EsriPress\DesktopAssociate\Chapter08\Woodside.gdb\Highways, from Esri Data & Maps, 2010; courtesy of DeLorme.

\EsriPress\DesktopAssociate\Chapter08\Woodside.gdb\SuggestedControlPoints, created by the author.

\EsriPress\DesktopAssociate\Chapter08\Georeferencing.mxd, created by the author.

\EsriPress\DesktopAssociate\Chapter08\Georeferencing_100.mxd, created by the author.
\EsriPress\DesktopAssociate\Chapter08\Roads.shp, from Esri Data & Maps, 2010; courtesy of DeLorme.
\EsriPress\DesktopAssociate\Chapter08\SpatialAdjustment.mxd, created by the author.
\EsriPress\DesktopAssociate\Chapter08\SpatialAdjustment_100.mxd, created by the author.

Chapter 9

\EsriPress\DesktopAssociate\Chapter09\Locators\Hamilton_DualRanges, courtesy of Hamilton County, Indiana.
\EsriPress\DesktopAssociate\Chapter09\Locators\Hamilton_DualRanges_100, courtesy of Hamilton County, Indiana.
\EsriPress\DesktopAssociate\Chapter09\Hamilton.gdb\Centerlines, courtesy of Hamilton County, Indiana.
\EsriPress\DesktopAssociate\Chapter09\Hamilton.gdb\CustomerInfo, courtesy of Hamilton County, Indiana.
\EsriPress\DesktopAssociate\Chapter09\Hamilton.gdb\CustomerInfoPlus, courtesy of Hamilton County, Indiana.
\EsriPress\DesktopAssociate\Chapter09\Hamilton.gdb\Zip_codes, courtesy of Esri.
\EsriPress\DesktopAssociate\Chapter09\Geocoding.mxd, created by the author.
\EsriPress\DesktopAssociate\Chapter09\Geocoding_100.mxd, created by the author.

Chapter 10

\EsriPress\DesktopAssociate\Chapter10\Imagery\AUSTIN_EAST-NEA3.JP2, courtesy of The City of Austin.
\EsriPress\DesktopAssociate\Chapter10\Imagery\AUSTIN_EAST-SED1.JP2, courtesy of The City of Austin.
\EsriPress\DesktopAssociate\Chapter10\Imagery\AUSTIN_EAST-SED2.JP2, courtesy of The City of Austin.
\EsriPress\DesktopAssociate\Chapter10\Imagery\AUSTIN_EAST-SED3.JP2, courtesy of The City of Austin.
\EsriPress\DesktopAssociate\Chapter10\Imagery\AUSTIN_EAST-SED4.JP2, courtesy of The City of Austin.
\EsriPress\DesktopAssociate\Chapter10\Austin.gdb\LandBase, courtesy of The City of Austin.
\EsriPress\DesktopAssociate\Chapter10\Austin.gdb\Parks, courtesy of The City of Austin.
\EsriPress\DesktopAssociate\Chapter10\Austin.gdb\RightOfWay, courtesy of The City of Austin.
\EsriPress\DesktopAssociate\Chapter10\Austin.gdb\Streets, courtesy of The City of Austin.
\EsriPress\DesktopAssociate\Chapter10\Austin.gdb\StreetsNortheast, courtesy of The City of Austin.
\EsriPress\DesktopAssociate\Chapter10\Austin.gdb\ZilkerPark, courtesy of The City of Austin.
\EsriPress\DesktopAssociate\Chapter10\ConstructFeatures.mxd, created by the author.
\EsriPress\DesktopAssociate\Chapter10\ConstructFeatures_100.mxd, created by the author.
\EsriPress\DesktopAssociate\Chapter10\ConstructionMethods.mxd, created by the author.
\EsriPress\DesktopAssociate\Chapter10\ConstructionMethods_100.mxd, created by the author.
\EsriPress\DesktopAssociate\Chapter10\ConstructionTools.mxd, created by the author.
\EsriPress\DesktopAssociate\Chapter10\ConstructionTools_100.mxd, created by the author.
\EsriPress\DesktopAssociate\Chapter10\FeatureTemplates.mxd, created by the author.
\EsriPress\DesktopAssociate\Chapter10\FeatureTemplates_100.mxd, created by the author.
\EsriPress\DesktopAssociate\Chapter10\Historical_Landmarks.shp, courtesy of The City of Austin.

Chapter 11

\EsriPress\DesktopAssociate\Chapter11\Imagery\AUSTIN_EAST-NEA1.jp2, courtesy of The City of Austin.
\EsriPress\DesktopAssociate\Chapter11\Imagery\AUSTIN_EAST-SWC1.jp2, courtesy of The City of Austin.
\EsriPress\DesktopAssociate\Chapter11\Imagery\AUSTIN_WEST-SED1.jp2, courtesy of The City of Austin.
\EsriPress\DesktopAssociate\Chapter11\Imagery\AUSTIN_WEST-SED2.jp2, courtesy of The City of Austin.
\EsriPress\DesktopAssociate\Chapter11\Imagery\AUSTIN_WEST-SED3.jp2, courtesy of The City of Austin.

\EsriPress\DesktopAssociate\Chapter11\Imagery\AUSTIN_WEST-SED4.jp2, courtesy of The City of Austin.
\EsriPress\DesktopAssociate\Chapter11\Austin.gdb\BartonSpringsPool, created by the author.
\EsriPress\DesktopAssociate\Chapter11\Austin.gdb\Parks, courtesy of The City of Austin.
\EsriPress\DesktopAssociate\Chapter11\Austin.gdb\Streets, courtesy of The City of Austin.
\EsriPress\DesktopAssociate\Chapter11\Austin.gdb\StreetsNortheast, courtesy of The City of Austin.
\EsriPress\DesktopAssociate\Chapter11\Extend_Trim.mxd, created by the author.
\EsriPress\DesktopAssociate\Chapter11\Extend_Trim_100.mxd, created by the author.
\EsriPress\DesktopAssociate\Chapter11\Reshape.mxd, created by the author.
\EsriPress\DesktopAssociate\Chapter11\Reshape_100.mxd, created by the author.
\EsriPress\DesktopAssociate\Chapter11\Split_Cut.mxd, created by the author.
\EsriPress\DesktopAssociate\Chapter11\Split_Cut_100.mxd, created by the author.

Chapter 12

EsriPress\DesktopAssociate\Chapter12\Austin.gdb\AddressPoints, courtesy of The City of Austin.
EsriPress\DesktopAssociate\Chapter12\Austin.gdb\Centerlines, courtesy of The City of Austin.
EsriPress\DesktopAssociate\Chapter12\EditingAttributes.mxd, created by the author.
EsriPress\DesktopAssociate\Chapter12\EditingAttributes_100.mxd, created by the author.

Chapter 13

EsriPress\DesktopAssociate\Chapter13\Austin.gdb\Centerlines, courtesy of The City of Austin.
EsriPress\DesktopAssociate\Chapter13\Austin.gdb\StreetTypes, courtesy of The City of Austin.

Chapter 14

EsriPress\DesktopAssociate\Chapter14\Noblesville.gdb\Hydrology\Hydrology_Topology, created by the author.
EsriPress\DesktopAssociate\Chapter14\Noblesville.gdb\Hydrology\Streams, courtesy of Hamilton County, Indiana.
EsriPress\DesktopAssociate\Chapter14\Noblesville.gdb\Hydrology\WaterSamples, created by the author.
EsriPress\DesktopAssociate\Chapter14\Noblesville.gdb\Landbase\Annexation, courtesy of Hamilton County, Indiana.
EsriPress\DesktopAssociate\Chapter14\Noblesville.gdb\Landbase\Landbase_Topology, created by the author.
EsriPress\DesktopAssociate\Chapter14\Noblesville.gdb\Landbase\Landtrust, created by the author.
EsriPress\DesktopAssociate\Chapter14\Noblesville.gdb\Landbase\Precincts, courtesy of Hamilton County, Indiana.
EsriPress\DesktopAssociate\Chapter14\Noblesville.gdb\ LandTrust_Hydro.mxd, created by the author.
EsriPress\DesktopAssociate\Chapter14\Noblesville.gdb\ LandTrust_Hydro_100.mxd, created by the author.
EsriPress\DesktopAssociate\Chapter14\Noblesville.gdb\ TopoEditing.mxd, created by the author.
EsriPress\DesktopAssociate\Chapter14\Noblesville.gdb\ TopoEditing_100.mxd, created by the author.

Chapter 15

EsriPress\DesktopAssociate\Chapter15\CityParkSiteSelection.gdb, created by the author.
EsriPress\DesktopAssociate\Chapter15\Fishers.gdb\Centerlines, courtesy of Hamilton County, Indiana.
EsriPress\DesktopAssociate\Chapter15\Fishers.gdb\Interstate, courtesy of Hamilton County, Indiana.
EsriPress\DesktopAssociate\Chapter15\Fishers.gdb\Parcels, courtesy of Hamilton County, Indiana.
EsriPress\DesktopAssociate\Chapter15\CityPark.tbx, created by the author.
EsriPress\DesktopAssociate\Chapter15\ FishersModel.mxd, created by the author.

EsriPress\DesktopAssociate\Chapter15\ FishersModel_100.mxd, created by the author.
EsriPress\DesktopAssociate\Chapter15\ FishersScript.mxd, created by the author.
EsriPress\DesktopAssociate\Chapter15\ FishersScript_100.mxd, created by the author.
EsriPress\DesktopAssociate\Chapter15\ FishersTool.mxd, created by the author.
EsriPress\DesktopAssociate\Chapter15\ FishersTool_100.mxd, created by the author.

Chapter 16

EsriPress\DesktopAssociate\Chapter16\Census.gdb\DaneCntyblk00, courtesy of the U.S. Census Bureau.
EsriPress\DesktopAssociate\Chapter16\Census.gdb\DaneCntyTrt00, courtesy of the U.S. Census Bureau.
EsriPress\DesktopAssociate\Chapter16\Census.gdb\TractPopulation, courtesy of the U.S. Census Bureau.
EsriPress\DesktopAssociate\Chapter16\Fishers.gdb\Centerlines, courtesy of Hamilton County, Indiana.
EsriPress\DesktopAssociate\Chapter16\Fishers.gdb\Interstate, courtesy of Hamilton County, Indiana.
EsriPress\DesktopAssociate\Chapter16\Fishers.gdb\Neighborhoods, courtesy of Hamilton County, Indiana.
EsriPress\DesktopAssociate\Chapter16\Fishers.gdb\Parcels, courtesy of Hamilton County, Indiana.
EsriPress\DesktopAssociate\Chapter16\Fishers.gdb\ParcelsforParks, courtesy of Hamilton County, Indiana.
EsriPress\DesktopAssociate\Chapter16\Attributes.mxd, created by the author.
EsriPress\DesktopAssociate\Chapter16\Attributes_100.mxd, created by the author.
EsriPress\DesktopAssociate\Chapter16\ParkSelection.mxd, created by the author.
EsriPress\DesktopAssociate\Chapter16\ParkSelection_100.mxd, created by the author.
EsriPress\DesktopAssociate\Chapter16\Selections.mxd, created by the author.
EsriPress\DesktopAssociate\Chapter16\Selections_100.mxd, created by the author.

Chapter 17

EsriPress\DesktopAssociate\Chapter17\Austin.gdb, DistCntr, courtesy of The City of Austin.
EsriPress\DesktopAssociate\Chapter17\Austin.gdb, PostOffice, courtesy of The City of Austin.
EsriPress\DesktopAssociate\Chapter17\Census.gdb, CensusBlk, courtesy of the U.S. Census Bureau.
EsriPress\DesktopAssociate\Chapter17\Overlay.mxd, created by the author.
EsriPress\DesktopAssociate\Chapter17\Overlay_100.mxd, created by the author.
EsriPress\DesktopAssociate\Chapter17\Proximity.mxd, created by the author.
EsriPress\DesktopAssociate\Chapter17\Proximity_100.mxd, created by the author.
EsriPress\DesktopAssociate\Chapter17\Statistics.mxd, created by the author.
EsriPress\DesktopAssociate\Chapter17\Statistics_100.mxd, created by the author.

Chapter 18

EsriPress\DesktopAssociate\Chapter18\SA_Rasters\SA_dem.tif, from Esri Data & Maps, 2007; courtesy of USGS EROS Data Center.
EsriPress\DesktopAssociate\Chapter18\SA_Rasters\SA_hillshade.tif, from Esri Data & Maps, 2007; courtesy of USGS EROS Data Center.
EsriPress\DesktopAssociate\Chapter18\South_America.gdb\Admin, from Esri Data & Maps, 2010; courtesy of ArcWorld Supplement.
EsriPress\DesktopAssociate\Chapter18\South_America.gdb\Cities, from Esri Data & Maps, 2010; courtesy of ArcWorld.
EsriPress\DesktopAssociate\Chapter18\South_America.gdb\Countries, from Esri Data & Maps, 2010; courtesy of ArcWorld Supplement.
EsriPress\DesktopAssociate\Chapter18\South_America.gdb\Lakes, from Esri Data & Maps, 2010; courtesy of U.S. Geological Survey, Esri.

EsriPress\DesktopAssociate\Chapter18\South_America.gdb\Railroads, from Esri Data & Maps, 2010; courtesy of DeLorme.

EsriPress\DesktopAssociate\Chapter18\South_America.gdb\Rivers, from Esri Data & Maps, 2010; courtesy of U.S. Geological Survey, Esri.

EsriPress\DesktopAssociate\Chapter18\South_America.gdb\Roads, from Esri Data & Maps, 2010; courtesy of DeLorme.

EsriPress\DesktopAssociate\Chapter18\South_America.gdb\World_Countries, from Esri Data & Maps, 2010; courtesy of ArcWorld Supplement.

EsriPress\DesktopAssociate\Chapter18\DataFrameProperties.mxd, created by the author.

EsriPress\DesktopAssociate\Chapter18\DataFrameProperties_100.mxd, created by the author.

EsriPress\DesktopAssociate\Chapter18\OrganizeLayers.mxd, created by the author.

EsriPress\DesktopAssociate\Chapter18\OrganizeLayers_100.mxd, created by the author.

EsriPress\DesktopAssociate\Chapter18\ReferenceFixedScale.mxd, created by the author.

EsriPress\DesktopAssociate\Chapter18\ReferenceFixedScale_100.mxd, created by the author.

Chapter 19

EsriPress\DesktopAssociate\Chapter19\SA_Rasters\SA_dem.tif, from Esri Data & Maps, 2007; courtesy of USGS EROS Data Center.

EsriPress\DesktopAssociate\Chapter19\SA_Rasters\SA_hillshade.tif, from Esri Data & Maps, 2007; courtesy of USGS EROS Data Center.

EsriPress\DesktopAssociate\Chapter19\South_America.gdb\Admin, from Esri Data & Maps, 2010; courtesy of ArcWorld Supplement.

EsriPress\DesktopAssociate\Chapter19\South_America.gdb\Cities, from Esri Data & Maps, 2010; courtesy of ArcWorld.

EsriPress\DesktopAssociate\Chapter19\South_America.gdb\Countries, from Esri Data & Maps, 2010; courtesy of ArcWorld Supplement.

EsriPress\DesktopAssociate\Chapter19\South_America.gdb\Lakes, from Esri Data & Maps, 2010; courtesy of U.S. Geological Survey, Esri. from Esri Data & Maps, 2010; courtesy of U.S. Geological Survey, Esri.EsriPress\DesktopAssociate\Chapter19\South_America.gdb\Railroads, from Esri Data & Maps, 2010; courtesy of DeLorme.

EsriPress\DesktopAssociate\Chapter19\South_America.gdb\Rivers, from Esri Data & Maps, 2010; courtesy of U.S. Geological Survey, Esri.

EsriPress\DesktopAssociate\Chapter19\South_America.gdb\Roads, from Esri Data & Maps, 2010; courtesy of DeLorme.

EsriPress\DesktopAssociate\Chapter19\South_America.gdb\World_Countries, from Esri Data & Maps, 2010; courtesy of ArcWorld Supplement.

EsriPress\DesktopAssociate\Chapter19\DefinitionQuery.mxd, created by the author.

EsriPress\DesktopAssociate\Chapter19\DefinitionQuery_100.mxd, created by the author.

EsriPress\DesktopAssociate\Chapter19\ScaleDependency.mxd, created by the author.

EsriPress\DesktopAssociate\Chapter19\ScaleDependency_100.mxd, created by the author.

EsriPress\DesktopAssociate\Chapter19\SymbolizeRasterLayers.mxd, created by the author.

EsriPress\DesktopAssociate\Chapter19\SymbolizeRasterLayers_100.mxd, created by the author.

EsriPress\DesktopAssociate\Chapter19\SymbolizeVectorLayers.mxd, created by the author.

EsriPress\DesktopAssociate\Chapter19\SymbolizeVectorLayers_100.mxd, created by the author.

Chapter 20

EsriPress\DesktopAssociate\Chapter20\SA_Rasters\SA_dem.tif, from Esri Data & Maps, 2007; courtesy of USGS EROS Data Center.

EsriPress\DesktopAssociate\Chapter20\SA_Rasters\SA_hillshade.tif, from Esri Data & Maps, 2007; courtesy of USGS EROS Data Center.

EsriPress\DesktopAssociate\Chapter20\South_America.gdb\Admin, from Esri Data & Maps, 2010; courtesy of ArcWorld Supplement.

EsriPress\DesktopAssociate\Chapter20\South_America.gdb\Cities, from Esri Data & Maps, 2010; courtesy of ArcWorld.

EsriPress\DesktopAssociate\Chapter20\South_America.gdb\Countries, from Esri Data & Maps, 2010; courtesy of ArcWorld Supplement.

EsriPress\DesktopAssociate\Chapter20\South_America.gdb\Lakes, from Esri Data & Maps, 2010; courtesy of U.S. Geological Survey, Esri.

EsriPress\DesktopAssociate\Chapter20\South_America.gdb\Railroads, from Esri Data & Maps, 2010; courtesy of DeLorme.

EsriPress\DesktopAssociate\Chapter20\South_America.gdb\Rivers, from Esri Data & Maps, 2010; courtesy of U.S. Geological Survey, Esri. from Esri Data & Maps, 2010; courtesy of U.S. Geological Survey, Esri.EsriPress\DesktopAssociate\Chapter20\South_America.gdb\Roads, from Esri Data & Maps, 2010; courtesy of DeLorme.

EsriPress\DesktopAssociate\Chapter20\South_America.gdb\World_Countries, from Esri Data & Maps, 2010; courtesy of ArcWorld Supplement.

EsriPress\DesktopAssociate\Chapter20\AddScalebar.mxd, created by the author.

EsriPress\DesktopAssociate\Chapter20\AddScalebar_100.mxd, created by the author.

EsriPress\DesktopAssociate\Chapter20\CreateGraph.mxd, created by the author.

EsriPress\DesktopAssociate\Chapter20\CreateGraph_100.mxd, created by the author.

EsriPress\DesktopAssociate\Chapter20\ExtentIndicator.mxd, created by the author.

EsriPress\DesktopAssociate\Chapter20\ExtentIndicator_100.mxd, created by the author.

EsriPress\DesktopAssociate\Chapter20\InsertLegend.mxd, created by the author.

EsriPress\DesktopAssociate\Chapter20\InsertLegend_100.mxd, created by the author.

EsriPress\DesktopAssociate\Chapter20\PopulationDensity.tee, created by the author.

Chapter 21

EsriPress\DesktopAssociate\Chapter21\South_America.gdb\Admin, from Esri Data & Maps, 2010; courtesy of ArcWorld Supplement.

EsriPress\DesktopAssociate\Chapter21\South_America.gdb\SA_Countries, from Esri Data & Maps, 2010; courtesy of ArcWorld Supplement.

EsriPress\DesktopAssociate\Chapter21\South_America.gdb\World_Countries, from Esri Data & Maps, 2010; courtesy of ArcWorld Supplement.

EsriPress\DesktopAssociate\Chapter21\CreateDDP.mxd, created by the author.

EsriPress\DesktopAssociate\Chapter21\CreateDDP_100.mxd, created by the author.

EsriPress\DesktopAssociate\Chapter21\PageDefinitionQuery.mxd, created by the author.

EsriPress\DesktopAssociate\Chapter21\PageDefinitionQuery_100.mxd, created by the author.

EsriPress\DesktopAssociate\Chapter21\RefineDDP.mxd, created by the author.

EsriPress\DesktopAssociate\Chapter21\RefineDDP_100.mxd, created by the author.

Chapter 22

EsriPress\DesktopAssociate\Chapter22\SouthAmerica.gdb\Anno_27_78, software-generated.

EsriPress\DesktopAssociate\Chapter22\SouthAmerica.gdb\PeruAdminAreas, from Esri Data & Maps, 2010; courtesy of ArcWorld Supplement.

EsriPress\DesktopAssociate\Chapter22\SouthAmerica.gdb\PeruCities, from Esri Data & Maps, 2010; courtesy of ArcWorld.

EsriPress\DesktopAssociate\Chapter22\SouthAmerica.gdb\PeruCitiesAnno, from Esri Data & Maps, 2010; courtesy of ArcWorld.

EsriPress\DesktopAssociate\Chapter22\SouthAmerica.gdb\PeruRoads, from Esri Data & Maps, 2010; courtesy of DeLorme.

EsriPress\DesktopAssociate\Chapter22\SouthAmerica.gdb\PeruSites, from Esri Data & Maps, 2010; courtesy of ArcWorld.

EsriPress\DesktopAssociate\Chapter22\SouthAmerica.gdb\SouthAmericaCountries, from Esri Data & Maps, 2010; courtesy of ArcWorld Supplement.

EsriPress\DesktopAssociate\Chapter22\SouthAmerica.gdb\ SouthAmericaCountriesAnno, from Esri Data & Maps, 2010; courtesy of ArcWorld Supplement.

EsriPress\DesktopAssociate\Chapter22\SouthAmerica.gdb\ SouthAmericaLakes, from Esri Data & Maps, 2010; courtesy of U.S. Geological Survey, Esri.

EsriPress\DesktopAssociate\Chapter22\SouthAmerica.gdb\ SouthAmericaRivers, from Esri Data & Maps, 2010; courtesy of U.S. Geological Survey, Esri.

EsriPress\DesktopAssociate\Chapter22\AnnoFeatures.mxd, created by the author.

EsriPress\DesktopAssociate\Chapter22\AnnoFeatures_100.mxd, created by the author.

EsriPress\DesktopAssociate\Chapter22\CustomMarkerSymbols.mxd, created by the author.

EsriPress\DesktopAssociate\Chapter22\CustomMarkerSymbols_100.mxd, created by the author.

EsriPress\DesktopAssociate\Chapter22\LabelClasses.mxd, created by the author.

EsriPress\DesktopAssociate\Chapter22\LabelClasses_100.mxd, created by the author.

EsriPress\DesktopAssociate\Chapter22\LabelExpressions.mxd, created by the author.

EsriPress\DesktopAssociate\Chapter22\LabelExpressions_100.mxd, created by the author.

Chapter 23

EsriPress\DesktopAssociate\Chapter23\World_U.N.gdb\Countries, from Esri Data & Maps, 2010; courtesy of ArcWorld Supplement.

EsriPress\DesktopAssociate\Chapter23\PrepareToPublish.mxd., created by the author.

EsriPress\DesktopAssociate\Chapter23\PrepareToPublish_100.mxd., created by the author.

EsriPress\DesktopAssociate\Chapter23\TimeAwareMap.mxd., created by the author.

EsriPress\DesktopAssociate\Chapter23\TimeAwareMap_100.mxd., created by the author.

Chapter 24

EsriPress\DesktopAssociate\Chapter24\SA_Rasters\SA_dem.tif, from Esri Data & Maps, 2007; courtesy of USGS EROS Data Center.

EsriPress\DesktopAssociate\Chapter24\SA_Rasters\SA_hillshade.tif, from Esri Data & Maps, 2007; courtesy of USGS EROS Data Center.

EsriPress\DesktopAssociate\Chapter24\World_UN.gdb\Countries, from Esri Data & Maps, 2010; courtesy of ArcWorld Supplement.

EsriPress\DesktopAssociate\Chapter24\World_UN.gdb\Oceans_Anno, created by the author.

EsriPress\DesktopAssociate\Chapter24\South_America.gdb.\Admin, from Esri Data & Maps, 2010; courtesy of ArcWorld Supplement.

EsriPress\DesktopAssociate\Chapter24\South_America.gdb.\Cities, from Esri Data & Maps, 2010; courtesy of ArcWorld.

EsriPress\DesktopAssociate\Chapter24\South_America.gdb.\Countries, from Esri Data & Maps, 2010; courtesy of ArcWorld Supplement.

EsriPress\DesktopAssociate\Chapter24\South_America.gdb.\World_Countries, from Esri Data & Maps, 2010; courtesy of ArcWorld Supplement.

EsriPress\DesktopAssociate\Chapter24\ExportMap.mxd, created by the author.

EsriPress\DesktopAssociate\Chapter24\ ExportMap_100.mxd, created by the author.
EsriPress\DesktopAssociate\Chapter24\PackageMap.mxd, created by the author.
EsriPress\DesktopAssociate\Chapter24\PackageMap_100.mxd, created by the author.
EsriPress\DesktopAssociate\Chapter24\PreparePrinting.mxd, created by the author.
EsriPress\DesktopAssociate\Chapter24\PreparePrinting_100.mxd, created by the author.
EsriPress\DesktopAssociate\Chapter24\RelativePaths.mxd, created by the author.
EsriPress\DesktopAssociate\Chapter24\RelativePaths_100.mxd, created by the author.
EsriPress\DesktopAssociate\Chapter24\Transportation.lpk, from Esri Data & Maps, 2010; courtesy of DeLorme.
EsriPress\DesktopAssociate\Chapter24\Transportation_100.lpk, from Esri Data & Maps, 2010; courtesy of DeLorme.

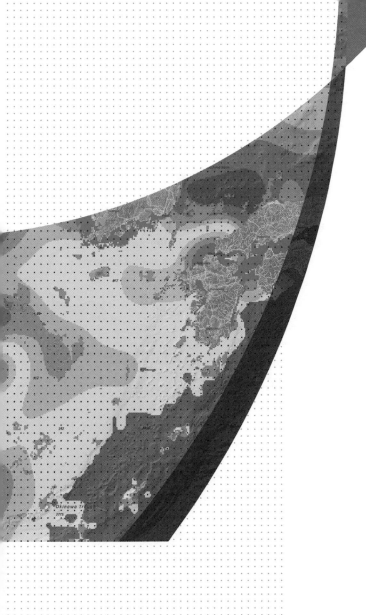

Installing the data and software

374 Installing the data and software

Esri ArcGIS Desktop Associate: Certification Study Guide includes a DVD containing data for working through the exercises. A free, fully functioning 180-day version of ArcGIS 10.1 for Desktop Advanced license level or ArcGIS Desktop 10 ArcEditor license level, can be downloaded at **www.esri.com/DesktopAssociateStudyGuide**. You will find an authorization number printed on the inside back cover of this book. You will use this number when you are ready to install the software.

If you already have a licensed copy of ArcGIS 10.1 for Desktop or ArcGIS Desktop 10 installed on your computer (or have access to the software through a network), do not install the trial software. Use your licensed software to do the exercises in this book. If you have an older version of ArcGIS installed on your computer, you must uninstall it before you can install the software that is provided with this book.

.NET Framework 3.5 SP1 must be installed on your computer before you install ArcGIS 10.1 for Desktop or ArcGIS Desktop 10. Some features of ArcGIS 10.1 for Desktop and ArcGIS Desktop 10 require Windows Internet Explorer version 8.0. If you do not have Internet Explorer version 8.0, you must install it before installing ArcGIS 10.1 for Desktop or ArcGIS Desktop 10.

Installing the exercise data

Follow the steps below to install the exercise data.

1. **Put the data DVD in your computer's DVD drive. A splash screen will appear.**

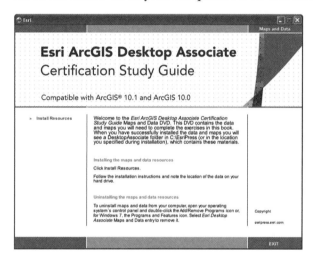

2. **Read the welcome, and then click the Install Data and Exercises link. This launches the InstallShield Wizard.**

3. **Click Next. Read and accept the license agreement terms, and then click Next.**

Installing the data and software **375**

4. Accept the default installation folder or click Browse and navigate to the drive or folder location where you want to install the data.

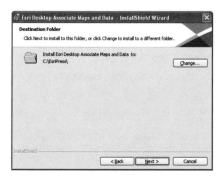

5. Click Next. The installation will take a few moments. When the installation is complete, you will see the following message.

6. Click Finish. The exercise data is installed on your computer in a folder called C:\EsriPress\DesktopAssociate, or in the location you chose in step 4.

Uninstalling the data and resources

To uninstall the data and resources from your computer, open your operating system's control panel and double-click the Add/Remove Programs icon. In the Add/Remove Programs dialog box, select the following entry and follow the prompts to remove it:

Desktop Associate Study Guide

Installing the software

Note: If you already have a licensed copy of ArcGIS 10.1 for Desktop or ArcGIS Desktop 10 installed on your computer or have access to the software through a network, use it to do the exercises in this book. If you need a copy of ArcGIS 10.1 for Desktop or ArcGIS Desktop 10, you can obtain a free 180-day trial version from Esri. This trial version is intended for educational purposes only and will expire 180 days after you install and register the software. The software cannot be reinstalled nor can the time limit be extended. It is recommended that you uninstall this software when it expires.

Installing ArcGIS 10.1 for Desktop Advanced license 180-day trial

Follow these steps to obtain the 180-day free trial of ArcGIS 10.1 for Desktop Advanced:

1. Uninstall any previous versions of ArcGIS Desktop that you already have on your computer.

2. Check the system requirements for ArcGIS to make sure your computer has the hardware and software required for the trial: **www.esri.com/ArcGIS101sysreq**.

3. Install the Microsoft .NET Framework 3.5 Service Pack 1 from the Download Center on the Microsoft website.

4. Go to www.esri.com/DesktopAssociateStudyGuide. Select the option that's right for you and follow the wizard instructions.

Note: You must have an Esri Global Account to receive your free trial software. Click "Create an Account" if you do not have one. When prompted, enter your 12-character authorization number (EVAxxxxxxxxx) printed on the inside back cover of this book, and then click Submit.

5. **Authorize your trial software.**

   - Once the ArcGIS Administrator Wizard dialog box opens, select Advanced (Single Use) in the ArcGIS for Desktop section, and then click Authorize Now.

   - Continue to click Next until you are prompted to enter your 12-character authorization number (EVAxxxxxxxxx), which you can find printed on the inside back cover of this book. Click Next.

   - Select "I do not want to authorize any extensions at this time." The trial software automatically authorizes the extensions.

   - On the next screen, verify that the extensions are listed on the left under Available Extensions. Click Next.

   - After the authorization process is complete, click Finish.

   - Click OK to close the ArcGIS Administrator dialog box.

6. **To start ArcGIS 10.1 for Desktop, on the taskbar, click the Start button. On the Start menu, click All Programs > ArcGIS > ArcMap 10.1.**

7. **Enable the extensions:**

   - Start ArcMap. Go to Customize > Extensions and make sure each extension is selected.

   - Click Close. The software is ready to use.

Assistance, FAQs, and support for your trial software are available on the online resources page at **www.esri.com/trialhelp**.

Installing ArcGIS Desktop 10 ArcEditor license 180-day trial

Follow the steps below to install the software.

1. Uninstall any previous versions of ArcGIS that you may already have on your machine. This includes any version of ArcGIS Desktop (ArcView, ArcEditor, ArcInfo), the ArcGIS 9.x License Manager, and ArcGIS Server. The only exception is ArcGIS Explorer.

2. Install an image (ISO) extractor program such as 7-Zip. Check the system requirements for ArcGIS to make sure your computer has the hardware and software required for the trial: **www.esri.com/AG10systemrequirements**.

3. Install the Microsoft .NET Framework 3.5 Service Pack 1.

4. Go to www.esri.com/DesktopAssociateStudyGuide. Select the option that's appropriate for you and click Next. Note: You must have an Esri Global Account to receive your free trial software. Click "Create an Account" if you do not have one. When prompted, enter your 12-character authorization number (EVAxxxxxxxxx), printed on the inside back cover of this book, and then click Submit.

5. Unzip the ArcGIS Desktop 10 image files that you downloaded using 7-Zip. Visit the online resources page for help with using 7-Zip at **esri.com/evalhelp**.

6. **Install ArcGIS Desktop 10.**

   To download the software

   - Extract (unzip) the ArcGIS Desktop 10 files once they have been downloaded to your computer.
   - Go to the location of the extracted files and run ESRI.exe.

   To install the software from a DVD

   - Insert the DVD into your computer.
   - The Quick Start Guide will run automatically. If not, go to My Computer and double-click your DVD drive to run ESRI.exe.
   - To the right of ArcGIS Desktop, click Setup. Click Next and accept the license agreement. Choose Complete Installation and click Next. Click Next to accept all default settings; then click Finish.

7. **Authorize your trial software**

   - Once the ArcGIS Administrator Wizard dialog box opens, select ArcEditor (Single Use) in the ArcGIS Desktop section, and then click Authorize Now.
   - Continue to click Next until you are prompted to enter your 12-character authorization number (EVAxxxxxxxxx), which you can find printed on the inside back cover of the book. Click Next.
   - Select "I do not want to authorize any extensions at this time." The trial software automatically authorizes the extensions.
   - On the next screen, verify that the extensions are listed on the left under Available Extensions. Click Next.
   - After the authorization process is complete, click Finish.
   - Click OK to close the ArcGIS Administrator dialog box.

8. **To start ArcGIS Desktop 10, on the taskbar, click the Start button. On the Start menu, click All Programs > ArcGIS and choose ArcMap 10.**

9. **Enable the extensions.**

   - Start ArcMap. Go to Customize > Extensions and make sure each extension is checked.
   - Click Close. The software is ready to use.

Assistance, FAQs, and support for your complimentary software are available on the online resources page, **esri.com/evalhelp**.

Uninstalling the software

To uninstall the software from your computer, open your operating system's control panel and double-click the Add/Remove Programs icon. In the Add/Remove Programs dialog box, select the following entry, depending on which version of the software you installed, and follow the prompts to remove it:

ArcGIS 10.1 for Desktop

or

ArcGIS Desktop 10

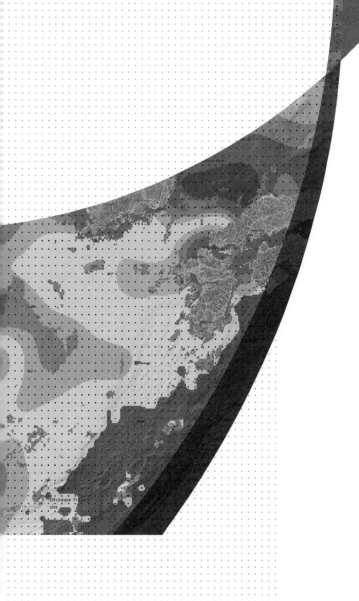

Data license agreement

Data license agreement

Important: Read carefully before opening the sealed media package.

Environmental Systems Research Institute, Inc. (Esri), is willing to license the enclosed data and related materials to you only upon the condition that you accept all of the terms and conditions contained in this license agreement. Please read the terms and conditions carefully before opening the sealed media package. By opening the sealed media package, you are indicating your acceptance of the Esri License Agreement. If you do not agree to the terms and conditions as stated, and then Esri is unwilling to license the data and related materials to you. In such event, you should return the media package with the seal unbroken and all other components to Esri.

Esri License Agreement

This is a license agreement, and not an agreement for sale, between you (Licensee) and Environmental Systems Research Institute, Inc. (Esri). This Esri License Agreement (Agreement) gives Licensee certain limited rights to use the data and related materials (Data and Related Materials). All rights not specifically granted in this Agreement are reserved to Esri and its Licensors.

Reservation of Ownership and Grant of License

Esri and its Licensors retain exclusive rights, title, and ownership to the copy of the Data and Related Materials licensed under this Agreement and, hereby, grant to Licensee a personal, nonexclusive, nontransferable, royalty-free, worldwide license to use the Data and Related Materials based on the terms and conditions of this Agreement. Licensee agrees to use reasonable effort to protect the Data and Related Materials from unauthorized use, reproduction, distribution, or publication.

Proprietary Rights and Copyright

Licensee acknowledges that the Data and Related Materials are proprietary and confidential property of Esri and its Licensors and are protected by United States copyright laws and applicable international copyright treaties and/or conventions.

Permitted Uses

Licensee may install the Data and Related Materials onto permanent storage device(s) for Licensee's own internal use. Licensee may make only one (1) copy of the original Data and Related Materials for archival purposes during the term of this Agreement unless the right to make additional copies is granted to Licensee in writing by Esri. Licensee may internally use the Data and Related Materials provided by Esri for the stated purpose of GIS training and education.

Uses Not Permitted

Licensee shall not sell, rent, lease, sublicense, lend, assign, time-share, or transfer, in whole or in part, or provide unlicensed Third Parties access to the Data and Related Materials or portions of the Data and Related Materials, any updates, or Licensee's rights under this Agreement.

Licensee shall not remove or obscure any copyright or trademark notices of Esri or its Licensors.

Term and Termination

The license granted to Licensee by this Agreement shall commence upon the acceptance of this Agreement and shall continue until such time that Licensee elects in writing to discontinue use of the Data or Related Materials and terminates this Agreement. The Agreement shall automatically terminate without notice if Licensee fails to comply with any provision of this Agreement. Licensee shall then return to Esri the Data and Related Materials. The parties hereby agree that all provisions that operate to protect the rights of Esri and its Licensors shall remain in force should breach occur.

Disclaimer of Warranty

The Data and Related Materials contained herein are provided "as-is," without warranty of any kind, either express or implied, including, but not limited to, the implied warranties of merchantability, fitness for a particular purpose, or noninfringement. Esri does not warrant that the Data and Related Materials will meet Licensee's needs or expectations, that the use of the Data and Related Materials will be uninterrupted, or that all nonconformities, defects, or errors can or will be corrected. Esri is not inviting reliance on the Data or Related Materials for commercial planning or analysis purposes, and Licensee should always check actual data.

Data Disclaimer

The Data used herein has been derived from actual spatial or tabular information. In some cases, Esri has manipulated and applied certain assumptions, analyses, and opinions to the Data solely for educational training purposes. Assumptions, analyses, opinions applied, and actual outcomes may vary. Again, Esri is not inviting reliance on this Data, and the Licensee should always verify actual Data and exercise their own professional judgment when interpreting any outcomes.

Limitation of Liability

Esri shall not be liable for direct, indirect, special, incidental, or consequential damages related to Licensee's use of the Data and Related Materials, even if Esri is advised of the possibility of such damage.

No Implied Waivers

No failure or delay by Esri or its Licensors in enforcing any right or remedy under this Agreement shall be construed as a waiver of any future or other exercise of such right or remedy by Esri or its Licensors.

Order for Precedence

Any conflict between the terms of this Agreement and any FAR, DFAR, purchase order, or other terms shall be resolved in favor of the terms expressed in this Agreement, subject to the government's minimum rights unless agreed otherwise.

Export Regulation

Licensee acknowledges that this Agreement and the performance thereof are subject to compliance with any and all applicable United States laws, regulations, or orders relating to the export of data thereto. Licensee agrees to comply with all laws, regulations, and orders of the United States in regard to any export of such technical data.

Severability

If any provision(s) of this Agreement shall be held to be invalid, illegal, or unenforceable by a court or other tribunal of competent jurisdiction, the validity, legality, and enforceability of the remaining provisions shall not in any way be affected or impaired thereby.

Governing Law

This Agreement, entered into in the County of San Bernardino, shall be construed and enforced in accordance with and be governed by the laws of the United States of America and the State of California without reference to conflict of laws principles. The parties hereby consent to the personal jurisdiction of the courts of this county and waive their rights to change venue.

Entire Agreement

The parties agree that this Agreement constitutes the sole and entire agreement of the parties as to the matter set forth herein and supersedes any previous agreements, understandings, and arrangements between the parties relating hereto.